Metabolic Phenotyping in Personalized and Public Healthcare

Metabolic Phenotyping in Personalized and Public Healthcare

Edited by

Elaine Holmes

Jeremy K. Nicholson

Ara W. Darzi

John C. Lindon

AMSTERDAM • BOSTON • HEIDELBERG • LONDON
NEW YORK • OXFORD • PARIS • SAN DIEGO
SAN FRANCISCO • SINGAPORE • SYDNEY • TOKYO
Academic Press is an imprint of Elsevier

Academic Press is an imprint of Elsevier
125, London Wall, EC2Y 5AS.
525 B Street, Suite 1800, San Diego, CA 92101-4495, USA
50 Hampshire Street, 5th Floor, Cambridge, MA 02139, USA
The Boulevard, Langford Lane, Kidlington, Oxford OX5 1GB, UK

Library of Congress Cataloging-in-Publication Data
A catalog record for this book is available from the Library of Congress

British Library Cataloguing-in-Publication Data
A catalogue record for this book is available from the British Library

ISBN: 978-0-12-800344-2

For information on all Academic Press publications
visit our website at http://store.elsevier.com

Working together
to grow libraries in
developing countries

www.elsevier.com • www.bookaid.org

Publisher: Mica Haley
Acquisition Editor: Catherine Van Der Laan
Editorial Project Manager: Lisa Eppich
Production Project Manager: Melissa Read
Designer: Maria Ines Cruz

Printed and bound in the United States of America

Contents

8. Handing on Health to the Next Generation: Early Life Exposures

Elaine Holmes, David MacIntyre, Neena Modi and Julian R. Marchesi

Foreword

Clinical medicine has always been driven forward by advances in technology, whether it be new approaches to drug design and discovery, new imaging modalities for improved diagnosis, or, more recently, the advent of the so-called "-omics" sciences. Huge amounts of data are now acquired from subjects, and these are interrogated in relation to disease or population risk factors to reveal biomarkers that can be used for exposure measures, susceptibility factors, improved diagnosis, and better prognosis. The human genome project proved to be the first of these new paradigms, and the improved understanding of genetic mutations has already yielded novel diagnostic and therapeutic combinations, particularly in the field of oncology.

It is clear that human beings are much more than the combination of their genes. The complex interactions between these genes and their environment and a person's lifestyle lead to the realization that the study of human metabolism might offer better clues to the malfunctions underlying diseases. Many diseases and conditions could have both genetic and environmental components, and there exists a wide spectrum, from the purely genetic (eg, Huntington's disease) to the predominantly environmental (eg, smoking-induced lung cancer). The complex nature of the microbial colonies that also share the human body, especially the gut bacteria, also play a vital part as they affect and impact the human metabolism.

The measurement of large amounts of data that describe a human being (the phenotype) is now known as *phenomics*, and the human *metabolic phenotype*, in particular, captures information on both human biochemistry and the effects of the microbiome, and this approach offers new insights into the perturbations caused by diseases or by exposure to external agents—be they air pollution, trace chemicals in the environment, lifestyle choices, or diet. Metabolic phenotyping is most often, but not exclusively, conducted by performing analysis of thousands of metabolites simultaneously, using advanced analytical chemistry technology on biofluids, such as urine or blood serum, or on tissue biopsy specimens.

Metabolic phenotyping is now developing to have an impact on both clinical medicine and on epidemiologic studies of disease risk, and this text provides a comprehensive survey and assessment of the field in both areas of application. It discusses unmet medical needs, provides an overview of the technologies used, and shows how the methodology can be used. It also offers a critical evaluation of the problems still to be overcome.

As Chief Medical Officer for England and Chair of the UK National Institute of Health Research (NIHR), which is part of the UK National Health Service, I am pleased to say that the NIHR, in conjunction with the UK Medical Research Council, awarded a grant to Imperial College London to set up the world's first phenome center. This is now fully functional and has carried out a number of successful studies in a wide range of disease risk areas, including cardiovascular disease, dementia, and diabetes. The template used has now been applied to setting up other phenome centers around the world. The approach promises to address major areas of risk in the populations of today. The topics that I have highlighted recently include the urgent need to tackle microbial resistance to currently available antibiotics, diabetes, and the associated rise in obesity, especially in females.

The editors of this book have brought together an impressive array of international experts who have provided a unified and comprehensive account of the history, development, and practice of metabolic phenotyping. This text should not only provide valuable information to the reader but also stimulate thinking about possible areas of application. I believe that this knowledge will be useful to research scientists, clinicians, epidemiologists, and other health service providers, as well as those involved in health economics. The state of the art is now such that real progress can be made quickly.

Professor Dame Sally Davies FRS FMedSci
Chief Medical Officer for England

Preface

Biomolecular research studies in the areas of clinical medicine and epidemiology are increasingly being directed by analyses of "big data" under the umbrella known as *phenotyping*. One major new and burgeoning aspect of phenotyping is the broad multi-analyte determination of a subject's metabolic phenotype. This can be based on analytes in biofluids or tissue samples, for example. This field has been previously known as *metabonomics* or *metabolomics*. The cohesion between metabolic phenotyping and clinical and population profiling studies has recently been shown, through the large published literature, to be highly fruitful in the search for biomarkers of diseases or population disease risks factors, and hence this leads to a better understanding of the underlying biochemical mechanisms of diseases and disease risks.

This book begins with a view on unmet medical needs, and this is followed by a chapter that reviews the history of metabolic phenotyping. Other chapters are devoted to the basis of the technological aspects of metabolic phenotyping, whether they are based on sample collection needs, the analytical chemistry of assays, or the subsequent comprehensive statistical analysis of data and the downstream biochemical interpretations. One chapter covers the principles and applications of predictive metabolic phenotyping, which can be used for prognostic studies, a process that is linked to the subject of another chapter on phenotyping the "patient journey" through diagnosis, therapy, and outcome. We have been fortunate to commission chapters from world experts describing various areas of application, from early life through to old age, including the effects of the microbiome, and another chapter covers the use of real-time metabolic phenotyping during surgery with the "intelligent knife." We have taken care to include the paradigm for studying the effects of exposure to environmental factors such as air pollution, exogenous chemicals, and so on. Finally, we have provided a chapter that describes the concept of specialized phenome centers for metabolic phenotyping, with the first now operational at Imperial College London and others coming on stream worldwide and a chapter on the problems and solutions associated with studies on "big data," and we conclude with a brief summary, a look at the state of the art, what problems still remain to be solved, and the exciting prospects for the future.

We hope that is text will become essential reading for academic, industrial, and clinical scientists who wish to gain a better understanding of the field and of the prospects of the metabolic phenotyping approach.

We are indebted to all of the very busy authors who agreed to write chapters for this book and thank them for their efforts. We would also like to thank the team at Elsevier and its associates (especially Lisa Eppich, who handled the difficult commissioning stage with great efficiency and Melissa Read who was most helpful during the production stage), who all contributed to bringing this book to publication.

We hope that this text will contribute to better understanding of how metabolic phenotyping can fit into clinical medicine and population screening, along with the other many advances that are paving the way to precision medicine and hence patient benefit.

Elaine Holmes
Jeremy K. Nicholson
Ara W. Darzi
John C. Lindon

List of Contributors

Hutan Ashrafian Department of Surgery and Cancer, Imperial College London, London, UK

Thanos Athanasiou Department of Surgery and Cancer, Imperial College London, London, UK

Seth Chitayat Department of Biomedical and Molecular Sciences, Queen's University, Kingston, ON, Canada; Department of Surgery, Queen's University, Kingston, ON, Canada

Ara W. Darzi Department of Surgery and Cancer, Imperial College London, London, UK; Institute of Global Health Innovation, Imperial College London, London, UK

Anthony C. Dona MRC-NIHR National Phenome Centre, Department of Surgery & Cancer, Imperial College London, London, UK; Kolling Institute of Medical Research, Northern Clinical School, University of Sydney, St Leonards, NSW, Australia

Jeremy R. Everett Medway Metabonomics Research Group, University of Greenwich, Kent, UK

Young-Mi Go Division of Pulmonary, Allergy and Critical Care Medicine, Emory University School of Medicine, Atlanta, GA, USA

Elaine Holmes Department of Surgery and Cancer, Imperial College London, London, UK

Dean P. Jones Division of Pulmonary, Allergy and Critical Care Medicine, Emory University School of Medicine, Atlanta, GA, USA

Sarah Kenderdine Expanded Perception and Interaction Centre, University of New South Wales, Sydney, New South Wales, Australia

James Kinross Section of Biosurgery and Surgical Technology, Department of Surgery and Cancer, Faculty of Medicine, Imperial College London, London, UK

Nadine Levin Institute for Society and Genetics, University of California, Los Angeles, CA, USA

Jia Li Department of Surgery and Cancer, Imperial College London, London, UK

John C. Lindon Department of Surgery and Cancer, Imperial College London, London, UK

Ken Liu Division of Pulmonary, Allergy and Critical Care Medicine, Emory University School of Medicine, Atlanta, GA, USA

David MacIntyre Institute of Reproductive and Developmental Biology, Department of Surgery and Cancer, Imperial College London, The Hammersmith Hospital, London, UK

Julian R. Marchesi Institute of Reproductive and Developmental Biology, Department of Surgery and Cancer, Imperial College London, The Hammersmith Hospital, London, UK; Centre for Digestive and Gut Health, Imperial College London, London, UK; School of Biosciences, Cardiff University, Cardiff, Wales, UK

Ingrid Mason Intersect Australia Pty Ltd, Sydney, New South Wales, Australia

Neena Modi Department of Medicine, Imperial College London, London, UK

Laura Muirhead Section of Biosurgery and Surgical Technology, Department of Surgery and Cancer, Faculty of Medicine, Imperial College London, London, UK

Jeremy K. Nicholson Department of Surgery and Cancer, Imperial College London, London, UK

Kurt D. Pennell Department of Civil and Environmental Engineering, Tufts University, Medford, MA, USA

John F. Rudan Department of Surgery, Queen's University, Kingston, ON, Canada; Human Mobility Research Centre, Queen's University and Kingston General Hospital, Kingston, ON, Canada

Reza M. Salek The European Bioinformatics Institute, Wellcome Genome Campus, Hinxton, Cambridge, UK

Christoph Steinbeck The European Bioinformatics Institute, Wellcome Genome Campus, Hinxton, Cambridge, UK

Zoltan Takats Department of Surgery and Cancer, Faculty of Medicine, Imperial College London, London, UK

Douglas I. Walker Division of Pulmonary, Allergy and Critical Care Medicine, Emory University School of Medicine, Atlanta, GA, USA; Department of Civil and Environmental Engineering, Tufts University, Medford, MA, USA

Ian D. Wilson Department of Surgery and Cancer, Imperial College London, London, UK

Chapter 1

Unmet Medical Needs

Hutan Ashrafian[1], Thanos Athanasiou[1], Jeremy K. Nicholson[1] and Ara W. Darzi[1,2]

[1]*Department of Surgery and Cancer, Imperial College London, London, UK* [2]*Institute of Global Health Innovation, Imperial College London, London, UK*

Chapter Outline

1.1 A HISTORICAL PERSPECTIVE

The 21st century has heralded dramatic changes in the global health care ecology. There have been significant fluxes in population dynamics, occupational shifts, environmental changes, drivers of health care economics, political forces, and a technologic explosion that can be considered as dramatic as the advances of both the agricultural and industrial revolutions. Despite an overall increase in the awareness of disease and a more unified approach to its management, health care remains a global problem that challenges society with a voluminous corpus of unmet medical needs.

As early as the 5th century BC, Hippocrates had clarified that disease pathology originated from both inherent patient factors and those of the patient's environment [1]. In subsequent eras, the increased scrutiny and understanding of disease mechanisms has offered a tentative breakdown of the relative contribution of disease from these two factors. These led to many of the very foundations of medicine as we know it today, including Edward Jenner's demonstration of controlled immunity, Louis Pasteur's germ theory of disease [2], and hybridization of Greek anatomy and eastern proto-pharmacotherapy described in Avicenna's *Canon of Medicine* [3].

E. Holmes, J.K. Nicholson, A.W. Darzi & J.C. Lindon (Eds): Metabolic Phenotyping in Personalized and Public Healthcare. DOI: http://dx.doi.org/10.1016/B978-0-12-800344-2.00001-X

While the work of the luminaries listed above added to a critical mass of health care expertise, the isolated work of John Gaunt in 1662 led to the birth of epidemiology [4]. In this work, for the first time, Gaunt was able to describe quantifiable trends in disease rates in a London population, and his description offered a wider interpretation of modifiable disease mechanism that could lead to a cure.

Three taxonomic questions arose then and continue to confront mankind:

1. What is the nature of diseases, and what are their trends?
2. What are the mechanisms that govern their initiation and progression?
3. How can we best treat them?

Twentieth century medicine, in turn, yielded a number of powerful scientific breakthroughs, including the discovery of the structure of deoxyribonucleic acid (DNA), characterization of the human genome, and the ability to differentiate the protein, transcription ribonucleic acid, and even metabolites of healthy as well as diseased individuals. Such discoveries were also augmented by treatment innovations, including the introduction of population-wide antibiotics following the birth of penicillin, and novel technologies that permit super-precise operations through robotic surgery.

1.2 UNMET MEDICAL NEEDS

Despite the developments described above, many current treatment strategies remain ineffectual and significant health care needs continue to challenge human society (Fig. 1.1). At a global level, the World Health Organization (WHO) [5] reported ischemic heart disease, stroke, lower respiratory infections, chronic obstructive lung disease, and diarrhea as the top five causes of death in 2011. Both ischemic heart disease and stroke originate from cardiovascular disease

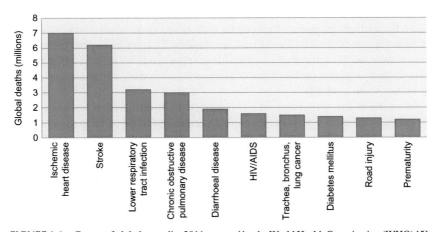

FIGURE 1.1   Causes of global mortality 2011 reported by the World Health Organization (WHO) [5].

and, together, account for 21.8% of all deaths. Adding deaths from diabetes mellitus, an associated condition, to this total results in just fewer than 25% of all deaths from cardiovascular disease (see Fig. 1.1).

The current rates of global mortality, nevertheless, represent a paradigm shift from the main drivers of mortality that afflicted mankind until the end of the 19th century. These included malnourishment, mycobacterial diseases (tuberculosis and leprosy), smallpox (which was eradicated in 1979), diphtheria, typhoid, and dysentery. Consequently, the majority of mankind's diseases originated from lack of food and exposure to communicable diseases (in the absence of antimicrobials), whereas more recently we have discovered vulnerability to noncommunicable diseases arising from obesity, smoking, aging, environmental change, and the persistence of global infectious diseases due to the expansive worldwide traveling footprint.

Disease trends as a result of industrialization also contribute distinct disease patterns. As an example, trends in mortality in the United States (Fig. 1.2) [6–9] demonstrate that in 1900 infectious diseases accounted for 37% of deaths, cardiovascular diseases (heart disease and stroke) were responsible for 17%, and cancer resulted in only 4.5% of deaths. By 1982, cardiovascular diseases accounted for over half of all deaths (58.3%) and cancer for approximately one quarter of mortality. At the beginning of the 21st century, Alzheimer disease entered the top 10 causes of death as the first example of mortality from neurodegenerative diseases (see Fig. 1.2).

These epidemiologic shifts in disease result from the balancing of excess calories, sedentary lifestyle, environmental pollution, and uncontrolled exposure

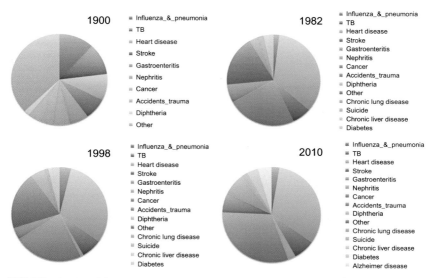

FIGURE 1.2   Trends in US mortality rates [6–8] in 1900–2010.

to toxins (such as smoking and alcohol) versus accessibility to advanced health care centers, health care economics, health policy and management, personal economic capital, efficacious pharmaceuticals, and well-trained medical expertise. This simple equilibrium is, in actuality, more complex than one of pathologic factors versus health-promoting elements. Factors such as advanced health systems also lead to longer life span, which, by itself, is associated with pathologic conditions such as frailty, increased cardiovascular risk, and a higher chance of tumorigenesis and neurodegenerative diseases.

At a higher level, based on our biomedical understanding of disease trends, are the roles of human networks and global health care policy. Obesity is a classic example of both a physical disease and a web-based communication network disease [10,11], whereas the global spread of viral diseases such as influenza and "bird flu" depends directly on global network hubs and travel dynamics. The eradication of smallpox [2] was achieved through successful international lobbying, at a political level, through the WHO, and conversely, increasing global rates of poliomyelitis have been linked to geopolitical rivalry. Also, the management of human immunodeficiency virus/acquired immunodeficiency syndrome (HIV/AIDS) has been less efficient in some nation states due to social nonacceptance of this viral illness as a disease. As a consequence, the 21st century medical community has recognized the need to address disease and health care management through a complete arsenal of (1) disease management, (2) socioeconomics, (3) health process, and (4) technology. Together, these constitute healthcare's unmet needs (Fig. 1.3).

The needs of disease management directly reflect the current status of disease trends. These comprise the major diseases that have been present for generations as well as newer diseases that represent current and future pathologies.

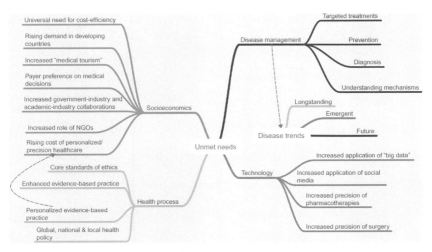

FIGURE 1.3   Unmet needs in health care.

For example, malaria has plagued mankind for millennia but is associated with our molecular evolution, as sickle cell disease may have come into existence because of a selection trait brought on by protection from this disease. Another example is obesity, which was considered an evolutionary advantage in the Ice Age, but its sudden and dramatic rise to a pandemic level disease has resulted in bariatric surgery rates exceeding those of traditional operations for common disorders such as gallbladder excision (cholecystectomy) in some parts of the Western world [12,13]. Population-based variances in disease trends also exist; for example, in Africa, autoimmunity and asthma are not well recognized, whereas these conditions are common in Western populations. Between 1990 and 2001, deaths due to communicable, maternal, perinatal, and nutritional diseases were reduced by 20% [14], but these have been replaced by new sources of mortality.

Globally, approximately 60–70 million people die each year, 15–20% of which are children. However, 99% of child mortality occurs in low-income and middle-income countries, and over 50% of these deaths result from infectious diseases, including acute respiratory infections, measles, diarrhea, malaria, and HIV/AIDS [14]. Analysis of global data has identified 19 risk factors [14] that account for 45% of deaths and 36% of disease burden worldwide. These include:

- *Nutritional*: underweight in children, high cholesterol, obesity (and overweight), low fruit and vegetable intake, zinc deficiency, iron deficiency anemia, vitamin A deficiency
- *Environmental*: unsafe water (as well as poor sanitation and hygiene), indoor air pollution (from solid fuels), urban air pollution
- *Behavioral and sociopathologic*: unsafe sex, smoking, alcohol use, physical inactivity, illicit drug use, unmet contraception need, sexual abuse of children
- *Iatrogenic*: contaminated injections
- *Physiologic*: hypertension.

Streptococcus pneumoniae is responsible for approximately 11% (8–12%) of deaths in children between 1 and 59 months, whereas *Haemophilus influenzae* type b (Hib) causes approximately 400,000 deaths in children of the same age group through bacterial meningitis, pneumonia, and other serious infections [15,16].

Other pertinent examples include Epstein–Barr virus, which, for half a century, has been associated with development of cancer in over 200,000, and, to date, the exact mechanism of viral tumorigenesis remains unknown [17]. Other well-known diseases with causes that are better understood, have yet to be addressed by the global society. In the developing world, the maternal mortality ratio is approximately 440 per 100,000 live births; approximately 34 times the rate in the developed world (13 per 100,000 live births) [18]. In terms of a predicted disease mortality, noncommunicable diseases are set to rise from 59% (2002) to 69% (2030). According to both baseline and worst-case scenarios, the three top global causes of death in 2030 have been predicted to be HIV/AIDS, unipolar depressive disorders, and ischemic heart disease [19], with road traffic accidents indicated as the fourth leading cause of death in some simulations.

The WHO currently estimates that HIV/AIDS represents the third most common cause of mortality in the developing world. The prevalence of a significant retroviral load in the populations of the developing world has led to deterioration in the fight against tuberculosis and its drug therapy, which is now heavily complicated by drug resistance [20]. These pathologic factors add to other elements of unmet health care, which include several socio-economic factors that are prominent contributors to early mortality in the developing world.

Although more prevalent in the developing world, chronic noncommunicable diseases are now emerging as a global health problem. Cardiovascular diseases, obesity, diabetes, cancer, chronic respiratory disease, and mental health now account for the majority of global mortality and disability [21]. Furthermore, it has been estimated that over a 10-year period, the financial loss caused by these disorders to low-income and middle-income countries is approximately $84 billion [22]. Despite this, noncommunicable diseases remain a concern of low priority in many health care systems; this situation has led to several calls to action on the global stage of health care, including a recent United Nations high-level meeting [21,23]. Finally, there is a second level of diseases not originating from nascent pathology but, rather, as a result of our unbalanced attempts to treat other diseases. One prominent example is the worsening threat of antimicrobial resistance (agents ineffective against all microbial organisms), specifically antibiotic resistance (agents ineffective against common bacterial infections). In this case, the WHO has identified 450,000 new cases of multidrug-resistant tuberculosis in 2012 and a further 64% higher risk of death from hospital-acquired methicillin-resistant *Staphylococcus aureus* infections compared with risk of death from nonresistant *S. aureus* infections [24]. This has resulted from incongruous overconsumption and overapplication of these therapies, which, however, demonstrate the highest efficacy when utilized in a targeted fashion [25].

1.3 ADDRESSING THE PROBLEMS

Both the developed and developing societies of the world recognize an escalation of health care costs, and there has been a call for targeted therapies that are cost efficient and yet provide care of quality and value. It is expected that the major part of global health care expenditure will come from countries with the fastest-growing markets, termed "BRIC" (Brazil, Russia, India, and China) and "MINT" (Mexico, Indonesia, Nigeria, and Turkey). As an example, the expenditure of the health care sector in India in 2012 [26,27] was estimated to be $40 million dollars.

With the changing disease burden, a shortage of qualified health care practitioners is emerging, even in the developed world. The Association of American Medical Colleges has estimated that in the United States, there will be a shortfall of 45,000 primary care physicians and of 46,000 surgeons and medical specialists by 2020 [27,28]. Staffing shortages in both developing and developed countries will likely result in an increased reliance on fast, accurate, and cost-effective diagnostic and treatment methods. Increased access to information with Web 2.0+ technologies, including social media, has empowered patients and increased public awareness of health care so that patient decisions will also contribute significantly to health care practice in the future. This development, coupled with increased global travel, has already resulted in the proliferation of "medical tourism" [27] (where patients visit countries specifically for the quality and availability of health care available there). This increased global travel has, however, also facilitated the spread of diseases, including pandemics of communicable diseases.

Increased requirements for cost-effectiveness in patient care will likely herald an era in which all stakeholders will join together to generate and deliver the next generation of treatments. This will include collaborations among patients, health care providers, industry, academia, governments, and nongovernmental organizations (NGOs), this, however, will require support from appropriate local and global policies to offer to patients treatments that are selected on the basis of the best evidence.

Obtaining the best evidence in patient care calls for the development of technologic tools that can offer cutting-edge treatments (eg, surgical robots with increased precision, pharmacotherapies with fewer side effects and more efficacy). The increased level of health care precision channeled through evidence-based practice, in turn, requires the interpretation of larger, more complex, and more accurate patient information or "big data" with the use of more powerful computation tools.

1.4 PERSONALIZED MEDICINE

Questions with regard to unmet medical needs can ultimately be answered through the recently developed concept of *Personalized Medicine*, which offers targeted health care provision that takes into consideration both inherent patient factors and environmental factors. This concept was developed in response to several advances in health care practice, including:

1. Increased awareness that not all treatments have the same effects on all patients. For example, not all patients respond to similar pharmacotherapies in the same way.
2. Increased precision in the characterization of patients and their diseases, for example, through the Human Genome Project and, more recently, the use of approaches known as Systems Biology (involving so-called omics

technologies—genomics, transcriptomics, proteomics, and metabolomics/
metabonomics, now often referred to as "metabolic phenotyping").
3. Increased ability to evaluate and compute the "big data" from these sciences
   to generate treatments.

The term *Personalized Medicine* has often been used synonymously with
"precision medicine" and "stratified medicine." It utilizes biomedical methodology, which can stratify patients within their subpopulations. As a result, it
can account for varying susceptibility of patients to different diseases and their
fluctuating responses to specific treatments. This powerful methodology can
thus maximize patient benefit and minimize patient harm. Numerous definitions
of Personalized Medicine have been proposed (Table 1.1), all of which demonstrate common elements [37] (eg, three definitions employ the term "tailored"
or "tailoring"). Here, we propose a single unifying definition:

> *Personalized Medicine is the tailored management and/or prevention of disease
> according to the specific characteristics of a stratified individual, subpopulation,
> or population to enhance patient care. These characteristics are derived from the
> integrated evaluation of phenotype, genotype, and treatment bioresponses real-
> ized through a systems biomedicine—the "-omics" approach. This employs com-
> plex multivariate, network, and hierarchical computation in the context of best
> evidence-based practice. It offers precision diagnosis and treatments in addition
> to the generation of targeted therapeutics.*

The birth of Personalized Medicine resulted from the coalescence of several underlying elements (Fig. 1.4). By the late 17th century, Archibald Pitcairne
had introduced the concept of applying complex mathematics to medicine
[38]. Approximately two and a half centuries later, two fundamental breakthroughs took place, in an unrelated but sequential fashion, in the late 1930s.
Alan Turing the renowned British mathematician and cryptographer addressed
the "Entscheidungsproblem" or "Decision Problem" in 1936 to initiate the era
of computation using computers [39]. Shortly after in 1937, Karl Ludwig von
Bertalanffy introduced the concept of General Systems Theory or "Allgemeine
Systemtheorie." This revolutionized our appraisal of biology, overcoming traditional basic models of bio-organization to offer an appreciation or real-life
complexity in terms of a hierarchical structure and an interactive, multifaceted
communication dialog of metabolic, energetic, and cellular intermediaries [40].
This paved the way for modern systems biology and the birth of the "-omics"
sciences (including genomics, transcriptomics, proteomics, and metabolomics/
metabonomics) in addition to contributing to cybernetics, management theory,
economics, and a massive range of allied specialties. The final component of
the Personalized Medicine coalescence was the renaissance of Avicenna's concept of treatment evidence discussed in his 11th century *Canon of Medicine* [3],
and subsequently re-adapted in the 1960s by pioneers such as Archibald Leman
Cochrane to promote the role of clinical evidence in decisions and was subsequently, in 1990, named "evidence-based medicine" by Gordon Guyatt.

TABLE 1.1 Definitions of Personalized Medicine

Organization	Year	Definition
President's Council of Advisors on Science and Technology [29]	2008	The tailoring of medical treatment to the specific characteristics of each patient. It does not literally mean the creation of drugs or medical devices that are unique to a patient. Rather, it involves the ability to classify individuals into subpopulations that are uniquely or disproportionately susceptible to a particular disease or responsive to a specific treatment.
PricewaterhouseCoopers (PWC) [30]	2009	As products and services that leverage the science of genomics and proteomics (directly or indirectly) and capitalize on the trends toward wellness and consumerism to enable tailored approaches to prevention and care.
Personalized Medicine Coalition (PMC) [31]	2011	Personalized Medicine is the tailoring of medical treatment to the individual characteristics of each patient. The approach relies on scientific breakthroughs in our understanding of how a person's unique molecular and genetic profile makes them susceptible to certain diseases. This same research is increasing our ability to predict which medical treatments will be safe and effective for each patient, and which ones will not be.
The National Academy of Sciences	2011	Precision medicine refers to the tailoring of medical treatment to the individual characteristics of each patient. It does not literally mean the creation of drugs or medical devices that are unique to a patient, but rather the ability to classify individuals into subpopulations that differ in their susceptibility to a particular disease, in the biology and/or prognosis of those diseases they may develop, or in their response to a specific treatment. Preventive or therapeutic interventions can then be concentrated on those who will benefit, sparing expense and side effects for those who will not. Although the term "Personalized Medicine" is also used to convey this meaning, that term is sometimes misinterpreted as implying that unique treatments can be designed for each individual.
National Cancer Institute [32]	2012	A form of medicine that uses information about a person's genes, proteins, and environment to prevent, diagnose, and treat disease. In cancer, Personalized Medicine uses specific information about a person's tumor to help diagnose, plan treatment, find out how well treatment is working, or make a prognosis. Examples of Personalized Medicine include using targeted therapies to treat specific types of cancer cells, such as HER2-positive breast cancer cells, or using tumor marker testing to help diagnose cancer. Also called precision medicine.

(Continued)

TABLE 1.1 Definitions of Personalized Medicine (Continued)

Organization	Year	Definition
National Human Genome Research Institute, NIH [33]	2012	Personalized Medicine is an emerging practice of medicine that uses an individual's genetic profile to guide decisions made in regard to the prevention, diagnosis, and treatment of disease. Knowledge of a patient's genetic profile can help doctors select the proper medication or therapy and administer it using the proper dose or regimen. Personalized Medicine is being advanced through data from the Human Genome Project.
The National Academy of Sciences [34]	2011	Precision Medicine refers to the tailoring of medical treatment to the individual characteristics of each patient. It does not literally mean the creation of drugs or medical devices that are unique to a patient, but rather the ability to classify individuals into subpopulations that differ in their susceptibility to a particular disease, in the biology and/or prognosis of those diseases they may develop, or in their response to a specific treatment. Preventive or therapeutic interventions can then be concentrated on those who will benefit, sparing expense and side effects for those who will not. Although the term "Personalized Medicine" is also used to convey this meaning, that term is sometimes misinterpreted as implying that unique treatments can be designed for each individual.
Medical Research Council (MRC) [35]	2012	Stratified medicine is based on identifying subgroups of patients with distinct mechanisms of disease, or particular responses to treatments. This allows us to identify and develop treatments that are effective for particular groups of patients. Ultimately stratified medicine will ensure that the right patient gets the right treatment at the right time.
Erasmus University [36]	2013	The use of combined knowledge (genetic or otherwise) about a person to predict disease susceptibility, disease prognosis, or treatment response and thereby improve that person's health.
Authors	2014	Personalized Medicine is the tailored management and/or prevention of disease according to the specific characteristics of a stratified individual, subpopulation, or population to enhance patient care. These characteristics are derived from the integrated evaluation of phenotype, genotype, and treatment bioresponses realized through a systems biomedicine "-omics" approach. This employs complex multivariate, network, and hierarchical computation in the context of best evidence-based practice. It offers precision diagnosis and treatments in addition to the generation of targeted therapeutics.

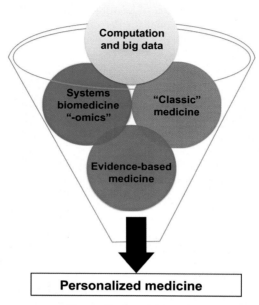

FIGURE 1.4   The components of Personalized Medicine.

The strength of Personalized Medicine can therefore offer a solution to health care's unmet needs through addressing six core areas [37,41], including:

- *Basic research*: disease mechanisms, molecular pathways, and drug metabolism
- *Drug discovery and development*: prediction of drug responders and identifying new roles for drugs already in use for treating other conditions (so-called repurposing)
- *Preclinical testing*: identifying drug candidates that can progress through the phases of drug testing
- *Clinical research*: enhanced selection of appropriate trial participants and enhanced analysis of outcomes
- *Clinical adoption*: better targeted therapies with higher efficacy for public health, medicine, and surgery
- *Health care (in general)*: better diagnostics, better treatments, better patient selection, cost-effectiveness of treatment strategies and enhanced patient outcomes, adoption of novel next-generation technologies such as super-high-precision intraoperative diagnostics.

The ultimate goal of Personalized Medicine is to improve patient outcomes by offering higher-precision treatments of the highest quality and value. It has been projected that reducing the burden of chronic disease by 2% a year over a 10-year period would result in the prevention of 36 million deaths and 500 million life-years saved. The majority of these benefits would result from the improved health of both young and older subjects in low-income and middle-income countries [42]. The potential for Personalized Medicine to offer solutions across all health care areas therefore renders it a powerful tool for current and future health care providers.

In summary, the global shift in the characters and trends of diseases continues to have a significant impact on the health care sector worldwide, which, in turn, is correlated to the scientific, political, and economic landscapes. Traditional medical practice has also expanded to meet the goals of next-generation targeted and precise health care. Personalized Medicine is one paradigm that offers an innovation tipping point, with the potential for achieving high-precision cost-effective treatments to augment medical care at all levels. The scene is now set for the development of this system and for incorporating it into the future of biomedicine, with the ultimate aim of achieving the highest quality treatments and best value-based health care outcomes.

1.5 PERSONALIZED MEDICINE: THE ROLE OF METABOLIC PHENOTYPING

Metabolic phenotyping uses a range of modern technologies in analytical chemistry to provide information on the small molecule metabolites present in biofluids such as urine or blood serum/plasma as well as in tissue extracts. Changes in the levels of such metabolites in patients can be used as biomarkers for disease diagnosis, therapy efficacy or side effects and also as prognostic markers of outcome. In population studies, the same approach can lead to epidemiologic biomarkers of disease risk.

The main techniques used are nuclear magnetic resonance (NMR) spectroscopy, and mass spectrometry, the latter detection method usually requiring a chromatographic separation stage first; currently, this is usually ultra-performance liquid chromatography, although other techniques such as capillary electrophoresis and gas chromatography are also used. NMR spectroscopy is also unique in that it can be used to obtain information on metabolites in intact tissue biopsy specimens by using a technique known as "magic-angle spinning."

For population screening, there are a number of published examples where novel metabolic biomarkers of disease risk have been discovered. These include markers of cardiovascular disease risk from metabolic profiling of serum [43] and blood pressure from markers in urine [44].

In the context of Personalized Medicine, the concept of the patient's metabolic trajectory has been suggested [45]. In this approach, biofluid (and stool samples) can be obtained prior to, during, and after a patient undergoes treatment. The metabolic information so obtained can then be used to aid diagnosis, to monitor the effects of any treatment, and to predict outcomes such as the need for Intensive Therapy/Care Unit (ITU) admission. Although the approach can be used for any therapeutic intervention, it has particular benefit for surgical procedures, where fast decision making is necessary and comprehensive biochemical information is usually not available. In surgery, the "intelligent knife" technology is already beginning to demonstrate its usefulness [46]. These and similar strategies offer a platform on which to build and integrate health care innovations from nanotechnology agents to robotic surgeons, which can, in turn, address pathology at a multitude of scales, ranging from the whole organism to nanoscale molecular targets. In this way, the promise of cutting-edge medical therapies in areas such as genetics, regenerative medicine, surgery, and pharmaceuticals may fulfill their long-term promises.

We believe that this approach holds tremendous potential for helping to fulfill the unmet medical needs of the 21st century.

REFERENCES

[1] Martin PM, Martin-Granel E. 2,500-year evolution of the term epidemic. Emerg Infect Dis 2006;12(6):976–80.

[2] de Quadros CA. History and prospects for viral disease eradication. Med Microbiol Immunol 2002;191(2):75–81.

[3] Ashrafian H. Avicenna (980–1037) and the first description of post-hepatic jaundice secondary to biliary obstruction. Liver Int 2014;34(3):479–80.

[4] Rothman KJ. Lessons from John Graunt. Lancet 1996;347(8993):37–9.

[5] World Health Organization. Mortality and global health estimates, <http://www.who.int/gho/mortality_burden_disease/en/>; 2011.

[6] Heron M. Deaths: leading causes for 2010 (U.S. Department of Health and Human Services, Centers for Disease Control and Prevention). Natl Vital Stat Rep 2013;62(6).

[7] Murphy SL, Xu J, Kochanek KD. Deaths: final data for 2010 (U.S. Department of Health and Human Services, Centers for Disease Control and Prevention). Natl Vital Stat Rep 2013;61(4).

[8] Hennekens CH, Buring JE. Epidemiology in medicine. Philadelphia, PA: Lippincott Williams and Wilkins; 1987.

[9] Murphy SL. Deaths: leading causes for 1998 (U.S. Department of Health and Human Services, Centers for Disease Control and Prevention). Natl Vital Stat Rep 2000;48(11).

[10] Christakis NA, Fowler JH. The spread of obesity in a large social network over 32 years. New Engl J Med 2007;357(4):370–9.

[11] Valente TW, Fujimoto K, Chou CP, Spruijt-Metz D. Adolescent affiliations and adiposity: a social network analysis of friendships and obesity. J Adolesc Health 2009;45(2):202–4.

[12] Ashrafian H, Darzi A, Athanasiou T. Bariatric surgery—can we afford to do it or deny doing it? Frontline Gastroenterol 2011;2(2):82–9.

[13] Li JV, Ashrafian H, Bueter M, Kinross J, Sands C, le Roux CW, et al. Metabolic surgery profoundly influences gut microbial-host metabolic cross-talk. Gut 2011;60(9):1214–23.

[14] Lopez AD, Mathers CD, Ezzati M, Jamison DT, Murray CJ. Global and regional burden of disease and risk factors, 2001: systematic analysis of population health data. Lancet 2006;367(9524):1747–57.

[15] O'Brien KL, Wolfson LJ, Watt JP, Henkle E, Deloria-Knoll M, McCall N, et al. Burden of disease caused by *Streptococcus pneumoniae* in children younger than 5 years: global estimates. Lancet 2009;374(9693):893–902.

[16] Watt JP, Wolfson LJ, O'Brien KL, Henkle E, Deloria-Knoll M, McCall N, et al. Burden of disease caused by *Haemophilus influenzae* type b in children younger than 5 years: global estimates. Lancet 2009;374(9693):903–11.

[17] Lieberman PM. Virology. Epstein-Barr virus turns 50. Science 2014;343(6177):1323–5.

[18] World_Health_Organization. The World Health Report (2005): make every mother and child count, <http://www.who.int/whr/2005/whr2005_en.pdf>; 2005.

[19] Mathers CD, Loncar D. Projections of global mortality and burden of disease from 2002 to 2030. PLoS Med 2006;3(11):e442.

[20] Anish T, Sreelakshmi P. Revisiting public health challenges in the new millennium. Ann Med Health Sci Res 2013;3(3):299–305.

[21] Beaglehole R, Ebrahim S, Reddy S, Voute J, Leeder S, Chronic Disease Action G Prevention of chronic diseases: a call to action. Lancet 2007;370(9605):2152–7.

[22] Abegunde DO, Mathers CD, Adam T, Ortegon M, Strong K. The burden and costs of chronic diseases in low-income and middle-income countries. Lancet 2007;370(9603):1929–38.

[23] Beaglehole R, Bonita R, Alleyne G, Horton R, Li L, Lincoln P, et al. UN High-Level meeting on Non-Communicable diseases: addressing four questions. Lancet 2011;378(9789):449–55.

[24] World_Health_Organization. Antimicrobial resistance, Fact sheet 194, <http://www.who.int/mediacentre/factsheets/fs194/en/>; 2014.

[25] Bell BG, Schellevis F, Stobberingh E, Goossens H, Pringle M. A systematic review and meta-analysis of the effects of antibiotic consumption on antibiotic resistance. BMC Infect Dis 2014;14:13.

[26] PWC_PriceWaterhouseCoopers. Healthcare in India: emerging market report, <http://www.pwc.com/en_gx/gx/healthcare/pdf/emerging-market-report-hc-in-india.pdf>; 2007.

[27] Dillon K, Prokesch S. Megatrends in global health care. Harvard Bus Rev, http://hbr.org/web/extras/insight-center/health-care/globaltrends/1-slide; 2010.

[28] AAMC. Physician shortages to worsen without increases in residency training, <https://http://www.aamc.org/download/153160/data/physician_shortages_to_worsen_without_increases_in_residency_tr.pdf>; 2010.

[29] President's Council of Advisors on Science and Technology (PCAST). Priorities for Personalized Medicine, <http://www.whitehouse.gov/files/documents/ostp/PCAST/pcast_report_v2.pdf>; 2008.

[30] PWC_PriceWaterhouseCoopers. The new science of personalized medicine, <http://www.pwc.com/us/en/healthcare/publications/personalized-medicine.jhtml>.

[31] PMC. The Age of Personalized Medicine, <http://www.ageofpersonalizedmedicine.org/objects/pdfs/Age_PM_factsheet.pdf>; 2012.

[32] NationalCancerInstitute. Personalized Medicine, <http://www.cancer.gov/dictionary?cdrid=561717>; 2012.

[33] NHGRI. Personalized Medicine, <http://www.genome.gov/Glossary/index.cfm?id=150>; 2012.

[34] NAS. Toward Precision Medicine: Building a Knowledge Network for Biomedical Research and a New Taxonomy of Disease, <http://www.ucsf.edu/sites/default/files/legacy_files/documents/new-taxonomy.pdf>; 2011.

[35] MRC. Stratified medicine, <http://www.mrc.ac.uk/Ourresearch/ResearchInitiatives/StratifiedMedicine/index.htm>; 2012.

[36] Redekop WK, Mladsi D. The faces of personalized medicine: a framework for understanding its meaning and scope. Value Health 2013;16(6 Suppl.):S4–S9.

[37] Branzén K. Personalized medicine: a new era for healthcare and industry, <http://ec.europa. eu/digital-agenda/futurium/sites/futurium/files/futurium/library/Branz%C3%A9n-2013-Personalized Medicine(2).pdf>. Lund, Sweden: Life Science Foresight Institute; 2013.

[38] Ashrafian H. Mathematics in medicine: the 300-year legacy of iatromathematics. Lancet 2013;382(9907):1780.

[39] Ashrafian H, Darzi A, Athanasiou T. A novel modification of the Turing test for artificial intelligence and robotics in healthcare. Int J Med Robot 2015;11(1):38–43.

[40] Ashrafian H, Darzi A, Athanasiou T. Autobionics: a new paradigm in regenerative medicine and surgery. Regen Med 2010;5(2):279–88.

[41] Battelle_TPP. Briefing report on framework for understanding personalized medicine opportunities for Arizona (prepared for Flinn Foundation), <http://www.flinnscholars.org/file/final_ arizona_personalized_framework_briefing.pdf>; 2010.

[42] Strong K, Mathers C, Leeder S, Beaglehole R. Preventing chronic diseases: how many lives can we save? Lancet 2005;366(9496):1578–82.

[43] Wurtz P, Havulinna AS, Soininen P, Tynkkynen T, Prieto-Merino D, Tillin T, et al. Metabolite profiling and cardiovascular event risk: a prospective study of three population-based cohorts. Circulation 2015;131(9):774–85.

[44] Holmes E, Loo RL, Stamler J, Bictash M, Yap IKS, Chan Q, et al. Human metabolic phenotype diversity and its association with diet and blood pressure. Nature 2008;453(7193):396–400.

[45] Nicholson JK, Holmes E, Kinross JM, Darzi AW, Takats Z, Lindon JC. Metabolic phenotyping in clinical and surgical environments. Nature 2012;491(7424):384–92.

[46] Balog J, Sasi-Szabo L, Kinross J, Lewis MR, Muirhead LJ, Veselkov K, et al. Intraoperative tissue identification using rapid evaporative ionization mass spectrometry. Sci Transl Med 2013;5(194):194ra93.

Chapter 2

The Development of Metabolic Phenotyping—A Historical Perspective

John C. Lindon and Ian D. Wilson

Department of Surgery and Cancer, Imperial College London, London, UK

Chapter Outline

2.1 INTRODUCTION

The idea of taking a holistic view of a body fluid for diagnosing diseases goes back into medieval times when so-called "pisse prophets" would examine a patient's urine and use color, smell, and even taste, to try to diagnose problems. The first documented examples of this may be the two "urine wheels" shown in Fig. 2.1. The left image is taken from a folio in the Royal Library of Copenhagen, Denmark, with the shelf mark NKS84b and an unlisted date of publication, and the right image is taken from a book, *Epyphanie Medicorum*, written by Ullrich Pinder and published in 1506. This approach was notoriously

E. Holmes, J.K. Nicholson, A.W. Darzi & J.C. Lindon (Eds): Metabolic Phenotyping in Personalized and Public Healthcare. DOI: http://dx.doi.org/10.1016/B978-0-12-800344-2.00002-1

FIGURE 2.1 Left, An image of a "urine wheel" taken from the book, identified as NKS 84 b folio, in the collection of the Royal Library, Copenhagen, Denmark. Right, An image of a "urine wheel" published in Ullrich Pinder's *Epyphanie Medicorum*, 1506. We acknowledge the Royal Library, Copenhagen, Denmark, for providing these images.

used by the doctors surrounding the British king George III, who was diagnosed with a form of madness, which was diagnosed partially on the basis of such a urine analysis. It is possible that he, in fact, had an inborn genetic error that resulted in the disease known as *porphyria*, which, among other symptoms, caused a dark bluish coloration of urine, although this postulation has recently been questioned and a psychiatric disorder proposed.

In the 1820s, a physician named Richard Bright, who was working at Guy's Hospital in London, first characterized a condition later named after him (Bright disease, now known to be acute or chronic nephritis). He worked with other pioneering physicians (such as Addison and Hodgkin), but it was his work on the causes and symptoms of renal diseases, as well as on the properties of urine, that led him to be considered the "father of nephrology" [1]. Indeed, even as clinical chemistry, based on qualitative and quantitative analyses, gained importance in the 19th century, consideration of factors such as the appearance (color) of urine, its odor, and the presence of turbidity for diagnostic purposes was still being discussed; for example, such a discussion is found in Neubauer and Vogel's book *On the Urine* published in 1863 [2].

2.2 THE 20TH CENTURY

As advances in analytical techniques gained speed in the 20th century, with the development of more advanced analytical chemistry methods, particularly the various forms of chromatography and electrophoresis, it became easier for scientists to identify specific substances in body fluids, and diagnosis then became more molecular-based. As a result, the presence of individual molecules in complex biofluids such as urine could then be characterized and a diagnosis could be made. One of the first examples of this multianalyte analysis came from the work of scientists such as Dent and Dalgliesh, who used two-dimensional paper chromatography to separate the metabolites in urine, which were visualized by using a variety of reagents, and produced a so-called "map of the spots" [3,4]. With the use of this technique, it was a relatively simple matter to discover the metabolites associated with a range of disorders (Fanconi syndrome, Hartnup disease, argininosuccinic aciduria, homocystinuria, cystinuria, etc.) [3,4].

By 1970, the technique of packed column gas chromatography (GC) was widely available and was being used to separate relatively volatile substances from complex mixtures. Linus Pauling et al. deployed this for analysis of the volatile components of urine and breath with a view to using the techniques for the detection of signatures of disease, noting that "sickness in human beings, whether it be from malfunctions of organs and tissues or from invasions of viruses or microorganisms, would be reflected in qualitative and quantitative changes in the metabolites emitted." They went on to say, "It seems that GC methods are now sensitive enough and have enough resolution and reliability to be useful as a diagnostic tool for indications of malfunctions in human beings" [5]. These ideas were developed in a number of publications, in which they showed GC profiles containing around 250 and 280 components for breath and urine volatiles, respectively [6]. The concept was further developed in a number of studies using a range of analytical techniques (GC and liquid chromatography (LC), mass spectrometry (MS)) and demonstrating that changes in the relative amounts of the various metabolites could be seen and, using a pattern recognition approach, be statistically associated with a range of endpoints such as diurnal variation, dietary intervention, and gender in normal individuals and disease profiles for patients with conditions such as multiple sclerosis, breast cancer, Duchenne muscular dystrophy, olivoponto cerebellar degeneration, Huntington disease, and then as so-called in the publication mental retardation, as well as use of oral contraceptives in healthy females [6]. These studies also employed amino acid analysis of urine via ion-exchange chromatography and used packed column GC for derivatized urine constituents as well as volatiles. The sample types analyzed in this research covered urine, breath, blood, and saliva and involved statistical analysis of data to find patterns of metabolites that are characteristic of the physiologic states under study [7]. However, despite its promise, the approach was not adopted by other groups and was largely forgotten when the approach was reinvented with newer technologies.

Around the same time as Pauling's study was being conducted, scientists at the Oak Ridge National Laboratory in the United States also developed what we would now refer to as methods for untargeted metabolic phenotyping using ion-exchange LC and were able to separate several hundred metabolites. Again, although these methods were technically advanced, the ideas were ahead of their time and did not become widely adopted [8–10]. The shortcoming of this early work using GC or LC was that, although the methods provided peak fingerprints, identification of potential "biomarkers" was complicated because robust interfaces with instruments capable of providing structural information, such as mass spectrometers, were yet to be developed (particularly for LC-based methods).

One of the earliest studies, even preceding Pauling's work in the early 1970s, was a study in 1967, in which nuclear magnetic resonance (NMR) and infra-red spectroscopies were used to identify a key urinary metabolite from an inborn error of metabolism [11]. In parallel, in 1974, GC-MS had been used to identify the metabolites in the urine of patients suffering from a whole range of inborn errors of metabolism [12]. Such studies on metabolic disorders mostly involved isolation of the endogenous compounds of interest prior to analysis of the purified component. Also, as early as 1968, NMR spectroscopy had found a niche in the area of identification of drug metabolites in synthetic solutions and of isolated metabolites in biofluids, predominantly urine [13].

However, by the early 1980s, pulse-Fourier transform proton NMR spectroscopy comprised a highly sophisticated set of techniques that was commonly used for the identification of small molecules by chemistry laboratories, and it was beginning to be used for the determination of the three-dimensional structures of proteins.

The possibility of using NMR spectroscopy for analysis of the multiple biochemicals present in biofluids such as urine and blood plasma was recognized by several people, including Jeremy Nicholson and Peter Sadler at Birkbeck College, University of London, [14–17]; Richard Iles at the London Hospital Medical School; and Richard Chalmers at the Medical Research Council (MRC) Clinical Research Centre [18,19]. This early work initiated a flow of reports on the identification of small molecule components in biofluid samples, noting the lipoprotein profiles in serum and plasma, observing altered metabolic profiles in certain diseases, and picking out drug metabolite peaks in urine and then identifying them. Spectroscopic profiling of inborn errors of metabolism was developed by Iles [18], who used both GC-MS and NMR platforms, as well as by Wevers [20] and Chalmers [21]; this profiling method was used to augment clinical diagnosis in many medical centers around the world. The report by Nicholson et al. in 1984 [16] described the first NMR spectroscopic profiling of the urine and serum of subjects with diabetes. An early example of an NMR spectrum of human urine measured at 400 MHz is shown in Fig. 2.2, which shows the relatively sparse information content in the spectrum compared with that obtained by a modern instrument.

In particular, Nicholson, often in collaboration with various pharmaceutical companies, carried out many studies of model toxins to elucidate their mechanisms of toxicity based on the altered urinary, bile, or plasma metabolite profiles, based on ^{1}H NMR spectra, and this area of effort was later reviewed [22]. This required the use of the rapidly developing two-dimensional NMR techniques to identify the metabolites [17]. Other groups in the second half of the

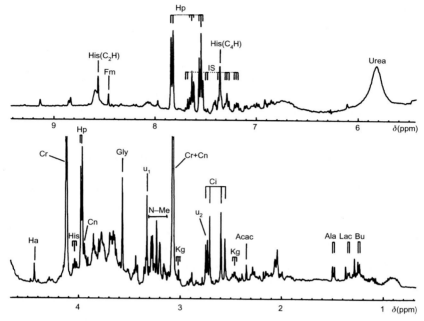

FIGURE 2.2 400 MHz ^{1}H NMR spectrum of human urine published in 1984. Water peak suppression was by continuous irradiation. A total of 232 FIDs were summed, and a 1 Hz line broadening factor was applied. Assignments taken from the paper are—*Cn*, creatinine; *Ci*, citrate; *Hp*, hippurate; *Gly*, glycine; *Cr*, creatine; *Bu*, 3-D-hydroxybutyrate; *Acac*, acetoacetate; *Ala*, alanine; *Lac*, lactate; *His*, histidine; *IS*, indoxyl sulfate; *Fm*, formate; *Ha*, dihydroxyacetone; *Kg*, α-ketoglutarate. *Reproduced with permission from Ref. [14].*

1980s were also studying other biofluids such as cerebrospinal fluid, seminal fluids, synovial fluid, and many others, and this work has been reviewed and summarized [23]. Of the fluids being studied using NMR spectroscopy, of note were cerebrospinal fluid [24], synovial fluid [25,26], aqueous humor [27], and seminal fluid [28,29].

In the description of early NMR work on biofluid analysis, the pioneering work of Iain Campbell et al. at Oxford University should not be forgotten; they studied red blood cells, the metabolites in them, and the transport of small molecules in and out of the cell by using paramagnetic reagents to edit out peaks from the inside or outside regions of the cells [30,31]. This mainly required the use of spectra edited on the basis of proton T2 spin relaxation times, giving spectra called *spin-echo spectra*. This spin-echo approach to editing NMR spectra was to have great usefulness in future studies of biofluids, especially blood plasma and serum, where it has been widely used to attenuate the interfering broad peaks from proteins and lipoproteins.

At the same time, others recognized the usefulness of such advanced approaches for analyzing the components of complex mixtures, and early studies on many natural mixtures such as wines [32], food extracts [33], and olive oils [34], for example, were published. In some cases, natural abundance ^{13}C NMR spectroscopy was also used.

With such cross-fertilization from other application areas, MS was also being employed to study complex mixtures. Jan van der Greef et al. at Leiden, The Netherlands, were among the first to obtain MS data on bacterial cultures and drug metabolites in biofluids [35,36], and profile the fluids themselves from controls and subjects with various conditions [37].

By the mid-1980s, the use of computers for analyzing spectroscopic data was becoming accepted. For example, the Dendral program was being developed to enable automatic interpretation of NMR spectra, which allowed identification of functional groups and molecular fragments [38]. Similar software was also being developed for molecular identification in MS. In parallel to such so-called artificial intelligence methods, approaches for performing multivariate statistical analysis on complex data sets had been developed, notably the use of principal components analysis and nonlinear, or Shannon, mapping pioneered by Svante Wold et al. in Sweden [39]. When such methods are applied to chemical data, the technique is termed *chemometrics*.

At the same time, John Lindon et al. in the pharmaceutical company Wellcome Foundation Ltd at their research laboratories in Beckenham, Kent, in the United Kingdom, were applying chemometrics methods to the large number of molecular parameters that could be calculated by using quantum mechanical calculations and similar calculations as part of the drug discovery effort to find those properties linked to the required biological activity.

In December 1987, these various strands came together in a serendipitous manner. John Lindon was asked to organize the NMR group meeting in London and invited Jeremy Nicholson to give a talk on his work on NMR spectroscopy of biofluids. Later, in a private discussion, it became clear that the methods being developed at Wellcome might be applicable to the complex NMR spectra to find out if chemometrics (or pattern recognition methods, as they were then called) would be useful for classifying biofluid spectra in terms of the toxicity of the animals from which they came. This started the collaboration between Nicholson and Lindon, and in order to investigate the approach, Wellcome (through Professor Sir Salvador Moncada, then Director of UK Research) funded a postdoctoral fellowship for 2 years. Kevan Gartland, who had just completed his doctoral studies with Nicholson, was appointed to the position and carried out the first studies; the success of the approach was proved, with the first report published in 1990 [40] followed quickly by other major publications [41–43]. The conjugation of chemometrics with spectroscopic profiling revolutionized the field, and most publications in the area now use some form of univariate or multivariate modeling. An early example of the use of chemometrics to classify samples according to their target organ of toxicity in the rat, based on ¹H NMR spectra of urine, is given in Fig. 2.3. The initial work was followed up by the work by Maria Anthony, a jointly supervised PhD student, and Elaine Holmes who was just completing her PhD work. This work showed that multivariate metabolic trajectories could be constructed on the basis of NMR spectra of rat urine, described the development and remission of kidney lesions [44], and elucidated the notion that the technology could be used to monitor organisms through dynamic metabolic processes, ultimately leading to the idea of the patient metabolic journey now coming to fruition.

In the 1990s, more examples followed, and the techniques used were greatly refined and developed. The term *metabonomics* was coined by Nicholson and

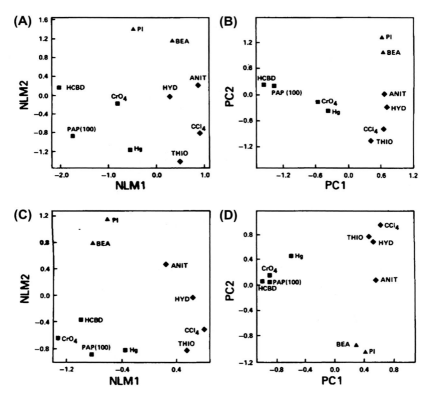

FIGURE 2.3   Nonlinear maps (A and C) and plots of the first two PCs (B and D) for NMR-derived data sets from urine, displaying renal and hepatic toxins. Values for the first three PCs are: PC1 = 41%, PC2 = 31%, and PC3 = 10% (B) and PC1 = 43%, PC2 = 40%, and PC3 = 7% of the total variance (D). Data used to generate (A) and (B) were autoscaled before analysis, whereas data used to generate (C) and (D) were not autoscaled. All parameters were equally weighted. The nephrotoxins are: *Hg*, mercuric chloride; *PAP*, p-aminophenol; *HCBD*, hexachlorobutadiene; *CrO₄*, chromate; *PI*, propyleneimine; *BEA*, bromoethanamine; and the hepatotoxins are *ANIT*, a-naphthyl-isothiocyanate; *CCl₄*, carbon tetrachloride; *HYD*, hydrazine; *THIO*, thioacetamide; plus the testicular toxin *Cd*, cadmium. *Reproduced with permission from Ref. [42].*

his collaborators, including Everett, then at Pfizer; the first public mention of the term was in the abstract of the plenary lecture given by Nicholson at the ISSX meeting at Hilton Head Island, South Carolina, in October 1997, and a paper setting out the definition of the field was subsequently published in 1999 [45]. An alternative term, *metabolomics*, was used by Oliver and Kell at the University of Manchester, initially in relation to metabolic studies of the consequences of modifying the yeast genome [46]. Both terms have become accepted, but with the formation of the Metabolomics Society and the *Metabolomics* journal, it is perhaps inevitable that *metabolomics* has become the more widely used term. However, it seems that both are now being replaced by the term *metabolic phenotyping*.

In parallel, a great deal of effort was expended on plant metabolism by several research groups, of note by Fiehn et al. in Germany. Fiehn later moved to the University of California (UC) Davis, and published an important paper in *Nature* in 2001 [47], describing the area that came to be known as functional genomics, connecting genes to endpoint metabolism. One technique that was especially widely used in plant metabolism studies was GC-MS, and a large database was established in Gölm, Germany, and later at UC Davis [48]. The application of LC-MS to plant metabolism was pioneered at the same time by several groups, including those of Wolfender [49] and Verpoorte [50].

A complementary technique for metabolic analysis, capillary electrophoresis, with and without MS detection, was developed in the early 2000s by several groups, including those of Britz-McKibbin at the Himeji Institute of Technology, Japan [51], Tomita at Keio University in Japan [52], and Barbas in Madrid, Spain [53].

Over the 15 years from 1990 to 2005, many papers describing metabolic profiling were published in a wide range of fields, including microbiology [54], invertebrates [55–57], food analysis [58], and animal biofluids for monitoring drug toxic effects [59,60], and small studies were conducted on human biofluids, mainly classifying different types of diseases and comparing patterns from those with diseases with those from control subjects [61,62]. Initially, these were mostly based on ^{1}H NMR spectroscopy, but in later years, more studies used MS, usually involving a high-performance liquid chromatography (HPLC) or GC separation stage first. In parallel, the use of magic angle spinning NMR spectroscopy of tissue samples was also developed and applied [63,64], and this has found several clinical applications, perhaps the most successful being in the analysis of tumors.

Essentially all of these studies were proof-of-concept investigations and were used as hypothesis-generating exercises. In fact, few of these studies tested the proposed hypotheses, and even fewer were repeated in an independently collected sample set for proper validation of any "biomarkers." Moreover, many published studies used chemometrics methods to classify samples (eg, according to disease vs health), but often the statistical models were not properly validated, and the models were overfitted and given overoptimistic predictive power. As a consequence, many of the biomarkers discovered have turned out to be less than rigorous.

By the early 2000s, a number of larger-scale investigations had begun to be undertaken in several areas, and these served to establish metabolic phenotyping as a mainstream approach. These included a large consortium project on drug toxicity known as the COMET project; the increasing use of metabolic profiling in epidemiologic studies also examined the links between metabolic biomarkers and disease risks in populations, initially with the INTERMAP project. The COMET project is described in more detail in the next section, and it provides a good example of what can be achieved in a productive academic–industrial collaboration.

In addition, several major projects were initiated with the aim of establishing a definitive metabolic composition of biofluids, mainly urine and serum or plasma, although some also encompassed cell systems. These included the HUSERMET consortium at the University of Manchester, with industrial sponsorship, to define the metabolite composition of human serum [65]; Lipid MAPS (Lipid Metabolites and Pathways Strategy) at the University of California to define the lipid components in the macrophage as well as other systems [66]; and efforts by Wishart et al. in Alberta, Canada, to define urine and serum "metabolomes" [67,68].

Several large industrial companies manufacturing nutritional products recognized the importance of metabolic profiling; it is worth noting the early participation of Unilever and Nestle in academic collaborations to assess the usefulness of the approach. Also, in the field of nutritional research, the Nutrigenomic Organization (NuGO) consortium has been heavily involved in metabolic profiling. NuGO is an association of academic organizations focusing on development of the molecular aspects of nutrition research, personalized nutrition, nutrigenomics, and nutritional systems biology [69].

2.3   THE COMET PROJECT FOR DRUG TOXICITY

The so-called attrition or failure of drug candidate compounds in pharmaceutical discovery pipelines has been, and still remains, a major problem. Although there are many reasons for such failures, one is the incidence of adverse effects (toxicity manifested in various target organs) in animal models prior to administration of the candidate into humans (in whom additional unexpected or idiosyncratic problems can be observed). The importance of postgenomic technologies for improving the understanding of the adverse effects of drugs was highlighted, and the inclusion of metabonomics in these approaches was recognized [70]. Although there was already a comprehensive literature on the use of metabonomics to investigate xenobiotic toxicity and this had been reviewed in 2004 [22], a rigorous and comprehensive evaluation was considered to be of considerable value. To this end, a consortium was formed to investigate the utility of NMR-based metabonomic approaches to the toxicologic assessment of drug candidates. The main aim of the consortium was to use ^{1}H NMR spectroscopy of biofluids (and, in selected cases, tissues), with the application of computer-based pattern recognition and expert system methods to classify the biofluids in terms of known pathologic effects caused by administration of substances causing toxic effects [71]. The project was hosted at Imperial College London, in the United Kingdom, and involved funding by six pharmaceutical companies, namely, Bristol-Myers-Squibb, Eli Lilly and Co., Hoffmann-La Roche, NovoNordisk, Pfizer Incorporated, and The Pharmacia Corporation (some of whom later ceased to exist as a result of corporate mergers and takeovers).

The main objectives of the project were to provide a detailed multivariate description of normal physiologic and biochemical variations of metabolites in urine, blood serum, and selected tissues, for primarily selected male rat and mouse strains, based on ^{1}H NMR spectra; the aim was (1) to produce a database of ^{1}H NMR spectra from animals dosed with model toxins, initially concentrating on liver and kidney effects, (2) to develop an expert system for the detection of the toxic effects of xenobiotics based on a chemometric analysis of their NMR-detected changes in biofluid metabolite profiles, (3) to identify combinations of biomarkers of the various defined classes, and (4) to test the methods to assess the ability of metabonomics to distinguish between toxic and nontoxic analogs and to assess the specificity of the predictive expert systems.

The classes of chemicals used and the types of toxicity investigated were selected to be as diverse as possible to assist the validation of NMR methods for use in early "broad" screening of candidates for toxicity. In all, a total of 147 in vivo 7-day toxicity studies were carried out in the partnering pharmaceutical companies, with all of them taking an equal part. The selection of the toxins for study was carried out at a Steering Committee level and a unanimous agreement was reached. The studies using a single strain of rat followed a highly detailed protocol, which was eventually agreed upon after much discussion by all companies. In certain cases, some studies were also repeated in a single mouse strain because of the known differences in toxicity that can be manifested by two species. In addition to each company running its own studies, all companies ran one compound, the liver toxin hydrazine, to check the reproducibility that could be achieved in animal handling, sample collection, and thus sample composition across all the sites in the United Kingdom, mainland Europe, and the United States.

The project arose from the recognition that single markers of toxicity are unlikely to be of value in the detection of subtle lesions that are the most problematic in primary toxicology studies on novel drugs. However, such lesions would cause multiple low-level disruptions of intermediary metabolites in biofluids which are indicators of the site and mechanisms of damage. Pattern recognition and multivariate statistical analysis of NMR spectra offered a realistic prospect of identifying these novel combination biomarkers of toxic effect for use in toxicologic screening at the discovery/development stage. In COMET, the scope of the NMR/pattern recognition approach was broadened, and an

NMR-based expert system suitable for "high-throughput" in vivo toxicology screening was devised. It was hoped that it would generate new markers of drug toxicity in vivo and further the understanding of the systemic toxic mechanisms and biochemical effects of novel drugs. The flowchart of the whole operation is shown in Fig. 2.4.

It is worth noting that at this time the US Food and Drug Administration (FDA) was publicizing on its web pages that, through its National Center for Toxicological Research, it was evaluating metabonomics as a tool for leveraging the development of new medicines and that the FDA's Vascular Injury Expert Working Group was exploring the use of biomarkers via genomics, proteomics, and metabonomics. Given both the increasing numbers of publications from different laboratories in which metabonomics had been used to probe xenobiotic toxicity and the interest shown by the pharmaceutical industry through the COMET project and by other means, it is clear that metabonomics could play an increasing role in drug safety evaluation and reduction in drug candidate attrition.

A useful supplementary result was an evaluation of the analytical and biological variations that might arise through the use of metabonomics, and a high degree of robustness was demonstrated. As an illustration, Fig. 2.5 shows a principal component scores plot based on the ^{1}H NMR spectra measured at both

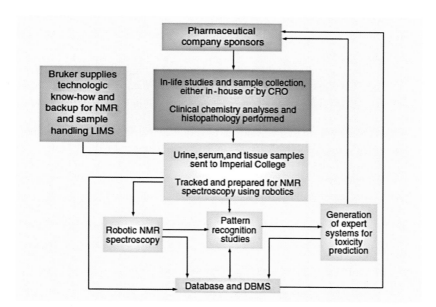

FIGURE 2.4 Operational flowchart for the COMET project. *COMET*, Consortium for Metabonomic Toxicology; *CRO*, contract research organization; *DBMS*, database management system; *LIMS*, laboratory information management system; *NMR*, nuclear magnetic resonance. *Reproduced with permission from Ref. [72].*

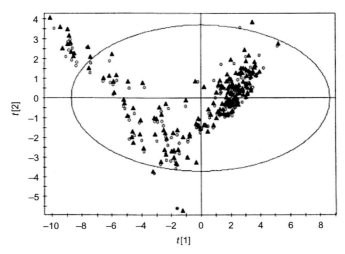

FIGURE 2.5 An example of the high degree of biochemical consistency between NMR spectra measured at Imperial College London and at Roche Pharmaceuticals on samples from the COMET hydrazine toxicity study. This is a plot of the first and second PC scores for all NMR spectra. Any given triangle and adjacent circle represent NMR spectra measured on the same samples at Imperial and Roche, respectively. The ellipse denotes the 95% significance limit. Samples appear as pairs showing similarity of biochemical profiles. The asterisk denotes a sample which gave an anomalous spectrum when measured automatically at the Imperial College London because of an instrumental artefact. *Reproduced with permission from Ref. [71].*

Imperial College London, and Hoffmann-La Roche Pharmaceuticals in Basel, Switzerland. The close pairing of the split samples measured at two different sites on two different NMR systems is clearly seen, indicating the robustness of the technology.

With the completion of 147 studies, the chief deliverables of a curated database of rodent biofluid NMR spectra and computer-based expert systems for the prediction of kidney or liver toxicity in rat and mouse, based on the spectral data, were generated and delivered to the sponsoring companies [72]. The project, with its relatively modest resources, met and exceeded all of its goals and was deemed a resounding success by the sponsoring companies, which made use of the acquired data and knowledge in their in-house studies.

After the COMET project was completed, a second phase denoted as COMET-2 was initiated. This involved a consortium of some of the pharmaceutical companies from COMET (Pfizer, BMS) as well as others that joined (Sanofi-Aventis, Servier), and the main aims of the project were to investigate in detail the biochemical mechanisms of drug toxicity for a model renal toxin and a hepatic toxin. The renal toxin was bromoethanamine, which attacks the renal medulla and had an unknown mechanism. The liver toxin was galactosamine, which also had an unknown mechanism of action and which had been shown previously to give highly variable levels of toxicity, from lethal to having no effect at all in some animals; in fact, so variable was the response in different laboratories and among different animals in the same laboratory that some reviewers suggested that the variation came from incompetence in dosing the animals.

The project discovered many new metabolites of bromoethanamine through the combined use of NMR and MS and gave new insight into the biochemical mechanism of this medullary toxin [73]. For galactose, the mechanism was elucidated [74], and the whole question of why some animals responded to the toxin and others did not was resolved by using a double dosing experiment with a washout period in between, in which it was shown that nonresponders after the first dosing turned into enhanced responders after the second dose. This showed that the effect was not genetic in nature but, in fact, arose from differences in the gut microbial flora [75].

2.4   ATTEMPTS TO DEFINE BIOFLUID COMPOSITION AND STANDARDIZE PROCEDURES

Over time it became clear that having a well-defined description of the metabolic composition of biofluids could, in some cases, be highly useful. To this end, a number of consortia were initiated, and several laboratories attempted to provide a complete biofluid metabolome.

Of note is the HUSERMET consortium, coordinated at the University of Manchester in the United Kingdom, with sponsorship from GlaxoSmithKline and AstraZeneca pharmaceuticals. Its aim was the development of methods for metabolic characterization and determination of the human serum metabolome. It has made all of its standard operating procedures available on its website and has used Fourier transform infrared spectroscopy, GC-MS, and LC-MS to define the serum metabolome. The definitive paper on this appeared in 2015 in the *Metabolomics* journal [76].

Also, the Lipid MAPS project, hosted at the University of California San Diego (UCSD), was a multi-institutional effort created in 2003 to identify and quantitate, using a systems biology approach and MS, all of the major and many minor lipid species in mammalian cells, as well as to quantify the changes in these species in response to perturbation [66].

Wishart et al. at the University of Alberta, Canada, have made great progress in defining the human serum metabolome by using a combination of analytical platforms, namely, NMR spectroscopy, GC-MS and LC-MS [77] and the human urine metabolome [78]. For the latter example, they employed NMR spectroscopy, GC-MS, direct flow injection MS, inductively coupled plasma MS, and HPLC experiments performed on multiple human urine samples. This multi-platform analysis enabled the identification of 445 urine metabolites and the quantification of 378 urine metabolites.

The first attempt to standardize the handling of biofluid samples, the collection of analytical chemistry data, and the reporting of results was initiated by Imperial College London, where a group was formed and named Standardised Metabonomics Reporting Structures. This was an informal collection of academics; researchers from pharmaceutical, analytical technology, and nutritional companies; and bioinformatics experts, who came together in two meetings to agree on minimum standards in these areas, analogous to the MIAME (Minimum Information About a Microarray Experiment) initiative in transcriptomics. This resulted in a set of guidelines that were published in *Nature Biotechnology* [79], and ultimately the process was taken up by the Metabolomics Society after initiation of working groups at a conference organized by the US National Institutes of Health (NIH). The output of this Metabolomics Standards Initiative (MSI) has been a series of papers on sample handling, sample analysis, and bioinformatics [80].

Finally, the continuing and increasing availability of online databases of metabolic information should be recognized. Of particular use are the Human Metabolome Database (HMBD) at the University of Alberta, Canada [81], which has a huge amount of background information on metabolites and, in many cases, NMR and MS data; and the METLIN database at the Scripps Center for Metabolomics, California [82], which contains MS data. In addition, the Biological Magnetic Resonance Bank (BMRB) database at the University of Wisconsin, Madison, contains one- and two-dimensional NMR spectra of many metabolites [83].

It now increasingly recognized that primary metabolic phenotyping data should be deposited in a database for general access so that other researchers can carry out their own analyses in their specialist fields. Two examples are the MetaboLights database hosted at the European Bioinformatics Institute in the United Kingdom [84] and the Metabolomics Workbench, funded as part of the NIH Core Center funding at UCSD [85].

2.5   EARLY EFFORTS AT INTEGRATING METABOLIC PHENOTYPING DATA WITH OTHER "-OMICS" DATA

One long-term aim in the field of metabolic phenotyping has been to integrate or cross-correlate data on subjects with results from proteomics, transcriptomics, or genomics in order to gain insight into the biochemical mechanisms underlying a disease process or other perturbation. In addition, integrating the metabonomics results with the metagenomics of gut bacteria has been a field of interest.

One of the first attempts was made in a preclinical toxicity study, where, in order to gain novel insight into the molecular mechanisms underlying hydrazine-induced hepatotoxicity, mRNAs, proteins, and endogenous metabolites were identified; these metabolites were altered in rats treated with hydrazine, compared with untreated controls, in a combined transcriptomics, proteomics, and metabonomics study. A single dose of hydrazine caused gene, protein, and metabolite changes that could be related to glucose metabolism, lipid metabolism, and oxidative stress [86].

Later, a novel approach was used to elucidate proteomic–metabonomic correlations in a human tumor xenograft mouse model of prostate cancer. Parallel 2D-DIGE proteomic and ^{1}H NMR metabolic profile data on blood plasma were collected from mice implanted with a prostate cancer xenograft and from matched control animals. To interpret the xenograft-induced differences in plasma profiles, multivariate statistical algorithms, including orthogonal projection to latent structure, were applied to generate models characterizing the disease profile [87].

Finally, the importance of the gut microbiome genes has been recognized, and changes in this metagenomics profile, together with metabolic phenotype changes in humans, have been assessed. Humans have evolved to develop symbiotic relationships with a consortium of gut microbes, and changes in the microbiome can influence host health and affect drug metabolism, toxicity, and efficacy. However, the molecular basis of these microbe–host interactions and the roles of individual bacterial species are still largely unknown. A combination of spectroscopic, microbiomic, and multivariate statistical tools was used to analyze fecal and urinary samples from seven Chinese individuals (sampled twice) and to model the microbe–host metabolic connectivities. At the species level, differences were found between the gut microbiomes in the Chinese subjects and those reported for American volunteers. As an example of the possible linkage, *Faecalibacterium prausnitzii* population variation was associated with modulation of eight urinary metabolites of diverse structures, indicating that this species is a highly functionally active member of the microbiome, influencing numerous host pathways [88].

Since then, many such studies of increasing complexity have been reported, culminating in the genome-wide association studies (GWASs), as described later.

2.6   POPULATION SCALE STUDIES AND BIOMARKERS OF DISEASE RISK

One huge challenge is how best to extract maximal and highly valuable information from thousands of samples taken from epidemiology and biobank repositories and use it to better understand population-based factors in health and disease. This process is complicated by the fact that humans are highly variable with regard to lifestyles and diets and often are not open or truthful when answering questionnaires. Another problem encountered in these studies is the presence of drugs and their metabolites as well as other environmentally present materials. A huge increase in complexity is caused by the significant variability in gut microflora among individuals and the complex, two-way biochemical relationship between the microflora and the host. Nevertheless, metabolic phenotyping of epidemiology cohort biofluid samples has been successful with the derivation of population-wide risk factors and the comparison of such outcomes with information from GWASs.

In the first large-scale investigations of human urine, two studies used metabolic profiling by ^{1}H NMR spectroscopy to study the metabolic variations within and among four distinct human populations—from China, Japan, the United States, and the United Kingdom; the information was collected as part of the INTERMAP project [89]. The population differences could be related to documented differences in diet and diet-related risk factors and also to cardiovascular disease [90]. The metabolites that discriminated between the populations were then also linked to data on blood pressure, and four metabolic biomarkers were determined [91]. Some of these could be related to diet and to gut microbial function.

Following this approach, a number of large cohort investigations used both untargeted and targeted metabolic assay methods involving both NMR spectroscopy and MS. It seems useful to summarize these briefly here because they indicate the scale of what is now possible and the desirability of establishing dedicated phenome centers.

A preliminary study using a large number of samples examined plasma specimens by using NMR spectroscopy to discover subjects at high risk of death in the short term [92]. To help identify biomarkers for all-cause mortality and to enhance risk prediction, over 100 candidate biomarkers were quantified by NMR spectroscopy of nonfasting plasma samples from a subset of the Estonian Biobank using nearly 10,000 samples. Significant biomarkers were validated and incremental predictive utility assessed in a separate 7500-strong population-based cohort from Finland. After adjusting for a range of conventional risk factors, four circulating biomarkers predicted the risk of all-cause mortality among the participants: (1) α-1-acid glycoprotein, (2) albumin, (3) very-low-density lipoprotein particle size, and (4) citrate. All four biomarkers were predictive of cardiovascular mortality, as well as death caused by cancer and other nonvascular diseases.

Similarly, GC-MS and LC-MS analyses of serum have been used to assess of risk of chronic kidney disease (CKD) in African Americans [93]. Many metabolites could be associated with kidney function, including glomerular flow rate, and two possible biomarkers were candidate risk factors for CKD, namely, 5-oxoproline and 1,5-anhydroglucitol.

Quantitative NMR-based metabolic phenotyping has been employed to identify the biomarkers for incident cardiovascular disease during long-term follow-up [94]. Biomarker discovery was conducted in the National Finnish FINRISK study on over 7000 samples. Replication of the discovered biomarkers and incremental risk prediction was assessed in approximately 6000 other samples from two other repositories. This study used a targeted analysis of 68 lipids and metabolites using MS, and 33 features were associated with incident cardiovascular events after adjusting for confounding parameters, that is, age, gender, blood pressure, smoking, diabetes mellitus, and medication. After further adjustment for routinely measured blood lipids, four metabolites were associated with future cardiovascular events, and higher quantities of serum phenylalanine and mono-unsaturated fatty acids were associated with increased cardiovascular risk, whereas higher ω-6 fatty acids and docosahexaenoic acid levels were associated with lower risk. The biomarker associations were further corroborated using MS in about 3000 samples from two further independent cohorts.

Another study was aimed at finding metabolic biomarkers of raised blood pressure, including the effects of lipid-lowering and antihypertensive drugs [95]. This reported a metabolome-wide association study with 295 metabolites in human serum from 1762 participants of the KORA F4 (Cooperative Health Research in the Region of Augsburg) study population. The intention was to find variations of metabolite concentrations related to the intake of the various drug classes and to generate new hypotheses about the expected and unexpected effects of these drugs. For β-blockers, 11 metabolic associations were discovered, whereas for other drugs, lesser numbers were found as follows: angiotensin-converting enzyme (ACE) inhibitors (4), diuretics (7), statins (10), and fibrates (9). For β-blockers, significant associations were observed with metabolite concentrations that are indicative of drug side effects, such as increased serotonin and decreased free fatty acid concentrations. Intake of ACE inhibitors and statins were linked to metabolites, which provide insight into the action of the drug itself on its target, such as an association of ACE inhibitors with des-Arg(9)-bradykinin and aspartylphenylalanine, a substrate and a product of the drug-inhibited ACE. The intake of statins that reduce blood cholesterol resulted in changes in the concentration of metabolites of the biosynthesis as well as of the degradation of cholesterol.

Finally, a novel approach of linking variations in the human genome with variations in metabolic profiles can yield information on the genes and hence the pathways associated with NMR or MS peaks that have significant correlation with chosen biological and clinical endpoints. Thus it may be possible to identify those unassigned metabolites by using the known functions of the connected genes. This approach has been applied to human serum. To test this hypothesis, the first GWAS with metabolic phenotyping data was based on the quantitative measurement of 363 targeted metabolites in the serum of 284 male participants of the KORA study [96]. Associations were found between common single nucleotide polymorphisms and the metabolite levels explaining up to 12% of the observed metabolic variance. Furthermore, using ratios of certain metabolite concentrations as a proxy for enzymatic activity, up to 28% of the variance could be explained. The study identified four genetic variants in genes coding for enzymes, in which the associated metabolites clearly matched the biochemical pathways where these enzymes are active. The results suggested that common genetic polymorphisms can induce considerable variation in the metabolic make-up of the human population. The authors proposed that this might lead to novel approaches to personalized health care based on a combination of genotyping and metabolic characterization.

The approach of studying large cohorts from epidemiologic collections or biobanks has led directly to the concept of dedicated centers for metabolic phenotyping analysis (phenome centers). This is discussed in more detail in chapter "Phenome Centers and Global Harmonization" of this book. The first such the center was set up at Imperial College London, following a £10 million grant from the MRC supported by the National Institute for Health Research (NIHR). To be successful, the approach has required strict operating procedures to ensure reproducibility [97] and access to state-of-the-art technologies.

In parallel to such innovations, the field of metabolic phenotyping has profited from major advances in analytical technologies. NMR spectroscopy is now routinely used for profiling at a frequency of 600 MHz, and access to higher frequency machines is available for metabolite identification. Although such instruments are not routinely used for profiling, Fig. 2.6 shows a 950 MHz ^{1}H NMR spectrum of human urine; this should be compared with that shown in Fig. 2.2, taken at 400 MHz, to see the vastly increased amount of information present. Similarly GC-MS is now widely used in metabolic profiling studies, but more usually in a targeted fashion, for example, to assay a wide range of bile acids. The use of HPLC-MS has largely been superseded by UPLC-MS with its improved chromatographic resolution and the possibility of shortened separation times, enabling more samples to be measured in a given time.

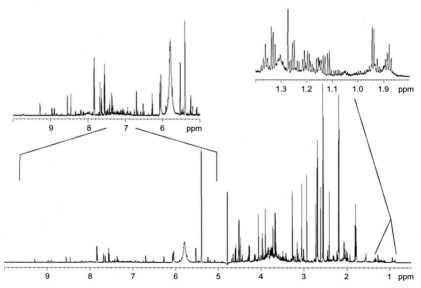

FIGURE 2.6 A 950 MHz ^{1}H NMR spectrum of human urine showing the dispersion of peaks and spectral resolution possible at this ultra-high operating frequency. The spectrum should be compared with that of a human urine sample shown in Fig. 2.2 and taken at an observation frequency of 400 MHz in the early 1980s. *Reproduced with permission from [112].*

Further details of studies in population screening and in the investigation of exposure effects are given in the chapter "Population Screening for Biological and Environmental Properties of the Human Metabolic Phenotype: Implications for Personalized Medicine" of this book.

2.7 PREDICTIVE METABOLIC PHENOTYPING AND ITS APPLICATION TO STRATIFIED MEDICINE

At the opposite end of the scale, one of the major aims of modern health care is to provide more effective therapies to minimize adverse effects and to maximize efficacy for a given individual. Although it has never really been practical to develop a drug targeted to be suitable to only a few individuals (truly personalized medicine), it is now clear that *stratified medicine*, that is, tailoring therapy to selected groups of patients based on some prognostic measure of their response, is a major goal of modern medicine and pharmaceutics. This requires systems to predict how a given patient will metabolize any administered drugs, whether there will be any adverse side effects, and whether the drugs will be efficacious. This approach extends beyond drug therapy and is also relevant to surgical and other interventions such as radiotherapy in cancer treatment, to nutritional and nutraceutical studies, or to sports training regimes. This requires a system of patient evaluation that would tell clinicians the correct drug, dose, or intervention for any individual before the start of therapy.

Most personalized approaches have so far been mainly based on measuring differences in genes related to susceptibility to cancer [98] or to polymorphisms in drug-metabolizing enzymes such as cytochrome P450 isoenzymes and *N*-acetyl transferases. There are now numerous examples of adverse drug reactions linked to specific enzymatic deficiencies or mutations, and thus precision medicine based on genetic knowledge has been the first line of attack, a paradigm known as *pharmacogenomics* [98]. Although highly successful in specific cases, mainly in oncology, the approach is limited because most major diseases involve a complex interplay between genetic and environmental influences.

An alternative approach has been developed at the level of small molecule metabolites, a method originally termed *pharmacometabonomics* (in analogy to *pharmacogenomics*) is now being used to understand and predict the outcomes (such as toxicity and xenobiotic metabolism in animal model systems) of interventional drugs, based on mathematical models derived from predose, biofluid metabolite profiles [99,100]. This was first applied to predicting variations in toxicity in an animal model [99] and then to prediction of variable drug metabolism in humans [100].

This concept has been used in a number of recent studies. Two early examples are given here, and others are summarized in a comprehensive chapter elsewhere in this book. One example is the investigation of the ability of pretreatment serum metabolic profiles to predict toxicity in patients with inoperable colorectal cancer who are treated with the single agent capecitabine [101].

Serum was collected from 54 patients with a diagnosis of locally advanced or metastatic colorectal cancer prior to treatment with capecitabine, and ^{1}H NMR spectroscopy was used to generate baseline metabolic profile data for each patient. After the chemotherapy, the patients' toxicities were graded according to National Cancer Institute Common Toxicity Criteria. Examination of the pretreatment serum metabolic profiles showed that higher levels of low-density lipoprotein–derived lipids, including polyunsaturated fatty acids and choline phospholipids predicted higher-grade toxicity over the treatment period, and more specifically, statistical analyses revealed a pharmacometabonomic lipid profile that correlated with severity of toxicity. This small study suggested that metabolic profiles could discriminate subpopulations susceptible to adverse events and that the profiles could have a potential role in the assessment of treatment viability for cancer patients prior to commencing chemotherapy.

Another small study evaluated the potential of a pretreatment metabolic phenotype to predict individual variation in the pharmacokinetics of the immunosuppressant drug tacrolimus [102]. LC-MS–based metabolic profiling was performed on 29 healthy volunteers by measuring the relative amounts of 1256 metabolite ions in predose urine samples. After oral administration of tacrolimus, its plasma concentration in these volunteers was measured for up to 72h, and the standard pharmacokinetics parameters were calculated. Chemometric modeling of the predose urine metabolic data was used to predict the postdose pharmacokinetic parameters of tacrolimus and also to identify the endogenous metabolites that contributed to that prediction. This metabolic profile allowed the generation of a clinically applicable index to predict the individualized pharmacokinetics of tacrolimus, making it a useful tool for personalized drug therapy.

A fuller explanation of the concept of pharmaco-metabonomics and a survey of all of the published literature is given in chapter "Pharmacometabonomics and Predictive Metabonomics: New Tools for Personalized Medicine" of this book.

2.8 MONITORING THE PATIENT JOURNEY

Given the state of current analytical technologies, it is now practical to screen, and follow up longitudinally, patients who present for a particular type of therapy such as cancer chemotherapy or radiotherapy. There is a long history of this type of longitudinal monitoring of a patient by using metabolic profiling [44], and possibly the first human study was a study of an unfortunate individual who fell into a vat of phenol and suffered severe renal toxicity [103]. A little later, time-course trajectories of metabolic changes observed after kidney transplantation were published [104].

Nevertheless, in a *Lancet* article, Kinross et al. proposed such a clinical phenotyping process for monitoring patient journeys before, during, and after surgery [105]. The concept is illustrated in Fig. 2.7. This indicates how patients can enter the diagnostic environment through community admission, electively, as

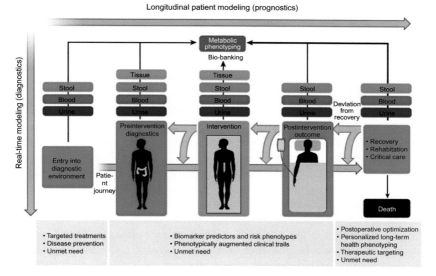

FIGURE 2.7   A scheme illustrating the paradigm of phenotyping the patient journey and of conducting phenotypically augmented clinical trials. *Initially published in Ref. [106].*

an acute case, or as an emergency. At any point in the patient journey, there are opportunities for collection of samples for metabolic phenotyping. Sections of these samples can also be stored in bio-banks prior to analysis for use in future research. Deviation from recovery can occur at any point in the patient journey, and samples can be taken again. This analysis can be used to enhance differential diagnosis, therapeutic responses, and long-term outcomes of therapy. There is also the prospect of real-time diagnosis and of prognosis to enhance clinical decision making. Before patient admission to hospital, this can be used for disease prevention and after intervention to optimize recovery. By modeling congruent longitudinal journeys by using pharmacometabonomic approaches, it is possible to derive prognostic biomarker predictors and risk phenotypes that allow patient stratification or can give mechanistic information relating to therapeutic responder or nonresponder status.

This paradigm leads to a separation between the original definition of pharmacometabonomics, which is the use of steady-state, predose metabolic profiling to predict drug metabolism, adverse effects, and efficacy, and longitudinal or dynamic pharmacometabonomics, which involves the measurement of a person's metabolic profiles prior to, during, and after some intervention in order to stratify the patient in terms of prediction of response to future treatments (eg, prognosis of the efficacy or adverse effects of chemotherapy) or the likelihood of any outcome (eg, need for intensive care). The differences in metabolic profiles for a given patient at different times can be considered a metabolic trajectory. From toxicity studies in animal models, it is known that characteristic biochemical changes are detectable in biofluids before any pathologic anatomical changes occur. As such data emerge, it is expected that the differences in metabolic trajectories among patients will provide useful information to help guide clinicians in their evaluation of different treatment options. This temporal extension to pharmacometabonomics holds the promise of significant benefits for patients in terms of optimizing efficacy and safety and for rapid responses in adjusting treatment; thus, it has implications for health service providers in terms of cost and efficiency [106]. It should also be noted that the same paradigm can be applied to drug clinical trials at both the subject/patient selection stage and during the trial itself.

The minimum numbers of samples required to build robust predictive models varies according to the application. For the human pharmacometabonomics study of paracetamol metabolism, 99 normal, healthy, nonsmoking, male subjects were used successfully [100]. For more diverse populations such as hospital in-patients, larger numbers of samples, formulated using appropriate statistical power calculations, will be required so that all of the normal and abnormal sources of human metabolic variations are well sampled [107]. In clinical longitudinal pharmacometabonomics, each patient can act as his or her own control, and it may not be necessary to provide a high level of predictive precision; often a simple traffic light system might be sufficient to alert clinicians to an unexpected event.

An extension of this approach is to monitor individual people (not necessarily patients) as a function of time and lifestyle so as to provide improved methods for personalized therapy. Although a number of pilot studies have shown promise, this approach still requires extensive validation. Here, individual diversity and environmental variability need to be factored in [108]. Comparing individuals against each other also represents a new avenue for exploration, and, indeed, research on twins has been carried out to study genetic and metabolic variations arising from heritability and environment [109].

One specific new technology that must be mentioned, since it is already having an impact in real-time metabolic analyses and is being tested in real clinical situations, is the intelligent knife, or i-Knife [110,111]. This is based on rapid evaporative ionization mass spectrometry, which provides real-time characterization of human tissue in vivo by analysis of the aerosol ("smoke") released during electrosurgical dissection. Recently [111], a study that validated the approach was published. Tissue samples from 302 patients were analyzed, resulting in data sets from 1624 cancerous samples and 1309 noncancerous samples, yielding a comprehensive database of tissue metabolic profiles. The technology was subsequently used in the operating theater, where the device was coupled to existing electrosurgical equipment to collect data during a total of 81 resections. The MS data were analyzed using multivariate statistical methods, and a spectral identification algorithm was developed. The approach differentiated accurately between distinct histologic and histopathologic tissue types, with malignant tissues yielding chemical characteristics specific to their histopathologic subtypes, as described in Fig. 2.8. Tissue identification via the intraoperative technique matched the postoperative histologic diagnosis in 100% (all 81) cases. The mass spectra reflected lipidomic profiles that varied among distinct histologic tumor types and also between primary and metastatic tumors.

Thus in addition to real-time diagnostic information, the spectra provided additional information on divergent tumor biochemistry, which might have mechanistic importance in cancer.

A separate chapter discusses the whole topic of surgical metabolic phenotyping in this book (see Chapter 4).

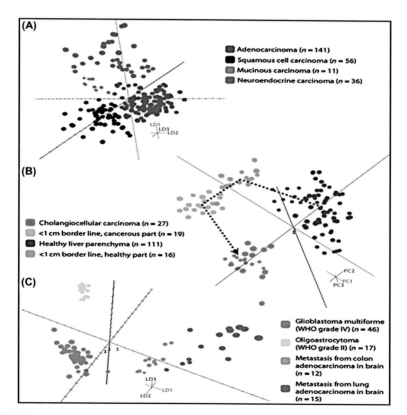

FIGURE 2.8 Multivariate statistical analysis of malignant tumor data obtained ex vivo. *n* represents the number of spectra in the training set. (A) Pseudo–three-dimensional (3D) linear discriminant analysis (LDA) plot of different human lung tumors. (B) 3D principal components scores plot of healthy liver parenchyma and hepatocellular carcinoma. Spectra obtained within 1 cm of the border lines are also depicted. The arrow demonstrates a shift from healthy liver parenchyma through tumor border into tumor tissue. (C) Pseudo-3D LDA plot of two different brain metastases and two different WHO-grade gliomas in the human brain. Numbers of patients were 40, 22, and 14 for (A), (B), and (C), respectively. *Reproduced with permission from AAAS from Ref. [111].*

2.9   PHENOME CENTER CONCEPT

The 2012 Olympic and Para-Olympic games were held in London, in the United Kingdom, and drug testing of athletes was conducted in a purpose-built laboratory. A strategy was developed to use this laboratory as part of the Olympics legacy to build a national facility for metabolic phenotyping. This vision was eventually achieved and funded by the MRC in conjunction with the NIHR, with major contributions from the NMR and MS instrument manufacturers Bruker and Waters.

The MRC-NIHR Phenome Centre, led by Imperial College London and King's College London, has the overall aim of delivering broad access to a world-class facility for metabolic phenotyping, which will benefit the whole translational medicine community in the United Kingdom. For logistical reasons, the facility was eventually located at the Hammersmith Hospital in west London (in laboratories that are part of Imperial College London) rather than at the original Olympic drug-testing site. The Centre offers a unique capability in targeted and exploratory high-throughput metabolic phenotyping and assay development, as well as a computational medicine capability for the provision of data, models, and interpretation. The Centre is set up to analyze both patient-based and population-based samples for biomarker discovery and validation. In the case of patients, the stated aim is to provide improved patient stratification through development of robust diagnostic and prognostic markers, as well as early identification of drug efficacy and safety and other responses to treatments. The Centre has state-of-the-art MS and NMR facilities and provides the means and infrastructure to advance the capabilities of researchers, both academic and industry-based, in terms of methodologies applied to the large-scale metabolic phenotyping of patient and epidemiologic samples. To ensure open and fair access to the Centre for the whole research community, it is controlled by an independent Access Committee. The Centre also has embedded in it a major international training center with world-class facilities. Many projects are already progressing through the Centre's system, including cohorts of samples used for the search for early biomarkers of diabetes, cardiovascular disease, hypertension, and Alzheimer disease.

This approach has served as a template for other phenome centers. Indeed, Imperial College London has already founded a second center, located at St Mary's Hospital, London, specifically for acute clinical studies focusing on patient journeys, and stratified medicine (see Section 2.7). These approaches are aimed at finding important translational applications for monitoring complex patient journeys in the UK NHS Hospital Trust environment such that "patient journey phenotyping" becomes a realistic prospect for improving patient care. Chapter "Phenome Centers and Global Harmonization" expands on the need for, and the philosophy of, operating such centralized phenome centers.

Another center has now been developed at the University of Birmingham in the United Kingdom as the second in a planned worldwide network of phenome

[49] Grata E, Boccard J, Glauser G, Carrupt PA, Farmer EE, Wolfender JL, et al. Development of a two-step screening ESI-TOF-MS method for rapid determination of significant stress-induced metabolome modifications in plant leaf extracts: the wound response in *Arabidopsis thaliana* as a case study. J Sep Sci 2007;30:2268–78.

[50] Choi HK, Choi YH, Verberne M, Lefeber AW, Erkelens C, Verpoorte R. Metabolic finger-printing of wild type and transgenic tobacco plants by ^{1}H NMR and multivariate analysis technique. Phytochemistry 2004;65:857–64.

[51] Britz-Mckibbin P, Nishioka T, Terabe S. Sensitive and high-throughput analyses of purine metabolites by dynamic pH junction multiplexed capillary electrophoresis: a new tool for metabolomic studies. Anal Sci 2003;19:99–104.

[52] Sato S, Soga T, Nishioka T, Tomita M. Simultaneous determination of the main metabolites in rice leaves using capillary electrophoresis mass spectrometry and capillary electrophoresis diode array detection. Plant J 2004;40:151–63.

[53] Barbas C, García A, de Miguel L, Simó C. Evaluation of filter paper collection of urine samples for detection and measurement of organic acidurias by capillary electrophoresis. J Chromatogr B Anal Technol Biomed Life Sci 2002;780:73–82.

[54] van der Werf MJ, Jellema RH, Hankemeier T. Microbial metabolomics: replacing trial-and-error by the unbiased selection and ranking of targets. J Ind Microbiol Biotechnol 2005; 32:234–52.

[55] Bundy JG, Osborn D, Weekes JM, Lindon JC, Nicholson JK. An NMR-based metabonomic approach to the investigation of coelomic fluid biochemistry in earthworms under toxic stress. FEBS Lett 2001;500:31–5.

[56] Bundy JG, Lenz EM, Bailey NJ, Gavaghan CL, Svendsen C, Spurgeon D, et al. Metabonomic investigation into the toxicity of 4-fluoroaniline, 3,5-difluoroaniline and 2-fluoro-4-methylaniline to the earthworm *Eisenia veneta* (Rosa): identification of novel endogenous biomarkers. Env Toxicol Chem 2002;21:1966–72.

[57] Phalaraksh C, Reynolds SE, Wilson ID, Lenz EM, Nicholson JK, Lindon JC. A metabonomic analysis of insect development: ^{1}H NMR spectroscopic characterisation of changes in the composition of the haemolymph of larvae and pupae of *Manduca sexta*. Science Asia 2008;34:279.

[58] Brennan L. NMR-based metabolomics: from sample preparation to applications in nutrition. Prog NMR Spectrosc 2014;83:42–9.

[59] Ament Z, Waterman CL, West JA, Waterfield C, Currie RA, Wright J, et al. A metabolomics investigation of non-genotoxic carcinogenicity in the rat. J Proteome Res 2013;12:5775–90.

[60] Li Y, Ju L, Hou Z, Deng H, Zhang Z, Wang L, et al. Screening, verification and optimization of biomarkers for early prediction of cardiotoxicity based on metabolomics.. J Proteome Res 2015;14:2437–45.

[61] Niewczas MA, Sirich T, Mathew AV, Skupien J, Mohney RP, Warram JH, et al. Uremic sol-utes and risk of end-stage renal disease in type-2 diabetes: metabolomics study. Kidney Int 2014;85:1214–24.

[62] Zordoky BN, Sung MM, Ezekowitz J, Mandal R, Han B, Bjorndahl TC, et al. Metabolomic fingerprint of heart failure with preserved ejection fraction. PLoS One 2015;10:e0124844.

[63] Cheng LL, Lean C, Bogdanova A, Wright Jr. SC, Ackerman JL, Brady TJ, et al. Enhanced resolution of proton NMR spectra of malignant lymph nodes using magic-angle spinning. Magn Reson Med 1996;36:653–58.

[64] Moka D, Vorreuther R, Schicha H, Spraul M, Humpfer E, Lipinski M, et al. Biochemical classification of kidney carcinoma biopsy samples using magic-angle-spinning ^{1}H nuclear magnetic resonance spectroscopy. J Pharm Biomed Anal 1998;17:125–32.

[65] Human Serum Metabolome in Health and Disease (HUSERMET) homepage. Available at: http://www.husermet.org/.

[66] LIPID MAPS® Lipidomics Gateway. Available at: http://www.lipidmaps.org/.

[67] Urine metabolome database. Available at: http://www.urinemetabolome.ca/.

[68] Serum metabolome database. Available at: http://www.serummetabolome.ca/.

[69] NuGo homepage. Available at: http://www.nugo.org/.

[70] Aardema MJ, MacGregor JT. Toxicology and genetic toxicology in the new era of "toxicogenomics": impact of "-omics" technologies. Mutat Res 2002;499:13–25.

[71] Lindon JC, Nicholson JK, Holmes E, Antti H, Bollard ME, Keun HC, et al. Contemporary issues in toxicology—the role of metabonomics in toxicology and its evaluation by the COMET project. Toxicol Appl Pharmacol 2003;187:137–46.

[72] Lindon JC, Keun HC, Ebbels TMD, Pearce JMT, Holmes E, Nicholson JK. The Consortium for Metabonomic Toxicology (COMET): aims, activities and achievements. Pharmacogenomics 2005;6:691–9.

[73] Shipkova P, Vassalo J, Aranibar N, Hnatyshyn S, Zhang H, Clayton AC, et al. Urinary metabolites of 2-bromoethanamine in the rat identified by stable isotope labeling: evidence for carbamoylation and glutathione conjugation. Xenobiotica 2011;41:144–54.

[74] Coen M, Hong YS, Clayton AC, Rohde CM, Pearce JT, Reily MD, et al. The mechanism of galactosamine toxicity revisited; a metabonomic study. J Proteome Res 2007;6:2711–19.

[75] Coen M, Want EJ, Clayton AC, Rhode CM, Hong YS, Keun HC, et al. Mechanistic aspects and novel biomarkers of responder and nonresponder phenotypes in galactosamine-induced hepatitis. J Proteome Res 2009;8:5175–87.

[76] Dunn WB, Lin W, Broadhurst D, Begley P, Brown M, Zelena E, et al. Molecular phenotyping of a UK population: defining the human serum metabolome. Metabolomics 2015;11:9–26.

[77] Psychogios N, Hau DD, Peng J, Guo AC, Mandal R, Bouatra S, et al. The human serum metabolome. PLoS One 2011;6:e16957.

[78] Bouatra S, Aziat F, Mandal R, Guo AC, Wilson MR, Knox C, et al. The human urine metabolome. PLoS One 2013;8:e73076.

[79] Lindon JC, Nicholson JK, Holmes E, Keun HC, Craig A, Pearce JT, et al. Summary recommendations for standardization and reporting of metabolic analyses. Nat Biotechnol 2005;23:833–8.

[80] MSI Board Members, Sansone SA, Fan T, Goodacre R, Griffin JL, Hardy NW, et al. The metabolomics standards initiative. Nat Biotechnol 2007;25:846.

[81] Human Metabolome Database. Available at: http://www.hmdb.ca/.

[82] METLIN metabolite database. Available at: https://metlin.scripps.edu/index.php.

[83] Biological Magnetic Resonance Data Bank. Available at: http://www.bmrb.wisc.edu/.

[84] MetaboLights database. Available at: http://www.ebi.ac.uk/metabolights/.

[85] UCSD Metabolomics Workbench. Available at: http://www.metabolomicsworkbench.org/.

[86] Klenø TG, Kiehr B, Baunsgaard D, Sidelmann UG. Combination of 'omics' data to investigate the mechanism(s) of hydrazine-induced hepatotoxicity in rats and to identify potential biomarkers. Biomarkers 2004;9:116–38.

[87] Rantalainen M, Cloarec O, Beckonert O, Wilson ID, Jackson D, Tonge R, et al. Statistically integrated metabonomic-proteomic studies on a human prostate cancer xenograft model in mice. J Proteome Res 2006;5:2642–55.

[88] Li M, Wang B, Zhang M, Rantalainen M, Wang S, Zhou H, et al. Symbiotic gut microbes modulate human metabolic phenotypes. Proc Natl Acad Sci USA 2008;105:2117–22.

[89] Stamler J, Elliott P, Dennis B, Dyer AR, Kesteloot H, Liu K, et al. INTERMAP: background, aims, designs, methods and descriptive statistics. J Hum Hypertens 2003;17:591–608.

[90] Yap IKS, Angley M, Veselkov KA, Holmes E, Lindon JC, Nicholson JK. Metabolome-wide association study identifies multiple biomarkers that discriminate north and south Chinese populations at differing risks of cardiovascular disease: intermap study. J Proteome Res 2010;3:6647–54.

[91] Holmes E, Loo RY, Stamler J, Bictash M, Yap IKS, Chan Q, et al. Human metabolic phenotype diversity and its association with diet and blood pressure. Nature 2008;453:396–400.

[92] Fischer K, Kettunen J, Würtz P, Haller T, Havulinna AS, Kangas AJ, et al. Biomarker profiling by nuclear magnetic resonance spectroscopy for the prediction of all-cause mortality: an observational study of 17,345 persons. PLoS One 2014;11:e1001606.

[93] Yu B, Zheng Y, Nettleton JA, Alexander D, Coresh J, Boerwinkle E. Serum metabolomics profiling and incident CKD among African americans. Clin J Am Soc Nephrol 2014;9:1410–17.

[94] Würtz P, Havulinna AS, Soininen P, Tynkkynen T, Prieto-Merino D, Tillin T, et al. Metabolite profiling and cardiovascular event risk: a prospective study of 3 population-based cohorts. Circulation 2015;131:774–85.

[95] Altmaier E, Fobo G, Heier M, Thorand B, Meisinger C, Römisch-Margl W, et al. Metabolomics approach reveals effects of antihypertensives and lipid-lowering drugs on the human metabolism. Eur J Epidemiol 2014;29:325–36.

[96] Gieger C, Geistlinger L, Altmaier E, de Angelis MH, Kronenberg F, Meitlinger T, et al. Genetics meets metabolomics: a Genome-Wide association study of metabolite profiles in human serum. PLoS Genet 2008;4:e1000282.

[97] Dona AC, Jimenez B, Schaefer H, Humpfer E, Spraul M, Lewis MR, et al. Precision high-throughput proton NMR spectroscopy of human urine, serum, and plasma for large-scale metabolic phenotyping. Anal Chem 2014;86:9887–94.

[98] Weng L, Zhang L, Peng Y, Huang RS. Pharmacogenetics and pharmacogenomics: a bridge to individualized cancer therapy. Pharmacogenomics 2013;14:15–24.

[99] Clayton TA, Lindon JC, Cloarec O, Antti H, Chareul C, Hanton G, et al. Pharmaco-metabonomic phenotyping and personalised drug treatment. Nature 2006;440:1073–7.

[100] Clayton TA, Baker D, Lindon JC, Everett JR, Nicholson JK. Pharmacometabonomic identification of a significant host-microbiome metabolic interaction affecting human drug metabolism. Proc Natl Acad Sci USA 2009;106:14728–33.

[101] Backshall A, Sharma R, Clarke SJ, Keun HC. Pharmacometabonomic profiling as a predictor of toxicity in patients with inoperable colorectal cancer treated with capecitabine. Clin Cancer Res 2011;17:3019–28.

[102] Phapale PB, Kim S-D, Lee HW, Lim M, Kale DD, Kim Y-L, et al. An integrative approach for identifying a metabolic phenotype predictive of pharmacokinetics of Tacrolimus. Clin Pharmacol Ther 2010;87:426–36.

[103] Foxall PJ, Bending MR, Gartland KP, Nicholson JK. Acute renal failure following accidental cutaneous absorption of phenol: application of NMR urinalysis to monitor the disease process. Hum Toxicol 1989;8:491–6.

[104] Foxall PJ, Mellotte GJ, Bending MR, Lindon JC, Nicholson JK. NMR spectroscopy as a novel approach to the monitoring of renal transplant function. Kidney Int 1993;43:234–45.

[105] Kinross JM, Holmes E, Darzi AW, Nicholson JK. Metabolic phenotyping for monitoring surgical patients. Lancet 2011;377:1817–19.

[106] Nicholson JK, Holmes E, Kinross JM, Darzi AW, Takats Z, Lindon JC. Metabolic phenotyping in clinical and surgical environments. Nature 2012;491:384–92.

[107] Broadhurst DI, Kell DB. Statistical strategies for avoiding false discoveries in metabolomics and related experiments. Metabolomics 2006;2:171–96.

[108] Robinette SL, Holmes E, Nicholson JK, Dumas ME. Genetic determinants of metabolism in health and disease: from biochemical genetics to genome-wide associations. Genome Med 2012;4:30.

[109] Menni C, Kastenmüller G, Petersen AK, Bell JT, Psatha M, Tsai PC, et al. Metabolomic markers reveal novel pathways of ageing and early development in human populations. Int J Epidemiol 2013;42:111–19.

[110] Guenther S, Schäfer K-C, Balog J, Dénes J, Majoros T, Albrecht K, et al. Electrospray post-ionization mass spectrometry of electrosurgical aerosols. J Am Soc Mass Spectrom 2011;22:2082–9.

[111] Balog J, Sasi-Szabó L, Kinross J, Lewis MR, Muirhead LJ, Veselkov K, et al. Intraoperative tissue identification using rapid evaporative ionization mass spectrometry. Sci Trans Med 2013;5 194ra93.

[112] Lindon J.C., Nicholson J.K., Analytical technologies for metabonomics and metabolomics, and multi-omic information recovery. Trends Anal Chem 2008;27:194–204.

Chapter 3

Phenotyping the Patient Journey

Elaine Holmes[1], Jeremy K. Nicholson[1], Jia Li[1] and Ara W. Darzi[1,2]

[1]*Department of Surgery and Cancer, Imperial College London, London, UK* [2]*Institute of Global Health Innovation, Imperial College London, London, UK*

Chapter Outline

Health care systems are among the most complex organizational and technical human constructs that are heavily regulated and thus institutionally resistant to change. The recognition of this intrinsic structural problem has contributed to the concept of "translational medicine," arising out of the need to harness the exponential growth of new technologies for real-world medical application. The aim of translational medicine is to improve the "bench to bedside" approach by developing and implementing new technologies and paradigms to improve the patient care pathway. There is also great enthusiasm to enable personalized health care programs through the use of stratified medicine approaches, where the characteristics of individual biological variations of the patients' disease (genetic and phenotypic) are used to tailor the therapeutic regimen for optimized treatments. If used effectively, personalized therapies should result

E. Holmes, J.K. Nicholson, A.W. Darzi & J.C. Lindon (Eds): Metabolic Phenotyping in Personalized and Public Healthcare. DOI: http://dx.doi.org/10.1016/B978-0-12-800344-2.00003-3
49

in improved care pathways, benefiting the individual patient and lowering health care costs by optimizing service utilization. There is also an increased emphasis on the involvement of patients in their own treatment, as this leads to improved patient experience and provides a psychological boost to aid recovery as well as improved doctor–patient relationships. Overall, the longitudinal hospital process is now viewed in terms of "the patient journey," mapping the progress from disease diagnosis and treatment to wellness, in an ideal case, and has been adopted as a model for improving health care by several governments and health authorities [1]. The patient journey begins when an individual presents at his or her general practitioner's (GP's) surgery or enters the hospital environment. The doctor or health care professional initiates a preliminary series of investigations, which may include verbal questioning about symptoms and medical history, a physical examination, and basic biochemical tests. For some patients, this triggers a second-stage investigation following referral to a specialized center for more sophisticated tests. Based on the outcome of these tests, a decision on treatment and management of the condition will be made and a course of action agreed upon. The endpoint of the patient journey may be resolution of a disease or condition with or without treatment, stabilization of a condition, or death [1,2]. The ultimate aim of the diagnostic tests applied in the patient journey is to define the nature and stage of disease; to define disease risk, where possible; and to select suitable therapies to improve the health of the patient. Ideally, the patient journey is a partnership between the clinician and the patient. Each patient journey is unique, since no two individuals are exposed to the same set of genetic and environmental factors. Moreover, the patient is the only person in the journey who provides continuity, since no other individual typically sees all parts of the journey [1], and parts of the journey may inevitably traverse horizontally through several hospital/health care departments, rather than tracking chronologically within a cohesive department. Thus a strategy that involves the patient as an interested partner is more likely to result in a successful outcome. However, clinician–patient interactions are all too often characterized by incomprehensible medical jargon, and information flow can be predominantly unidirectional, going only from the doctor to the patient. The patient, thus, experiences a "journey" over which he or she has little control; and particularly in the case of more serious disease conditions, the patient often finds the journey confusing and frightening. The anxiety and confusion experienced by the patient is compounded by lack of information about his or her particular condition and treatment options.

The model whereby GPs and highly specialized clinical experts form the building blocks of the health care system is very much a product of the modern philosophy of Western medicine. Other cultures have medical practices that are more deeply rooted in holistic treatment but lack the scientific rigor of Western medicine. For example, traditional Chinese medicine and other traditional medicine systems are based on a set of empirical observations and intrinsically adopt a personalized approach to treatment, using a series of observations and noninvasive measurements to stratify individuals into subgroups and typically allow for continuous monitoring of individuals, and the patient typically participates strongly in that journey [3]. However, since the mechanism and synergistic action of many of the active compounds in traditional medicines are not understood and because rigorous scientific evaluation of adverse patient responses to treatment have not been conducted at a population level, these practices are ultimately unsuitable for incorporation into Western medical practices. In Western medicine, there is a tendency to treat the disease rather than the patient, and the efficiency of the treatment process is constrained by both physical resource limitations as well as effective communication of information and data. In the true sense of the patient journey, the patient, rather than the disease or condition, must become the focal point, with capacity for processing multiple dynamic measurements reporting on a range of metabolic and physiologic processes and functions.

The concept of the patient journey and the appropriate mapping of that journey offer a solution to tackling some of the bottlenecks in the health care system and to improving data and information flow such that the patient experiences and perceives better and more holistic care. The term *patient journey* is often seen as being synonymous with process mapping of the events encountered by patients upon entering the hospital environment and is concerned with using the method of process mapping to identify and alleviate bottlenecks in the clinical pathway. The process of charting patient journeys also allows interactive troubleshooting and encourages better use of existing databases, which can be augmented to improve diagnosis and provide a framework for stratified or personalized health care. For example, the process can be made more efficient by identifying the diagnostic procedures that take the longest time to deliver results and batching the requests for tests or the subsequent delivery of results. Charting the patient journey is, in itself, a whole field of research, and its implementation has been taken seriously by several governments and health care providers. One of the best examples of this is the Scottish government's support for and focus on the patient journey, which culminated in a range of published guidelines and training documents [4]. *Patient journey* is also a term adopted by brand makers in the pharmaceutical industry to identify points of leverage for introducing new drugs. Here, the focus is less on the mechanical process of mapping the patient journey, but, rather, on the exploration of the potential of metabolic profiling to accompany and inform the diagnostic and therapeutic aspects of the patient journey. Many of the diagnostic tests employed by health care providers involve measurement of single or relatively few chemicals in biofluids (eg, urine or blood), obtaining images (eg, ultrasound, magnetic resonance imaging), or making slides of tissue biopsy specimens for histopathologic staining. These processes are often time consuming and costly and deliver limited information. One tool that is becoming available to the clinician is *metabolic profiling*, which utilizes high-resolution spectroscopic analyses of biofluid or tissue samples to deliver a simultaneous readout of hundreds of chemicals and can be accomplished in a short time frame. It is of note that the metabolic phenotype of an individual expressed in readily sampled compartments or liquid biopsy specimens is closely related to the site, mechanism, and severity of disease. Thus generating metabolic fingerprints of biological samples allows the metabolic phenotypes of individuals to be followed through their journeys and provides a systems-based approach to monitoring the patient pathway. Metabolic phenotyping, therefore, could be a valuable addition to current frameworks for monitoring and shaping the patient journey. It is now time to explore not only the potential but also the barriers to adopting metabolic phenotyping as a tool for enhancing the patient journey and to establish the necessary validation phases required to implement this platform in clinical settings.

3.1 BOTTLENECKS IN THE PATIENT JOURNEY

Currently, many health care systems rely on a pool of generalists such as GPs and community nurses, as well as a few physicians and surgeons who specialize in certain focus areas (eg, endocrinology, neuropathology, or cardiology) and are accustomed to working autonomously. In fact, there is an increasing trend toward medical specialization, and patients may have to see and interact with many clinicians in the course of their journey. This structuring of the medical community encourages excellence in specialist disciplines but fails to align itself with the fact that, in reality, the human body functions as a network of interconnected tissues and biological processes and seldom does a single organ or system fail in isolation. Indeed, there is strong genetic and systems biology evidence of the molecular connections between a wide range of diseases that are physically disparate in terms of signs and symptoms [5]. The deep biological lessons of systems biology have yet to be translated to medical practice. Most of the common diseases have significant environmental components. Furthermore, as individuals age and evidence of wear and tear on various tissues, organs, or metabolic networks becomes apparent, multiple pathologic conditions, which may or may not be connected, tend to be manifested. Therefore, the specialist who examines each patient with a view to correcting one aspect of a syndrome or dysregulated process can often be unaware of the totality of the medical needs of that patient. Thus the current tendency to direct patients to specialists who treat each condition in isolation, rather than taking a systemic approach and considering the consequences of interacting physiologic and pathologic processes in their totality, leads to suboptimal health care provision.

The lack of human, medical, or financial resources, typically dependent on geographic distribution, is a major contributory factor to the poor management of patient journeys. For example, bed availability in the accident and emergency (A&E) departments of hospitals is one of the biggest problems in the United Kingdom and imposes a constraint on the overall time for the in-patient component of the patient journey. The time taken to send patients to specialist wards or centers severely impacts on the quality of the patient journey. There is evidence to suggest that a shorter length of stay is beneficial to patient outcomes, as it can signal that the patient journey was managed efficiently [6]. Length of stay in hospitals is often artificially increased by (1) length of time between ward rounds, particularly where hospitals run on a Monday-to-Friday timetable; (2) in-hospital access to support services such as physiotherapy or dietitians; (3) delayed processing or batching of key diagnostic tests; (4) junior staff carrying out the main care, but still requiring advice from a consultant or senior doctor; and (5) handover to support staff in the community in cases where the patient requires continued nonacute physical or social support [7]. Thus prolonged stay in a hospital is associated with poorer patient outcomes [8] and also adds to the financial burden on the health care system.

Standard clinical assays and diagnostics can constitute bottlenecks in the patient journey. The majority of clinical assays are automated, but there are many that still require manual input, either at the sample preparation stage or at the interpretation stage. If the assays depend on trained technical staff, their workload can be the limiting factor in the sample analysis, resulting in delays before information is transferred back to the consultant and the patient. Metabolic phenotyping can deliver comprehensive readouts of the metabolic status of an individual at any given point in time and can do so on a rapid timescale with little cost, once the analytical infrastructure is in place. The spectral profiles obtained from biofluids and tissues carry information relating to potentially hundreds of metabolic pathways. The knowledge of how these pathways are behaving at a global level can help direct clinicians in making rapid and more accurate diagnoses. Recently, we have advanced the concept of the "Phenotypically Augmented Patient Journey" [9,10], in which an array of metabolic profiling technologies are aligned to follow the patient journey through routine longitudinal biofluid sampling to enable improved diagnosis, stratification, and prediction of outcomes. The focus of this chapter is to explore how metabolic profiling can contribute to the conventional clinical assay flow and to deliver information to improve the patient journey.

3.2   WHAT DOES A GOOD PATIENT JOURNEY LOOK LIKE?

An optimized patient journey is one in which the patient experiences the best and most efficient treatment, qualitative and quantitative assessment is possible at each stage, and the progress of treatment is not delayed by bottlenecks in the system. Ideally, the patient journey would begin with unaffected "healthy" individuals prior to onset of any pathology. This would have particular benefit for individuals at high risk of developing disease, for example, those with a genetic predisposition for breast cancer or dyslipidemia. However, more practically, in the short to medium term, one could envision the patient journey beginning with the first visit to the GP or the hospital emergency department.

In order to facilitate a system in which patient journeys can be implemented, monitored, and improved, the presence of the following features is a requirement:

- Ability to map and integrate patient information and test results in an interactive and iterative manner
- Application of multiple diagnostic tests in parallel fashion, where possible, to avoid delays
- Systems for interaction of clinicians, other health care staff, and scientists involved in patient care to allow for efficient transfer of information
- Potential to identify, report, and address bottlenecks in the system to improve system efficiency
- Construction of relational databases to hold the multimodal data streams characterizing the patient journey
- Mathematical modeling tools for analysis and integration of highly multivariate data that accommodate the complexity of human metabolism
- Visualization of clinical metadata and systems -omics or metabolic data in a way that allows for "clinical actionability."

Several health care systems have embraced the patient journey with enthusiasm and have contributed to the array of tools available for mapping it and improving it. For example, the Scottish health care system piloted the use of process maps to chart the patient journey and to identify bottlenecks and is also using software to provide flow analysis tools for identifying process and functional bottlenecks in the patient journey [11]. New multilayer visualization tools have been implemented in New South Wales, Australia, to identify whether changes in public health policy are impacting on patient flow in a beneficial manner or an adverse manner [12].

Internet search results for "patient journey" are dominated by cancer journeys, focusing on end-of-life care and palliative medicine [13,14]. However, we propose the patient journey concept in a much broader context. The patient journey for any procedure or disease can be charted, and any process in which the patient embarks on can be planned and recorded. This has been effectively implemented in many and diverse areas of the health care system. Examples include online programs for patients experiencing mild hearing impairment, analysis of bottlenecks in the detection and treatment of melanoma, end-of-life care, and colon cancer, to name but a few. For patients with early hearing impairment, a four-stage Internet-based pilot program was tested in Swansea to explore the journey through hearing loss, with a view to increasing the numbers of patients who seek help or use audiologic devices at early stages of the condition [15]. For melanoma, typical patient journeys were mapped for different countries across Europe. Skin cancer is the second most common cancer in individuals aged 15–34 years in the United Kingdom, and prognosis, as in most cancers, relies on early detection and treatment. Steps in the patient journey, including screening and management of individuals with skin cancer, differed across Europe, leading to disparities in clinical outcomes [16,17]. Bottlenecks in the patient journey in the United Kingdom include time from detection to referral to a specialist (2 weeks in contrast to the next day in Greece, Germany, Poland, Romania, and Malta) and ratio of dermatologists to GPs (United Kingdom 1:57, Poland 1:6, Greece 1:1) [16]. The patient journey is influenced by many factors, including medium of entry (GP vs emergency care) [18]; the quality of electronic health records [19], accessibility of diagnostics and treatment driven by geographic location, and, finally, patient perception and opinions, which are difficult to model given their subjective nature and diverse contextualization [20]. The Canadian health system has embraced the patient journey and has explored and implemented new means of health care system–patient communication throughout patient journeys, particularly in the case of cancer. One particularly innovative approach was to form hub-and-spoke models to enhance communication and continuity of care, where treatment for various aspects of disease requires input from multiple clinical centers [21]. The UK National Health Services (NHS) Institute for Innovation and Improvement has also responded to the need to focus on the patient journey as an entity recognizing that "the efficiency of the whole patient journey is more important than the individual team's efficiencies" [22].

Clear planning of a good patient journey not only in terms of clinical outcomes but also patient experience involves establishing the logistics of patient care in an integrated manner, from bed and consultant availability to storage and interpretation of test results. For the purposes of this chapter, we will focus on implementation of clinical diagnostic and prognostic procedures in the context of the patient journey and explore how rich phenotypical information, measured by rapid and minimally invasive metabolic profiling, can be used to augment clinical decision making.

Diagnostic assays are critical to the identification and characterization of a disease process and can consist of conventional clinical assays, histopathologic evaluation of tumor biopsy specimens, physical or surgical examination, imaging, and genetic tests. More recently, "-omics" technologies such as proteomics and metabonomics have been tentatively explored with respect to their potential for clinical translation as point-of-care assays. Metabolic phenotyping of biofluids or tissues, by high-resolution nuclear magnetic resonance (NMR) spectroscopy, mass spectrometry (MS), or, in special cases, spectroscopic imaging methods, offers a means of generating metabolic signatures and capturing information simultaneously for multiple metabolites. Its application in clinical studies was initially confined to offline exploration of disease mechanisms, but the evolution of analytical and computational technology has opened the door to the implementation of this technology within the time frame and scope of the patient journey.

These advanced metabolic phenotyping ("metabotyping") methods, described more extensively in chapters "The Development of Metabolic Phenotyping—A Historical Perspective" and "High-Throughput Metabolic

Screening", can be used to create a phenotypically enhanced patient journey, building on classic clinical diagnostic criteria to improve sensitivity and/or specificity of diagnosis [23,24]. Where multiple technology platforms are deployed throughout the patient handling pipeline, for example, combining NMR spectroscopy with ultra-performance liquid chromatography coupled with mass spectrometry (UPLC-MS) or with MS-directed chemical imaging of tissue sections at appropriate stages of a patient's journey through an illness, this technology has the ability to significantly and beneficially influence the capacity for diagnosis [25–27]. Sampling of multiple biological matrices will enable a comprehensive categorization of dysregulated metabolic processes. Blood (plasma or serum) and urine are the two most widely used clinical diagnostic fluids because of the minimally invasive nature of sampling and the potential for longitudinal sampling. Biofluids such as serum and urine represent snapshots of the metabolic state of a patient and reflect biochemical processes occurring in multiple organs and tissues. The theory is that early diagnostic or prognostic signatures of disease will be superimposed on the background of physiologic variability. Complementing biofluid signatures with directly sampled tissue from areas related to disease processes can help obtain more detailed information of the etiology or mechanism of a disease. There are three questions we can ask with regard to metabolic profiles: (1) Do they carry diagnostic or prognostic information? (2) Do the changes in metabolic profiles over time reflect the progression of the disease progress? (3) Can we use the dynamic metabolic profiles to map and monitor a patient's response to therapeutic intervention? In reality, sampling of a biological system will be incomplete—low frequency with missing data—and in the case of the patient journey, we are bound by practical and ethical considerations. Thus in order to maximize the value of metabolic profiling in the patient journey, a compromise needs to be made with regard to sampling frequency, patient comfort, cost, and analysis load. The challenge is to use computational modeling, which is constrained by sampling frequency and topographic or pathway coverage, to bridge and predict the gaps between the phenotypical information islands generated by spectroscopic analysis of biological samples.

3.3 THE ROLE OF METABOLIC PROFILING IN THE PATIENT JOURNEY

Since a journey is, by definition, a dynamic process, and since each patient experiences an individual journey, longitudinal sampling of the patient generates a metabolic pattern or trajectory, which carries much more information on the site, severity, and potential mechanism of damage compared with a single snapshot. As we have discussed, spectroscopic profiling generates molecular fingerprints of high information density reflecting the patient's metabolic status; it is described in more detail in chapters "The Development of Metabolic Phenotyping—A Historical Perspective" and "High-Throughput Metabolic

Screening". These molecular fingerprints can provide rich phenotypical information, which can be followed through time, and indicate metabolic stability, progression of a pathologic process, or response to a surgical or pharmaceutical intervention [9]. Variation in metabolic phenotypes has been explored in laboratory species and in humans in relation to a wide range of physiologic and pathologic conditions [28,29]. Gender, age, diet, physical activity, and hormonal state all have a systemic influence on spectral biofluid profiles, and these effects have been characterized by a range of clinical and population studies (see chapters: Precision Surgery and Surgical Spectroscopy and Population Screening for Biological and Environmental Properties of the Human Metabolic Phenotype: Implications for Personalized Medicine, for descriptions of metabolic signatures associated with these variables). However, beyond reporting on physiologic status, metabolic profiles of biological samples can contain disease-specific information. Every patient will have a different journey through a particular illness through to eventual recovery or, in some cases, adverse outcomes.

Analysis of samples in the patient journey has real costs both in the time and effort to acquire and interpret data, and in real financial budgets. However, metabolic phenotyping of biofluids is both rapid and relatively inexpensive compared with other clinical assays (as described in detail in chapters: The Development of Metabolic Phenotyping—A Historical Perspective and High-Throughput Metabolic Screening), so the quality of phenotyping does not usually need to be constrained.

Urine and plasma are generally chosen as biofluid samples, although feces, saliva, synovial fluid, breath condensate, and other biofluids might also be included. As discussed in chapter "The Development of Metabolic Phenotyping—A Historical Perspective", urine and plasma samples carry different and complementary information sets. Plasma or serum provides a snapshot of metabolism and a combined profile of low-molecular-weight molecules, superimposed on a background of lipoproteins and fatty acids [9]. Urine, on the other hand, is time averaged because of collection and storage in the bladder, and since it is an excretory fluid, it is not under tight homeostatic control as is plasma. Other biofluids can be obtained at specific stages of the patient journey, if needed, for more specific pathologic conditions [30], for example, cerebrospinal fluid (neurologic disorders), ulcer fluid (diabetic ulcers or varicose veins), synovial fluid (arthritis of the knee joints), bronchoalveolar lavage fluid (lung disease).

We propose that the patient journey can be represented as a process comprising six phases. At each stage, information can be gathered and modeled (Fig. 3.1), and metabolic phenotyping can be applied to augment decision-making processes.

Stage 1: involves initial patient awareness of a disease or condition, after which the patient will normally seek medical advice from a GP or a clinic.

Stage 2: the health care professional will take or access detailed patient medical history, supplemented with a physical examination. Often, blood or urine

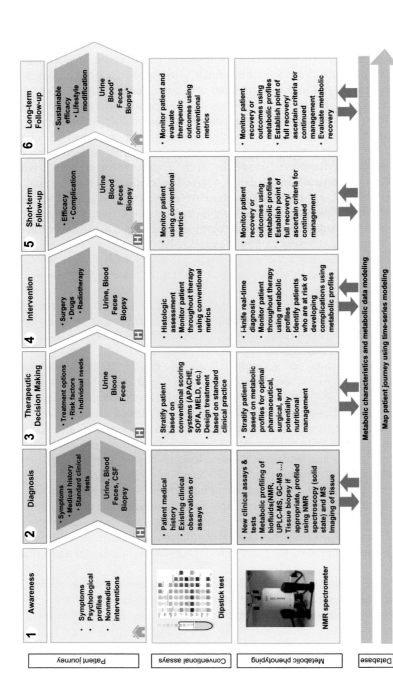

FIGURE 3.1 Schematic of the patient journey indicating the nodes at which metabolic phenotyping can contribute to shaping clinical opinions.

samples are obtained at this stage to diagnose the condition. For example, if diabetes is suspected, then the fasting plasma glucose level and sometimes the glycated hemoglobin level will be measured. These are conventional clinical assays, and one analyte is measured per sample. Metabolic phenotyping allows for obtaining a range of biochemical parameters simultaneously, and this profile can be used to enrich the existing clinical information. A recent MS-based study suggested that glyoxylate levels become elevated up to 3 years before diabetes is manifested, and therefore they may be an earlier biomarker of type 2 diabetes compared with blood glucose levels [31]. Another example where metabolic profiling was shown to augment existing clinical metrics is in the stratification of liver fibrosis induced by chronic hepatitis. Plasma profiles, reflecting altered amino acid and lipoprotein metabolism, were found to provide a robust correlation to the more invasively obtained enhanced liver fibrosis and METAVIR scores [32]. Where diagnosis involves obtaining a tissue biopsy, new metabolic profiling methods such as magic angle spinning (MAS)–NMR, conventional NMR, MS-based profiling of extracted tissue, or MS-based imaging (MSI) of tissue slices can be used to augment the traditional histopathologic analysis, which is relatively slow and often requires the use of multiple staining procedures. MAS-NMR spectroscopy uses intact tissue biopsy specimens, and since it is intrinsically a nondestructive technique, the tissue can be used for other analytical or histologic purposes after measurement [33]. This technology has been applied to mapping the tumor stage in colorectal cancer [34], where increased concentrations of taurine, isoglutamine, choline, lactate, phenylalanine, and tyrosine and decreased triglyceride concentrations were characteristic of tumor tissue, with several of these metabolites being reflective of the T- or N-stage of cancer. New microcoil probes, with the advantage of requiring considerably less sample (nanoliter volumes, as opposed to 10 mg for standard MAS-NMR analysis [35]), are being developed. MSI is also useful for delivering detailed characterization of chemically heterogeneous tissue and can define specific features such as focal disease plaques. This technology can use matrix-assisted laser-desorption ionization [36], desorption electrospray ionization [37,38] or secondary ion mass spectrometry [39].

Stage 3: involves stratification of patients with respect to treatment, be it surgical, pharmaceutical, or lifestyle modification. The potential for predicting a patient's response to an intervention, based on the metabolic phenotype of a baseline profile, is perhaps the ultimate achievement in Personalized Medicine, and for this reason, we have dedicated a separate chapter to the subject (chapter: Pharmacometabonomics and Predictive Metabonomics: New Tools for Personalized Medicine). However, it deserves special mention here since this is a tool that will help select appropriate treatment strategies for patients based on their unique set of metabolic and genetic cards during the patient journey. In pharmacometabonomic studies, preinterventional profiles of biofluids such as urine or plasma are utilized to create mathematical models by using the outcome of therapeutic interventions as variables against which spectral variables are regressed. The concept of pharmacometabonomics originates from the knowledge that not all patients respond equally to pharmaceutical interventions and that it is possible to be a slow responder, fast responder, nonresponder, or even adverse responder to a particular drug. The same philosophy holds true when considering response to surgical procedures. It has been clearly shown in humans that the metabolic fate of certain drugs such as paracetamol [40] and atorvastatin [41] can be predicted from baseline samples. More importantly, patient outcomes in response to pharmaceuticals can also be predicted from pre-interventional metabolic profiles for some pathologies, for example, mortality prediction in patients with septic shock treated with L-carnitine [42], the association of preintervention fatty acid levels with response to ketamine treatment for bipolar depression [43], and prediction of capecitabine toxicity in patients with inoperable colon cancer [44].

Stage 4: involves monitoring of a patient through an intervention. Intelligent surgical devices such as the "intelligent knife" (iKnife), which allow for continuous collection of metabolic profiles throughout an operation are an effective means of delivering near real-time chemical information [9] (chapter: Precision Surgery and Surgical Spectroscopy). MAS-NMR can also be used to inform clinicians about chemical composition and may aid in deciding whether enough tissue has been cut from a tumor, or in deciding when tissue is viable in situations where removal of necrotic tissue is necessary. If the procedure is pharmaceutical rather than surgical, then dynamic monitoring of biofluids is useful.

Stage 5: involves initiation of short-term follow-up after an intervention, or after commencing on a particular line of drug therapy. This may include ensuring the efficacy of a drug, surveillance for adverse drug reactions, or monitoring the patient's clinical signs after an operation. Metabolic phenotyping lends itself to the critical care scenario, where biofluid samples are easily obtained from hospitalized patients at short intervals. In a recent study, children undergoing surgery for congenital heart disease were monitored for 48 hours postoperatively via metabolic profiling of plasma samples (Fig. 3.2) [45]. In most patients, the metabolic profiles were found to be most deranged 6 hours after the operation (see Fig. 3.2A), and this mirrored the inflammatory response (see Fig. 3.2B). The metabolic trajectories are illustrated in this figure for two patients: patient A, who underwent major surgery with a prolonged postoperative recovery based on length of invasive ventilation, pediatric intensive care unit (PICU) free days, and risk adjusted congenital heart surgery (RACHS) score; and patient B, whose surgery was less severe and who recovered relatively quickly.

Stage 6: in situations where there is a need for longer-term follow-up, metabolic profiling can facilitate detection of disease remission or can be used to determine whether patients are becoming resistant to ongoing therapies. In one case study, an infant with ornithine transcarbamylase deficiency was followed up through a series of hepatocyte transplantations until eventual corrective liver transplantation at age 215 days, when the operation was less dangerous. In this

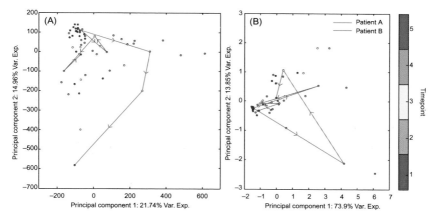

FIGURE 3.2   Principal component scores plots formed from (A) plasma NMR spectra, and (B) cytokine data from children undergoing surgery for congenital heart disease. Samples were collected before the operation (red), at time of operation (orange), 6 h after the operation (yellow), 24 h after the operation (green), and 48 h after the operation (purple). The trajectories are shown for two patients: patient A, who underwent a more serious operation with a prolonged recovery period; and patient B, who had a less severe condition and a shorter operation and recovery period. The child with the more serious condition and prolonged recovery period had the larger trajectory. *Reproduced from Ref. [45].*

single-patient study, NMR spectroscopy of urine and plasma samples showed that the urea cycle could be normalized by hepatocyte infusion, providing early indication of adverse metabolic events [46]. Infection is another setting in which prolonged metabolic profiling can be useful. In a study by Cui and colleagues, adults infected with Dengue fever were followed up using a combination of liquid chromatography coupled with mass spectrometry (LC-MS) and gas chromatography coupled with mass spectrometry (GC-MS) profiling over the early febrile stage (visit 1), which typically occurs within the first 72 hours after infection as the viremic load peaks, the defervescence stage (visit 2; 7–14 days after the infection), and the recovery stage (3–4 weeks after the infection (Fig. 3.3) [47]). Multivariate discriminant analysis of patients and matched noninfected controls showed clear separation of all three visits, with visit 1, corresponding to the acute infection stage, being least similar to the control group. The plot also showed that at visit 3, the profiles remained metabolically distinct from the control group. Altered serum lipid concentrations were key components of the differential biochemical signature of Dengue in addition to metabolites of proinflammatory cytokines such as arachidonic acid.

Despite the utility of metabolic profiling as an adjunct tool for clinicians, and the relative ease of noninvasively obtaining samples, there are surprisingly few examples of metabolic profiling time series data in human studies. Those studies that do exist tend to have sparse sampling and low numbers of study participants. One of the first metabolic profiling studies to introduce dynamic

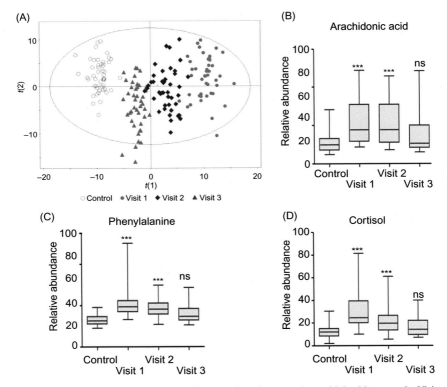

FIGURE 3.3 Analysis of three stages of Dengue fever in comparison with healthy controls: Visit 1 (febrile stage); Visit 2 (defervescence stage); Visit 3 (recovery phase). (A) partial least squares discriminant analysis scores plot, showing progressive recovery of patients over a 3–4-week period; (B) plasma arachidonic acid levels; (C) plasma phenylalanine levels; and (D) plasma cortisol levels. *Adapted from Ref. [47].*

monitoring of patient journeys was a study in which patients were followed up through renal transplantation in order to predict clinical outcome with respect to graft rejection or toxic reaction to cyclosporine A, which, at the time of analysis, was the immunosuppressant of choice [48]. Renal function, as assessed by the NMR spectroscopic profiles of urine and plasma, was compared with traditional clinical assessment by using a plasma creatinine concentration of 300 μmol/L as the cutoff point, below which the graft was assumed to be functioning. Graft dysfunction was clearly indicated by changes in renal osmolytes such as trimethylamine-*N*-oxide, dimethylamine, and myoinositol. Moreover, creatine and creatinine concentrations appeared to be different in discriminating poor and good graft function after transplantation (Fig. 3.4). Other groups have shown that the combination of technologies such as NMR and GC-MS allows for a richer coverage for mapping the metabolome of patients receiving transplants [49].

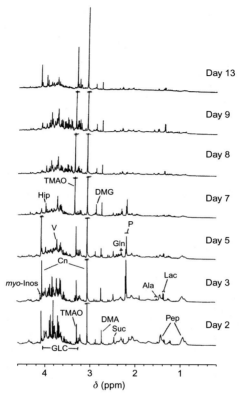

FIGURE 3.4 A series of ^{1}H NMR spectra of urine obtained from a patient who received a kidney transplant for acute tubular necrosis and who subsequently underwent graft rejection. High levels of trimethylamine-*N*-oxide and dimethylamine prior to day 7 are characteristic of graft dysfunction. Methylprednisolone was administered between days 7 and 9 after the transplantation. Key: *Ala*, alanine; *Cn*, creatinine; *DMA*, dimethylamine; *DMG*, dimethylglycine; *Glc*, glucose; *Gln*, glutamine; *Hip*, hippurate; *Lac*, lactate; *myo-inos*, myoinositol; *P*, paracetamol; *Suc*, succinate; *TMAO*, trimethylamine-*N*-oxide; *V*, drug carrier vehicle. *Reproduced from Ref. [48].*

The concept of Personalized Medicine is further illustrated in Fig. 3.5, where metabolic profiles of urine following bariatric surgery have been monitored over 12 months using ^{1}H NMR spectroscopy; these observed metabolic shifts were comapped with surgical outcomes and, in this case, related to weight loss to assess patients' metabolic status and evaluate the metabolic benefits of the surgery. Following earlier studies, multivariate modeling methods for mapping dynamic metabolic trajectories have been developed to deal with dynamic processes for various academic disciplines. They include chronologic connection of mean, median, or individual samples over consecutive time points using standard multivariate analyses [50], batch processing based on applications in

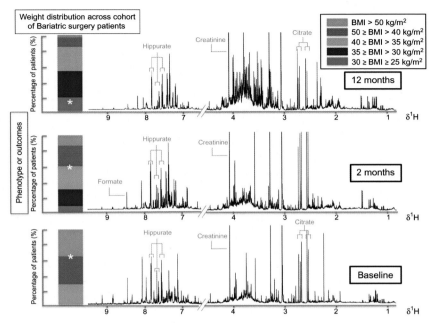

FIGURE 3.5 Metabolic characterization of a patient journey for a 32-year-old woman undergoing surgery. The left hand panel demonstrates the change in weight for a cohort of 40 patients after bariatric surgery through to 12 months after the operation. The white asterisk indicates the position of the individual patient within these weight bands. The right hand panel shows the NMR spectrum of urine samples from this patient, showing metabolic changes over the period of weight loss.

manufacturing to capture deviation form a normal trajectory [51], time series analysis [52], multivariate input–output analysis [53], time-varying Fourier analysis [54], and space–state analysis [55]. Since metabonomic data tend to be sparsely sampled and require processing methods to correct for issues such as retention time drift, there is much scope for improvement in the dynamic modeling of data, which is essential for mapping and interpretation of patient journeys.

Two other properties of metabolic profiling that are useful in mapping the patient journey remain to be discussed. First, drugs and their metabolites carry their own chemical signatures, which appear in the metabolic biofluid profile alongside the endogenous signature. These signals can confound the analysis of spectral profiles if they are not accounted for, and even drug excipients can have a biological effect in their own right [56]. However, the information on drug metabolism, which is carried in the metabolic profiles, can be useful in assessing interpatient differences in response to therapeutic intervention [57]. Several statistical correlation methods have been developed to differentiate endogenous metabolism from drug metabolism, largely on the basis of calculation of correlation matrices, driven by a signal known to arise from a particular chemical [58].

This process works especially well in the case of NMR data, since the technique is intrinsically quantitative, and the notion of statistical total correlation spectroscopy (STOCSY) has since been developed to include variants that make use of iterative correlation procedures, enlist multiple nuclei to profile solely drug-related molecules, or focus on subsets of individuals with distinct phenotypes within a patient cohort. Many of these correlation methods are described in more detail in chapters "The Development of Metabolic Phenotyping—A Historical Perspective" and "High-Throughput Metabolic Screening" and have been summarized, along with other statistical tools for spectral classification and structural elucidation, by Robinette et al. [59]. The second of the additional sources of information that metabolic profiling can deliver that is of direct relevance to health is about other exogenous chemicals originating either from the diet or from the metabolic activity of the gut bacteria (microbiota). Much attention has been given recently to the role of the intestinal microbiome in human health and disease, and dysregulation of the microbiota has been associated with intestinal bowel disorders, colon and other cancers, allergies, metabolic syndrome, cardiovascular disease, and even neurobehavioral disorders such as autism [60–62]. The gut microbiome is highly metabolically active and has coevolved with its human host such that it synthesizes the chemicals that humans require. In equilibrium, the gut microbiome makes beneficial contributions to our immune system such as harvesting of calories from the diet, chemical signaling within the gut–brain axis, and many other biological functions [61]. If the natural equilibrium is perturbed, then opportunistic pathogens can negatively impact on human health. The role of the microbiome in maintaining health is particularly prominent at the beginning and end of life, and further illustrations of the importance of the microbiome during those periods are provided in chapters "Handing on Health to the Next Generation: Early Life Exposures" and "The Ageing Superorganism", respectively. With regard to the patient journey, each individual has a unique microbiome, and this represents an additional body of cells and chemical communication channels that can be targeted pharmacologically with the aim of tailoring and optimizing personalized health care.

Although the patient journey generally describes the patient's response to disease and subsequent treatment, pregnancy represents a special case. As discussed in detail in chapter "Handing on Health to the Next Generation: Early Life Exposures", complications in pregnancy not only impact on the health of the infant but can also be predictive of disease in the mother later in her life. For example, women who get gestational diabetes are more at risk of developing type 2 diabetes in later life, and similarly, those who develop preeclampsia are at increased risk of hypertension. In the Western world, women are monitored at regular intervals throughout their pregnancy, and this provides an opportunity for prognostic application of metabolic profiling. Several studies have described the metabolic journey through healthy pregnancy, and some have successfully profiled pregnancy complications. For example, Diaz and colleagues showed systematic modulation of 21 urinary metabolites reflecting changes in choline, methionine, and ketone body metabolism in women with gestational diabetes compared with healthy nonpregnant women at the same time [63].

In summary, metabolic profiling provides a cost-effective and time-efficient means of characterizing the patient journey. Spectroscopic profiles generated from biological samples are information-rich, as described in detail in chapter "High-Throughput Metabolic Screening", allowing for simultaneous mapping of multiple biochemical pathways and processes that enable extraction of diagnostic and mechanistic information. Data can be visualized in high resolution, and unlike other systems biology "-omics" tools, the metabolic phenotype exposes metabolically driven processes that are triggered by external factors such as drugs or the gut microbiota.

3.4 STRATIFIED AND PERSONALIZED HEALTH CARE

The ideal of personalized or stratified health care is an attractive one and promises to improve health outcomes and patient experience and, if applied correctly, to reduce the financial burden of major acute and chronic diseases. The cost per day for intensive care has been estimated at €1425± €520 in a recent study in France [64] and a range of €1168 to €2025 in a study across Germany, Italy, the United Kingdom, and The Netherlands [65]. Thus by reducing the time a patient spends in an intensive care unit (ICU) by just 1 day, significant savings will be achieved; the statistics for clinical outcome suggest that reducing length of stay in the ICU improves the prognosis. The key question is, how can we maximize the use of the Personalized Medicine approach to improve patient care, reduce the length of hospital stay, and improve the economic burden on the health care system? With respect to metabolic phenotyping, there are five core areas that can be developed or modified to maximize the potential of the technology: (1) data capture; (2) sample collection and processing; (3) spectral acquisition of data; (4) data processing and modeling; and (5) data storage, query, and integration.

3.4.1 Data Capture

The first step in capturing the patient journey is to define the parameters that are required to generate a comprehensive phenotype. These parameters, of necessity, should not only include disease-related metrics but should also capture medical history, patient sociodemographics, and, where possible, information about diet, medication, and lifestyle. The process of data capture should be fluid, and the system requires enough flexibility to update the metadata when new information is brought to light. Without solid patient histories, for example, information on medication, it is impossible to interpret the spectral data with any degree of accuracy. Moreover, although we are primarily concerned here with mapping individual patients, drawing inferences from population-based studies can deliver new knowledge about candidate biomarkers, stratification of individuals according to subtypes of disease, and new information about associations between environmental factors and disease risk (see chapter: Population Screening for Biological and Environmental Properties of the Human Metabolic Phenotype: Implications for Personalized Medicine).

3.4.2   Sample Collection and Processing

The old adage "garbage in, garbage out" is particularly pertinent to sample collection and processing. This is the primary building block on which all subsequent metabolic profiling analysis is based. Clinicians and research nurses are often acting under tight time constraints and stressful conditions and so require unambiguous protocols for material required. Although considerable emphasis has been placed on sample handling and acquisition, as evidenced by a series of publications [66–70], much less action has been taken to define sample collection procedures [71–73]. Rigorous standard operating procedures and protocols are essential for effectively mapping patient journeys, particularly where there is no one individual responsible for obtaining samples across multiple sampling points.

3.4.3   Spectroscopic Profiling of Biofluids, Cells, and Tissues

As with sample collection, the acquisition of spectra has to adhere to standards and guidelines to make comparisons across patients and through time. The starting point for any spectroscopic diagnostic is, at least partially, agnostic, given that every individual holds a unique set of genetic and environmental cards, and therefore it is appropriate to begin by conducting global profiling assays where the metabolites are not selected a priori. These global NMR and MS screening profiles can then be supplemented with quantitative and targeted assays. For example, inflammation can be effectively profiled by using global lipidomic analysis [74] that is supplemented by a quantitative analysis of eicosanoids [75]. Spectroscopic profiling methods are described in greater detail in chapter "High-Throughput Metabolic Screening".

3.4.4   Data Processing and Modeling

Data structures vary across different biological matrices and analytical platforms. Since it is not possible yet to automate annotation because of the ambiguity of discrete signals in the NMR spectra and m/z ratios in MS data (unless the MS assay is targeted for pre-identified compounds), first-line analysis should not rely on annotated data but rather on patterns or fingerprints of metabolites. Molecular identification can be carried out at a later stage in analysis when signals relevant to particular clinical correlates are discovered. A myriad of multivariate algorithms are available for interrogating data, and it can be confusing as to which methods should be used. Methods fall into two distinct groups: (1) unsupervised methods, where disease- or intervention-related information is not used to model the data; and (2) supervised methods, where classification information is used effectively to find the difference between a disease state and the appropriate control. An overview of these methods is given in chapter "The Development of Metabolic Phenotyping—A Historical Perspective".

For longitudinal clinical samples, time series analysis is the obvious choice. Various multivariate methods have been used to analyze spectral data collected for the same individual over time, as described above, in relation to stage 6 or long-term follow-up of patients. Some of these methods require complete datasets (ie, they will not accommodate missing samples), whereas others are more forgiving of missing samples and time-displaced collection series. The timing of sample collection for a cohort of patients tends not to be uniform for practical reasons such as surgeons and research nurses not working on nonurgent cases at the weekends or, in the case where sampling is undertaken by patients themselves, forgetting to bring the sample to a particular clinic visit. In instances where the disease does not follow a linear process and metabolically "jumps" from one stage to another, then linear methods may not accommodate this behavior well, and nonlinear methods (eg, neural networks) or Bayesian probabilistic methods (eg, CLOUDS analysis) may be required to overcome the noncontiguous data structure [76].

One of the greatest challenges in data modeling is integrating disparate datasets to gain a holistic view of the patient. Spectral datasets collected across different spectroscopic platforms and different biofluids; for example, global NMR profiling of urine and UPLC-MS plasma lipidome spectra are combined with clinical metrics and metadata. In order to integrate different datasets in a medically meaningful way intelligent pre-processing and data fusion are necessary. There is a growing body of literature describing existing methods [77,78] and studies adopting a multi-omic approach are being used to address unmet clinical needs, for example, early biomarkers for Alzheimer's disease [79].

3.4.5 Data Storage and Query

In order to provide a vehicle for mapping and interrogating dynamic phenotypical data that reflect the metabolic progression of a patient through a disease episode, it is necessary to have a standardized data storage and curation system that can be iteratively incremented. The detailed architecture and design of such a system is covered in chapter "From Databases to Big Data"; basic requirements for such a system include intuitive organization, flexibility for storage of different data types, modular arrangement of biologically relevant subunits, architecture for data sharing, and relational linkages to archives where raw data can be stored.

3.5 DATA VISUALIZATION FOR THE PHYSICIAN

If the technology is to be useful in a clinical environment, it must be presented in an easily accessible format to the clinician such that no specialized training is required to interpret the output from the spectroscopic diagnostics. Raw or even annotated spectra and images are of little use. Therefore, the output needs to consist of a classification for a given disease or condition and a measure of confidence in that classification or diagnosis. For the data analyst, the models used to derive the classification should be transparent such that these models can augment existing databases as new data are added and identify any anomalies in the samples. Finally, innovative modes of data visualization can play a strong role in teaching, thus providing an interactive learning environment.

3.6   COMMUNICATION WITH THE PATIENT AND FAMILY: IMPROVING THE PATIENT EXPERIENCE

Perhaps one of the most important aspects of the patient journey is how the journey is presented to the patient and his or her family and caregivers. Lack of information, misinformation, or overcomplicated explanations cause anxiety and fear in the patient. In order to achieve a successful model, the patient journey should be accompanied at every stage by transparent and, where possible, unambiguous information, providing the patient with the tools to make more informed decisions about his or her treatment and journey. Data visualization and informatics technology can be employed here in the context of providing a clear and interactive interface, through which the patient can explore his or her own journey. This would include a well-documented history of procedures, access to results, and their interpretation in a patient-specific context.

3.7   IMPLEMENTING CLINICAL METABOLIC PHENOTYPING IN THE PATIENT JOURNEY: THE CLINICAL PHENOME CENTER CONCEPT

A primary requirement of a clinical metabolic phenotyping facility is the deployment of well-validated, high-throughput, exploratory and targeted analytical procedures operating in proximity to the patients and the clinical decision-making environment to provide a framework for mapping surgical and pharmaceutical patient journeys [80]. In reality, this means physical placement of spectroscopic instrumentation within a hospital, closely linked to the routine clinical chemistry and chemical pathology laboratories that will be handling the biofluid samples. Although at present a few operational facilities exist around the world (eg, the Imperial National Institute of Health Research Clinical Phenome Centre (CPC) at St Mary's Hospital, London; The Chung Guang Hospital, Taiwan), these are still mainly operating in an experimental medicine mode, where studies are carried out to help understand human disease rather than performing routine diagnosis of patients or following their journeys from the point of view of clinical decision making. Of course, both NMR spectroscopy and MS have been employed for screening of newborns for inborn errors of metabolism, but this use, although increasingly frequent, is still not widespread and is not connected to the patient journey per se. An important feature of CPC operations is the scalability and translatability of the technology: analytical platforms can be applied to multiple disease classes and biomedical problems. The philosophy behind developing CPCs and the necessary tools for their implementation are described in chapter "Phenome Centers and Global Harmonization". However, CPCs have to be operationally congruent with standard diagnostic procedures, the functional objective being augmentation of decision making rather than replacement of conventional chemical pathology. This, of course, means that analytical procedures and data interpretation have to be on a timescale that matches the crucial decision-making events that mark the patient journey for a particular disease type. This is in contrast to national-level phenome centers, which deal with biobanking or epidemiologic samples, in which case urgent diagnostics are not required. The level of validation of methodology, regulatory requirements, and the volume of analysis make the CPC proposition more challenging than an epidemiologic phenotyping center. We are still many years away from the widespread use of these approaches for routine patient pathway analysis in a general hospital environment, but the same can be said of genomics, epigenetics, proteomics, and metagenomics. The pathway to the clinical application of any new technology or approach is always long and complex. However, the overwhelming need for personalized health care tools means that eventually these technologies *will* be refined and find their appropriate place in the management of the patient journey. The challenge is to find which combinations of technologies will serve the many different patient journeys to optimize the path to recovery in the most cost-effective way. This challenge is now being addressed in many academic and medical institutions around the world. The challenge will ultimately be met, and platforms such as metabolic phenotyping will find their rightful place in health care delivery.

REFERENCES

[1] Ben-Tovim DI, Dougherty ML, O'Connell TJ, McGrath KM. Patient journeys: the process of clinical redesign. Med J Aust 2008;188(Suppl. 6):S14–7.

[2] The Free Dictionary. Definition: patient journey. <http://medical-dictionary.thefreedictionary.com/patient+journey>.

[3] Zhao L, Nicholson JK, Lu A, Wang Z, Tang H, Holmes E, et al. Targeting the human genome-microbiome axis for drug discovery: inspirations from global systems biology and traditional Chinese medicine. J Proteome Res 2012;11(7):3509–19.

[4] Anon. Understanding the patient journey - process mapping. <http://www.gov.scot/resource/doc/141079/0036023.pdf>.

[5] Goh KI, Cusick ME, Valle D, Childs B, Vidal M, Barabási AL. The human disease network. Proc Natl Acad Sci USA 2007;104(21):8685–90. Epub 2007 May 14.

[6] Gilligan S, Walters M. Quality improvements in hospital flow may lead to a reduction in mortality. Clin Gov Int J 2008;13:26–34.

[7] Allder S, Silvester K, Walley P. Managing capacity and demand across the patient journey. Clin Med 2010;10(1):13–15.

[8] Mahesh B, Choong CK, Goldsmith K, Gerrard C, Nashef SA, Vuylsteke A. Prolonged stay in intensive care unit is a powerful predictor of adverse outcomes after cardiac operations. Ann Thorac Surg 2012;94(1):109–16.

[9] Nicholson JK, Holmes E, Kinross JM, Darzi AW, Takats Z, Lindon JC. Metabolic phenotyping in clinical and surgical environments. Nature 2012;491(7424):384–92.

[10] Kinross JM, Holmes E, Darzi AW, Nicholson JK. Metabolic phenotyping for monitoring surgical patients. Lancet 2011;377(9780):1817–9.

[11] Wales A. A national health knowledge network to support the patient journey. Health Info Libr J 2008;25(Suppl. 1):99–102.

[12] Sayvong R, Curry J. Using patient journey modelling to visualise the impact of policy change on patient flow in community care. Stud Health Technol Inform 2015;216:429–33.

[13] Tulbah A, Chaudhri N, Al Dayel F, Akhtar M. The journey toward personalized cancer therapy. Adv Anat Pathol 2014;21(1):36–43.

[14] Mirnezami R, Jiménez B, Li JV, Kinross JM, Veselkov K, Goldin RD, et al. Rapid diagnosis and staging of colorectal cancer via high-resolution magic angle spinning nuclear magnetic resonance (HR-MAS NMR)spectroscopy of intact tissue biopsies. Ann Surg 2014;259(6):1138–49.

[15] Manchaiah VK, Stephens D, Andersson G, Rönnberg J, Lunner T. Use of the 'patient journey' model in the internet-based pre-fitting counseling of a person with hearing disability: study protocol for a randomized controlled trial. Trials 2013;14:25.

[16] Trakatelli M, Siskou S, Proby C, Tiplica GS, Hinrichs B, Altsitsiadis E, et al. EPIDERM. The patient journey: a report of skin cancer care across Europe. Br J Dermatol 2012;167(Suppl. 2): 43–52.

[17] Forsea AM, Del Marmol V, de Vries E, Bailey EE, Geller AC. Melanoma incidence and mortality in Europe: new estimates, persistent disparities. Br J Dermatol 2012;167(5):1124–30.

[18] Gerrand C, Francis M, Dennis N, Charman J, Lawrence G, Evans T, et al. Routes to diagnosis for sarcoma—describing the sarcoma patient journey. Eur J Surg Oncol 2015.

[19] Thompson CA, Kurian AW, Luft HS. Linking electronic health records to better understand breast cancer patient pathways within and between two health systems. EGEMS 2015;3(1):1127.

[20] Sun V, Grant M, McMullen CK, Altschuler A, Mohler MJ, Hornbrook MC, et al. From diagnosis through survivorship: health-care experiences of colorectal cancer survivors with ostomies. Support Care Cancer 2014;22(6):1563–70.

[21] Fung-Kee-Fung M, Boushey RP, Watters J, Morash R, Smylie J, Morash C, et al. Piloting a regional collaborative in cancer surgery using a "community of practice" model. Curr Oncol 2014;21(1):27–34.

[22] NHS website. <http://www.institute.nhs.uk/quality_and_service_improvement_tools/quality_and_service_improvement_tools/a_focus_on_the_whole_patient_journey.html#sthash.DfvRqt5O.dpuf>.

[23] Ladep NG, Dona AC, Lewis MR, Crossey MM, Lemoine M, Okeke E, et al. Discovery and validation of urinary metabotypes for the diagnosis of hepatocellular carcinoma in West Africans. Hepatology 2014;60(4):1291–301.

[24] Mohamad N, Ismet RI, Rofiee M, Bannur Z, Hennessy T, Selvaraj M, et al. Metabolomics and partial least square discriminant analysis to predict history of myocardial infarction of self-claimed healthy subjects: validity and feasibility for clinical practice. J Clin Bioinforma 2015;5(3).

[25] Al Awam K, Haußleiter IS, Dudley E, Donev R, Brüne M, Juckel G, et al. Multiplatform metabolome and proteome profiling identifies serum metabolite and protein signatures as prospective biomarkers for schizophrenia. J Neural Transm 2015;122(Suppl. 1):111–22.

[26] Ashrafian H, Li JV, Spagou K, Harling L, Masson P, Darzi A, et al. Bariatric surgery modulates circulating and cardiac metabolites. J Proteome Res 2014;13(2):570–80.

[27] Suhre K, Meisinger C, Döring A, Altmaier E, Belcredi P, Gieger C, et al. Metabolic footprint of diabetes: a multiplatform metabolomics study in an epidemiological setting. PLoS One 2010;5(11):e13953.

[28] Dunn WB, Lin W, Broadhurst D, Begley P, Brown M, Zelena E, et al. Molecular phenotyping of a UK population: defining the human serum metabolome. Metabolomics 2015;11:9–26.

[29] Bollard ME, Stanley EG, Lindon JC, Nicholson JK, Holmes E. NMR-based metabonomic approaches for evaluating physiological influences on biofluid composition. NMR Biomed 2005;18(3):143–62.

[30] Zhang A, Sun H, Wang P, Han Y, Wang X. Recent and potential developments of biofluid analyses in metabolomics. J Proteomics 2012;75(4):1079–88.

[31] Nikiforova VJ, Giesbertz P, Wiemer J, Bethan B, Looser R, Liebenberg V, et al. Glyoxylate, a new marker metabolite of type 2 diabetes. J Diabetes Res 2014;2014:685204.

[32] Sands CJ, Guha IN, Kyriakides M, Wright M, Beckonert O, Holmes E, et al. Metabolic phenotyping for enhanced mechanistic stratification of chronic hepatitis C-induced liver fibrosis. Am J Gastroenterol 2015;110(1):159–69.

[33] Santos CF, Kurhanewicz J, Tabatabai ZL, Simko JP, Keshari KR, Gbegnon A, et al. Metabolic, pathologic, and genetic analysis of prostate tissues: quantitative evaluation of histopathologic and mRNA integrity after HR-MAS spectroscopy. NMR Biomed 2010;23(4):391–8.

[34] Jiménez B, Mirnezami R, Kinross J, Cloarec O, Keun HC, Holmes E, et al. 1H HR-MAS NMR spectroscopy of tumor-induced local metabolic "field-effects" enables colorectal cancer staging and prognostication. J Proteome Res 2013;12(2):959–68.

[35] Wong A, Jiménez B, Li X, Holmes E, Nicholson JK, Lindon JC, et al. Evaluation of high resolution magic-angle coil spinning NMR spectroscopy for metabolic profiling of nanoliter tissue biopsies. Anal Chem 2012;84(8):3843–8.

[36] Schwamborn K, Caprioli RM. Molecular imaging by mass spectrometry—looking beyond classical histology. Nat Rev Cancer 2010;10(9):639–46.

[37] Takats Z, Wiseman JM, Gologan B, Cooks RG. Mass spectrometry sampling under ambient conditions with desorption electrospray ionization. Science 2004;306(5695):471–3.

[38] Eberlin LS, Norton I, Orringer D, Dunn IF, Liu X, Ide JL, et al. Ambient mass spectrometry for the intraoperative molecular diagnosis of human brain tumors. Proc Natl Acad Sci USA 2013;110(5):1611–6.

[39] Cillero-Pastor B, Eijkel G, Kiss A, Blanco FJ, Heeren RM. Time-of-flight secondary ion mass spectrometry-based molecular distribution distinguishing healthy and osteoarthritic human cartilage. Anal Chem 2012;84(21):8909–16.

[40] Clayton TA, Baker D, Lindon JC, Everett JR, Nicholson JK. Pharmacometabonomic identification of a significant host-microbiome metabolic interaction affecting human drug metabolism. Proc Natl Acad Sci USA 2009;106(34):14728–33. Epub 2009 Aug 10.

[41] Huang Q, Aa J, Jia H, Xin X, Tao C, Liu L, et al. A pharmacometabonomic approach to predicting metabolic phenotypes and pharmacokinetic parameters of atorvastatin in healthy volunteers. J Proteome Res 2015 [Epub ahead of print].

[42] Puskarich MA, Finkel MA, Karnovsky A, Jones AE, Trexel J, Harris BN, et al. Pharmacometabolomics of L-carnitine treatment response phenotypes in patients with septic shock. Ann Am Thorac Soc 2015;12(1):46–56.

[43] Villaseñor A, Ramamoorthy A, Silva dos Santos M, Lorenzo MP, Laje G, Zarate Jr C, et al. A pilot study of plasma metabolomic patterns from patients treated with ketamine for bipolar depression: evidence for a response-related difference in mitochondrial networks. Br J Pharmacol 2014;171(8):2230–42.

[44] Backshall A, Sharma R, Clarke SJ, Keun HC. Pharmacometabonomic profiling as a predictor of toxicity in patients with inoperable colorectal cancer treated with capecitabine. Clin Cancer Res 2011;17(9):3019–28.

the bowel) may influence drug metabolism, the inflammatory response, microbiome dysbiosis, or even other critical functions such as the prediction of cancer outcomes or survival [3]. This chapter explores these themes and reviews the current evidence to support this theory.

4.2 SURGERY AS A MODEL FOR STUDYING METABOLISM

Surgery provides unique opportunities for exploring the functional and symbiotic biochemical relationship between humans and microbiota. Numerous surgical and minimally invasive procedures provide access to normally unreachable anatomical targets (hepatic portal venous system, foregut, or even the brain). However, surgery is particularly pertinent in the analysis of systems biology, where the microbiome contributes significantly to disease and clinical outcomes. Therefore where surgery is responsible for deviating the anatomical flow of biofluids (eg, urine or feces), useful information about systems function can be gained by studying preintervention and postintervention physiologic functions. A good example of this is provided in gastric bypass of the foregut, which is increasingly being performed for the treatment of obesity and its metabolic complications such as type II diabetes.

There is now convincing animal and human data to suggest that surgical bypass fundamentally disrupts the distal gut microbiome and that this may play a role in the resolution of the metabolic complications of obesity. The subsequent disruption to the metabolome can be objectively measured by using approaches based on nuclear resonance imaging (NMR) - as well as mass spectrometry (MS) and, in turn, be correlated with metagenomic data sets from feces [4,5]. Following bariatric surgery, the gut microbiota move from a Firmicute- and Bacteroidetes-dominated ecosystem to one dominated by proteobacteria [6]. Metabolic profiles reflect this shift in the microbial community with a persistent alteration in the urinary, serum, and fecal levels of cresols, indoles, and biogenic amines. This suggests that personalized biomarkers for the prediction of long-term weight loss will have to account for the gut microbiome and that the long-term health consequences of permanently altered enteric flow have yet to be fully defined.

4.3 MOLECULAR PHENOTYPING AND SAMPLE PHENOTYPING

Over 800 million pathology tests are performed in the United Kingdom alone each year, and over 95% of all clinical patient pathways rely on access to effective pathology services [7]. The "gold standard" for diagnostics is tissue biopsy and histopathologic assessment, which is deployed in diagnosis, screening, and monitoring of chronic disease. Specifically, there is an increasing demand for cancer diagnostics with improved quality, safety, efficiency, and lower costs.

If cancer outcomes are to be improved, there is also a critical demand for augmented histopathology technologies based on a deeper molecular interrogation of cancer biology. Current univariate diagnostic and prognostic biomarkers are unlikely to deliver this vision, as they do not reflect cancer systems biology, which is modulated by multiple complex interacting molecular components. In surgical oncology, specifically, they also fail to account for intra- and intertumor heterogeneity and the increasing number of genomic molecular oncologic phenotypes described for solid tumors [8].

Complementary metabonomic analytical platforms may therefore provide data that will not only influence which incision should be made and when but also the long-term nutritional, pharmacologic, or environmental changes required to maximize recovery and long-term health. The precedent for phenome augmented decision making comes from the theory of "pharmacometabonomics," which proposes that predose drug toxicity or efficacy may be feasible through a systems metabolic analysis of predose biofluids. In this instance however, surgical systems biology will be able to support clinical decision making, enable targeted therapy, and predict and identify complications earlier.

For example, MS targeted analysis of 69 serum metabolites in a large population of patients undergoing cardiac surgery ($n = 478$) was able to accurately predict a poor operative outcome over a mean follow-up period of 4.3 ± 2.4 years [9]. Short-chain dicarboxylacylcarnitines, ketone-related metabolites, and short-chain acylcarnitines were all independently predictive of an adverse outcome after multivariate adjustment. However, personalized or stratified approaches to health care must not focus only on mammalian biology. Initial work also suggests that a metabonomic strategy is able to predict early-onset systemic inflammatory response in those patients exposed to major trauma. Partial least squares discriminant analysis was also able to clearly discriminate the systemic inflammatory response syndrome and multiorgan dysfunction syndrome, according to variations in carbohydrate, amino acid, glucose, lactate, glutamine signals, acyl chains, and lipids [10]. This is a critical area of research as sepsis is a major complication of surgery. Similar approaches are experimental models of renal transplantation, where it is being assessed for its ability to predict graft failure [11] and end-organ drug toxicity [12] and to assess hypoxic injury in cadaveric specimens. It has also been widely used in experimental models of liver and gut transplantation [13]. Metabonomics may have a significant impact in transplantations, where there is a particular demand for rapid molecular diagnostics within the operating room to predict graft suitability and survival.

4.4 CURRENT CHALLENGES IN SURGICAL CANCER BIOMARKER DISCOVERY

The World Health Organization reports that 8.2 million people each year die of cancer, and this figure is set to increase dramatically as the population ages [14]. There has, therefore, been a global drive on cancer prevention through the deployment of national screening programs in common cancers, where Wilson's criteria are met. For example, in the United Kingdom, the national Bowel Cancer Screening Program screens all patients between the age of 60 and 74 years with a fecal occult blood test and, if positive, a colonoscopy. This has, in turn, led to an increase in detection rates of early rectal cancers. Of the 1772 cancers detected by the Bowel Cancer Screening Program (Jun. 2006–Oct. 2008), 71% were in the early stage and 29% were rectal cancers [1]. This creates a surgical challenge, as there are very few biological data to help stratify the management of early cancer, and the merits of local resection versus radical colorectal surgery is still being debated. Clearly, experimental diagnostic and prognostic biomarkers based on metabonomic strategies have been described for numerous cancers, but relatively few "-omics" biomarkers have translated into routine clinical use for guiding surgical therapy. For example, some genomic biomarkers are used for targeting adjuvant therapy in colorectal cancer (CRC) (Table 4.1); however, none is based on "-omics" or systems-level analytical platforms.

This is because largescale prospective randomized control trials are currently lacking. However, genomic sequencing has also now identified the true heterogeneity within cancers and created new insights into the evolution of cancer biology during treatment. Therefore personalized biomarkers for therapeutic efficacy must be dynamic and highly personalized. Thus, there is a considerable opportunity for novel technologies based on a systems approach, as they are able to help plan and define surgical strategy, predict response to treatment, or predict survival. The ability of metabonomic strategies to track downstream cellular changes with great sensitivity facilitates a deeper understanding of the biochemical processes underlying breast cancer etiopathology and response to therapy. Over the last two decades, such strategies have been employed in breast cancer research in the development of molecular-based tissue diagnostics [41,42,43]; biofluid-based biomarker discovery [44,45]; the prediction of metastasis, recurrence [46], prognosis, and outcome [47].

4.5 CHEMICAL BIOPSY AND CHEMISTRY IN THE CLINIC

4.5.1 Magic Angle Spinning NMR

Chemical biopsy is feasible through numerous metabonomic platforms for augmenting near real time decision making in clinic or at the patient's bedside. Magic angle spinning (MAS) ^{1}H NMR spectroscopy has been used for some time for nondestructive metabolic analysis of both oncologic and nononcologic conditions with some effect. It has been extensively used in the analysis of brain tumors, and it is able to differentiate between malignant tumor types [29]; these detailed biochemical profiles can then be related to the lower-resolution spectra obtained in vivo, using magnetic resonance spectroscopy (MRS) [30]. This significantly improves MRI-based characterization of grade IV glioblastomas,

TABLE 4.1 Biomarkers Commonly Deployed in the Clinical Management of Colorectal Cancer

Biomarker	Use	Details
CEA (carcinoembryonic antigen)	Serum biomarker used clinically in surveillance for disease recurrence	A glycoprotein involved in cell adhesion. It has poor specificity as a diagnostic or screening biomarker.
MisMatch Repair (MMR) genes—MLH1, MSH2, MSH6, or PSM2	Tissue-based molecular biomarkers of Lynch syndrome	Lynch syndrome is an autosomal dominant condition with variable penetrance, which leads to a lifetime risk of CRC of 80% if not screened and treated [15].
KRAS and BRAF	Tissue molecular markers of predicted response to anti-EGFR chemotherapy	KRAS encodes a g-protein, and BRAF encodes a protein kinase. Both are part of the ras/raf/mitogen-activated protein kinase intracellular pathway [15]. The KRAS mutation predicts a complete lack of response to anti-EGFR therapy [16]. BRAF mutation also has a predictive role in the response to therapy with anti-EGFR in cancers in patients with wild-type KRAS [17].
Ki 67	Tissue-based predictive marker of the benefit of adjuvant chemotherapy in stage III colon cancer	Ki 67 is a monoclonal antibody that recognizes an antigen present in the nuclei of cells in all phases of the cell cycle except G_0. A marker of tumor proliferation—high expression is associated with an increased disease-free survival after adjuvant chemotherapy in stage III colon cancer.
p53	Tissue-based potential prognostic marker of 5-year survival and predictor of response to adjuvant chemotherapy	p53 is a key regulator of cell growth control and plays a central role in the induction of genes that are important in cell cycle arrest and apoptosis following DNA damage [18]. There is no general consensus on the prognostic or predictive role of p53 in colorectal cancer yet, but p53 mutation has been associated with significantly worse 5-year survival [19].

p21	Tissue-based potential predictive marker of response to chemotherapy	p21, a cell cycle inhibitor, is transcriptionally regulated by p53 and mediates p53-dependent growth arrest. Additionally, p21 acts as an effector of multiple tumor-suppressor pathways independent of the p53 tumor-suppressor pathway [20]. Some evidence that high expression is associated with poor response to chemotherapy in rectal cancer [21].
EMVI (Extramural venous invasion)	MRI and tissue-based prognostic marker and predictive marker of response to neoadjuvant therapy	Evidence of tumor cells is found in the vasculature outside the muscularis propria. It is a prognostic indicator associated with decreased disease-free survival. Evidence of EMVI regression on MRI is also a predictive marker of response to neoadjuvant therapy.
High levels of microsatellite instability (MSI-H)	Tissue-based prognostic biomarker	Associated with improved prognosis, even though MSI tends to be associated with poorly differentiated tumors [22].
Loss of heterozygosity of chromosome 18 (18qLOH)	Tissue-based prognostic biomarker	Chromosome 18 contains several important genes involved in carcinogenesis, and 18qLOH is associated with chromosomal instability. It is also an independent poor prognostic indicator in patients with stage II disease [23].
DYPD mutations	Serum-based marker of increased risk of 5-fluorouracil (5-FU) toxicity	Dihydropyridimine dehydrogenase (DPD) is a hepatic enzyme involved in the catabolism of uracil and thymine and is the initial rate-limiting enzyme in the metabolism of 5-FU in the liver; therefore deficiency is associated with increased toxicity secondary to impaired metabolism [24]. Currently, its predictive adequacy is not considered accurate enough for screening all patients prior to therapy [25].
Hypermethylation of the *Septin 9* gene	Plasma screening test	Hypermethylated in specimens of colorectal cancer tissue and can be detected in plasma in circulating DNA fractions of colorectal cancer cells. As a screening tool, the test has a sensitivity of 72% and specificity of 90% for the detection of CRC [26]. A multicenter clinical trial (PRESEPT trial) is currently underway.

(Continued)

TABLE 4.1 (Continued)

Biomarker	Use	Details
Panels of biomarkers that include p53, KRAS, APC, BAT-26, and long DNA	Stool-based DNA screening test for colorectal cancer and dysplastic adenomas	Panels of stool-based markers have been proposed as better screening tools than individual markers. Sensitivity for the detection of cancers and adenomas of 52–91% and 15–82%, respectively, and specificities of 93–100% [15].
		These tests are expensive, and currently the cost for population screening would be too high [27].
M2-PK (M2 pyruvate kinase) levels	Stool DNA screening test	Tumor M2-PK, an isoform of the glycolytic enzyme pyruvate kinase, is found in proliferating tissues with high nucleic acid synthesis such as tumor cells. Detection in stool samples as a screening test for colorectal cancer had a sensitivity and specificity of 72–91% and 78–79% [15].
Circulating free tumor mRNA	Plasma biomarker for surveillance of disease recurrence	Detection of free cancer cell mRNA in the plasma of patients with colorectal cancer is an independent predictor of postoperative recurrence [28].
		Currently, this approach requires mutation profiling of each patient's primary tumor prior to detection and therefore is costly and time consuming [15].

Despite rapid advances in "-omics" technologies, none is currently routinely deployed at the patient bedside.

metastases, medulloblastomas, lymphomas, and glial tumors. Low concentrations of citrate and high concentrations of choline-containing compounds (ChoCC) are metabolic characteristics observed by MRS of prostate cancer tissue. A similar approach has, therefore, also been used in prostatic cancer, where this technique demonstrated a 93–97% overall accuracy for detecting the presence of prostate cancer lesions [31]. Bertilsson et al. investigated the gene expression changes underlying these metabolic aberrations to find regulatory genes with potential for targeted therapies using 133 fresh-frozen samples from 41 patients undergoing radical prostatectomy. Gene products predicting a significantly reduced citrate level were acetylcitrate lyase (ACLY, $P = 0.003$) and m-aconitase (ACON, $P < 0.001$). The two genes whose expression most closely accompanied the increase in ChoCC were those of phospholipase A2 group VII (PLA2G7, $P < 0.001$) and choline kinase alpha (CHKA, $P = 0.002$) [32]. Thus MAS-NMR spectroscopy when incorporated into a systems-level analysis provides new insights into cancer disease mechanisms, and it is inconceivable that MAS-NMR spectroscopy can be performed within a 10- to 20-min time frame if facilities are co-located near or within clinical environments. It thus has translational potential as a clinical resource for rapid diagnostics within either the outpatient clinic or operating room environment.

More than 30 metabolites in breast cancer tissue have been identified and assigned by using MAS-NMR of intact breast tissue and extracts from breast cancer tissue (Table 4.2). Levels of metabolites, including choline and its metabolites, taurine, glycine, lactate, myo-inositol, and UDP-hexose, as well as relative intensities of glycerophosphocholine (GPC), phosphocholine (PCho), and choline, have been shown to be relevant in the differentiation of malignant and nonmalignant breast tissue. Similar shifts in metabolite profiles (relative intensities of taurine, GPC, lactate, glycine, scyllo-inositol and PCho) have been demonstrated in the differentiation of node-positive disease and increased levels of lactate, and glycine have been shown to be associated with a reduced 5-year survival in estrogen receptor (ER)–positive cancers (see Table 4.2). Using desorption electrospray ionization and mass spectrometry (DESI-MS), metabolic profiles of breast cancer tissue were sufficient for the accurate differentiation of grade 2 and grade 3, and ER-positive and ER-negative cancers (area under the curve (AUC) 0.97 and 0.96, respectively) and a further high-performance liquid chromatography–electrospray-MS study of breast cancer tissue extracts demonstrated that increasing tumor grade and negative hormone receptor status are both individually correlated with an increase in concentration of the same lipids [33].

MAS-NMR has been shown to differentiate between colorectal adenocarcinoma and normal adjacent mucosa with an AUC of 0.98. Interestingly, it has also demonstrated, in prospective analyses, that off-tumor chemistry has diagnostic as well as prognostic value, which may exceed that of the biopsy specimen. This may represent tumor heterogeneity; however, it also suggests that there is a metabolic field effect around tumors that may be of clinical significance [43].

TABLE 4.2 Summary of Metabolic Data From MAS-NMR Studies of Breast Cancer

Study	Normal Breast Tissue	Breast Cancer Tissue	Clinical Features
Cheng et al. [34]	↓Phosphocholine ↑Choline	↑Phosphocholine	**Higher grade** ↑PCho:Ch ratio ↑Lactate:choline ratio
Beckonert et al. [35] (tissue extract ^{1}H NMR)	↑Myoinositol	↑Phosphocholine (higher in higher grade) ↓Glycerophosphocholine ↑Taurine ↓Glucose	**Higher grade** ↑Phosphocholine, ↑phosphoethanolamine, ↑alanine, ↑taurine, ↑fatty acids, ↓inositol, ↓glucose, ↑UDP-hexose
Sitter et al. [36]	↑Glycerophosphocholine Ratio of PC:Cho high	↑Phosphocholine	**Tumor >2 cm** ↑Choline ↓Glycine **Lymph node involvement** ↑Glycine, ↓taurine
Bathen et al. [37]			Lymph node involvement ↑Taurine, ↑GPCh, ↓lactate, ↓PCHo, ↓glycine, ↓scyllo-inositol
Sitter et al. [38]			**Poor prognosis** ↑Glycine Altered ratio of Taurine:glycine GPC:glycine Total choline:glycine
Li et al. [39]		↑Phosphocholine ↑Taurine	**Lymph node involvement** ↑Taurine, ↑GPCh, ↓lactate, ↓PCHo, ↓glycine, ↓scyllo-inositol

Study		Findings
Cao et al. [40]		**Survivors** ↑tCho and ↓lactate pretreatment compared with nonsurvivors before treatment Survivors ↓GPC, ↓glycine, ↓lactate after treatment compared with nonsurvivors after treatment ↓Taurine after treatment best predictor of clinical response
Cao et al. [40]		**Nonsurvivors** ↓GPC and taurine after NAC compared with before NAC ↑lactate and glycine **Survivors** ↓PCh and choline-containing compounds, ↓glycine and ↑glucose after NAC compared with before NAC
Giskeødegård et al. [41]		**Poor prognosis** ↑Phosphocholine ↑Lactate, ↑glycine, ↓glucose, ↓creatine, ↓taurine
Bathen et al. [42] MAS-NMR	↑Ascorbate, ↑lactate, ↑creatine, ↑glycine, ↑taurine, ↑choline-containing metabolites (GPC, PCho, Cho) ↓glucose	

For example, this is able to classify patients according to lymph node involvement and even predict 5-year cancer-related mortality. Moreover, these analyses consistently demonstrate a fall in lipoprotein expression and a rise in fatty acid, phosphocholine, and choline expression on tumor. As the tumor stage progresses, MAS-NMR analysis in rectal cancer is able to provide accurate staging data; that is, it is able to differentiate between T1/2, T3, and T4 tumors, as lipid expression consistently changes as the tumor progresses [44].

4.5.2 Mass Spectrometry Imaging in Surgery

MS imaging (MSI) enables the simultaneous measurement of hundreds to thousands of molecular species from intact tissue sections in a spatially resolved manner, and it represents a compelling and transformative technology for the management of cancer. MSI has the potential to deliver a paradigm shift in digital pathology services and cancer diagnostics because it augments cellular morphologic analysis with highly accurate and robust data on cellular metabolic and proteomic molecular content. Moreover, it provides completely novel quantitative data on gene–environment interactions, pharmacology, microbiology, and the cancer–exposome axis. It is high throughput, reproducible, and it can describe novel companion prognostic and diagnostic markers for the stratification of cancer therapy. The advancements in the field of MSI have focused primarily on instrument development and validation of the technology. However, MSI studies have demonstrated the histologic specificity of certain lipids [45] and that individual tissue types can be identified on the basis of the presence of these lipid markers [33,46,47].

MSI characterization of intact biological tissues has, therefore, been pursued for more than 30 years. The three most commonly used imaging techniques in biological research are secondary ion mass spectrometry (SIMS), matrix-associated laser desorption ionization–mass spectrometry (MALDI-MS), and DESI-MS. MSI studies have revealed the molecular fingerprint of tissues featuring metabolic constituents, lipids, and proteins and show a high histologic specificity, and a number of prognostic markers (both single and complex) have been identified by using this approach.

4.5.2.1 *Matrix-Associated Laser Desorption Ionization–Mass Spectrometry*

MALDI is the most widely used MSI technology deployed in the analysis of surgical specimens. It is capable of detection of low and high masses to charge values ranging over 50,000 Da with a resolution of approximately 1 μm. Developed in 1997, MALDI is based on the application of an energy-absorbing matrix and uses a laser beam to cause desorption and ionization [48]. The resulting ions can then be analyzed by the mass spectrometer. The matrix is composed

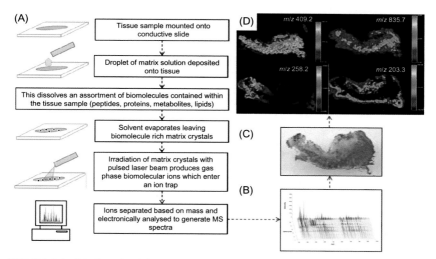

FIGURE 4.1   Matrix-assisted laser desorption ionization–mass spectrometry (MALDI-MS) for chemical mapping of intact tissue sections. (A) Sample analysis workflow; (B) spectral acquisition; (C) coregistration of features with corresponding H&E stained tissue; (D) topographical localization of different molecular compounds. *Courtesy of Dr Rezza Mirnezami.*

of compounds that can absorb energy at the laser's wavelength and normally consists of small, acidic, and aromatic molecules, which crystallize molecular structures prior to ionization (Fig. 4.1). Commonly employed solutions for this purpose include α-cyano-4-hydroxycinnamic acid (CHCA) and 2,5-dihydroxybenzoic acid (DHB). The properties of the matrix and the quality of the matrix application can have a significant effect on the mass and spatial resolution and sensitivity of the experiment [49]. This requirement for application of matrix means that MALDI measurements can only be done on an ex vivo sample.

MALDI has been used for analysis of the tumor margins and the tumor microenvironment in fresh-frozen soft tissue sarcoma and kidney samples [50]. In addition, proteins, peptide fragments, or protein and lipid fingerprints for tumor and normal tissue have been identified at least in gastrointestinal [51], respiratory [52], bladder [53], breast [54], and prostate [55] cancers and acute myeloid leukemia [56]. In ovarian cancer, it has identified candidate biomarkers such as 11S proteasome activator complex Reg-Alpha fragment, oviductin (mucin-9), and orosomucoid. Reg-Alpha, or PA28, is an antigen-processing protein, whose increased expression may allow presentation of self-peptides on tumor cells and, subsequently, immune evasion. Oviductin is a marker of oviductal epithelium and tubal differentiation marker, and orosomucoid is an acute-phase protein that had previously been evaluated as a marker of ovarian cancer and a possible immune suppressor through its action on T lymphocytes

[57]. One of the characteristics of this work is that the proteins and peptides that differentiate tissue types may be different in each type of cancer.

The successful application of MALDI-based methods to profiling biomolecules at the lower end of the m/z range (m/z values of <1000 arise from metabolites and lipids) is challenging because matrix and/or matrix-analyte cluster peaks (generated as a consequence of matrix deposition) can interfere with the detection of target low-molecular-weight compounds. Despite this, it has been deployed with success in CRC, where is provides spatially resolved data on lipid deposition, and it has confirmed the presence of cancer-adjacent metaboplasia, as initially proposed in studies by MAS-NMR analysis [58]. Findings across all tumors suggest significantly increased levels of cancer-related monounsaturated fatty acids and mono-unsaturated phosphatidylcholines relative to polyunsaturated fatty acids and polyunsaturated phosphatidylcholines in the cancer microenvironment, compared with adjacent normal tissue associated with de novo lipogenesis, with increased levels of FASN, CKα, and SCD1, further consolidating evidence of altered lipid and choline metabolism in the oncologic process [59].

A major disadvantage of the technique is the time demand of the analysis, which is currently in the range of 2–20 h for a single sample, even at coarse (100 μm) resolution; that is, it not necessarily suitable for real-time deployment in an operating room environment. However, Bregeon et al. have published a technique that could provide results within 30 min from acquiring tissue, which is subsequently homogenized [60]. An additional advantage of the technique is its ability to detect proteins from formalin-fixed tissue, which is a tissue processing technique used routinely in histopathologic workflows [61].

4.5.2.2   Desorption Electrospray Ionization–Mass Spectrometry

DESI is a novel method that allows MSI to be used to record spectra of condensed-phase samples under ambient conditions without sample preparation; it is applicable to solid samples, including complex biological materials, but it can also be applied to liquids, frozen solutions, and adsorbed gases [62]. The technology is based on charged droplets, produced by pneumatically assisted electrospray, which collide with a sample surface, causing both desorption and ionization. Critically for surgical analysis, the resulting spectra can be observed individually or as an image, similar to MALDI-MSI. This approach has specific use in surgery, however, as the ambient "soft ionization" technique causes minimal tissue architectural disruption during analysis and requires no tissue desiccation (as it operates at atmospheric pressure) or matrix-solution application; that is, it allows the same tissue section to be first subjected to MSI and then hematoxylin and eosin (H&E) staining, allowing precise feature correlation (Fig. 4.2).

Ambient ionization of tissue samples most frequently results in detection of an abundance of lipid species, including phospholipids, triglycerides,

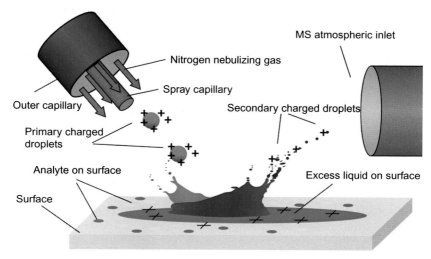

FIGURE 4.2   Illustration of desorption electrospray ionization–mass spectrometry imaging and its application in tissue analysis.

sphingolipids, and fatty acids. The role of proteins in the cancer mechanisms is well established; however, lipidomic profiling is newer, and the importance of lipid metabolism is less well defined. Metastatic liver adenocarcinoma samples were the first human specimens to be analyzed with DESI-MS using line scans. This analysis identified unique spectral differences among the tumor region, the transition regions between the tumor and normal tissue, and the normal region [61]. In stark contrast to MALDI-MS, the spectra were obtained within 5 s, although this was achieved on fresh-frozen slides [63].

DESI is now being widely deployed in the analysis of numerous tumors, including brain [64], breast [65], and gastric tumors [66]. It has been able to identify cancer-specific metabolic targets of descriptive importance, such as 2-hydroxyglutarate [64]. This is derived from α-ketoglutarate and accumulates in leukemia and specifically in grade II and III gliomas due to a mutation in isocitrate dehydrogenase 1 and 2 [67,68]. An intraoperative translational application was been developed for the detection of the biomarker during intraoperative sampling of gliomas containing a mutation in IDH1. With the use of DESI-MS installed in the operating theater, it was possible to detect the tumor metabolite 2-HG within minutes from intraoperative glioma specimens containing mutant variants of IDH1. DESI-MS is also able to differentiate among gliomas (oligodendrogliomas, astrocytomas, and oligoastrocytomas) representing a variety of grades by using lipid species such as fatty acids, glycerophosphoinositols, glycerophoserines, plasmenyl phosphoethanolamines, and sulfatides [47]. Classifiers constructed from multivariate statistical analysis of spectral signatures and correlated with histopathology revealed a high cross-validated predictive accuracy of 97% for tumor type, grade, and cellular concentration. A similar approach has also been used in breast cancer margin assessment [65].

Further attempts have been made to deploy this technology into minimally invasive operating room environments. For example, the sampling probe has been deployed via an endoscope for direct tissue analysis [69]. However, significant technical issues must be overcome with this type of analysis because the tubing is long, and it is not possible to use organic solvents for electrospray ionization.

Despite the significant advances in these areas, the major barrier to the wider adoption and transition of chemically augmented imaging into mainstream medicine and industry has been the challenge of interpreting and managing the huge metabolic and proteomic data sets that MSI generates. A stepwise increase in the computational interpretation of these highly complex data sets, is now required to ensure clinical translation. However, advances in desktop and mobile computing are making this possible. Recently, a spectral interconversion algorithm has been developed, and it bridges the MS and i-Knife approaches to fully utilize the capabilities of information system technologies [70]. Specifically, this enabled optimized data preprocessing for information recovery, precise image coregistration, and efficient extraction of tissue-specific molecular technology combined with the richness of generated molecular ion signatures for enhanced biochemical distinction of different pieces of information. This brings us closer to a reality of chemically augmented histopathology, where this process may be fully automated with the potential for virtual histochemistry.

4.6 LIPID METABOLISM IN CANCER

The influence of lipid metabolism on cancer initiation and progression has yet to be fully defined. However, these molecules are detected in abundance by ambient MS techniques, and therefore a brief summary of some key functions is provided here. The most well-known metabolic change in cancer cells is the Warburg effect. In contrast to normal differentiated cells, tumor cells rely heavily on aerobic glycolysis for the generation of the energy required for cellular processes, rather than the more efficient mitochondrial oxidative phosphorylation. Rapidly growing tumor cells may have glycolytic rates significantly higher than their tissues of origin, even in the presence of normal oxygen levels, requiring a high rate of glucose uptake to meet energy requirements and support tumor progression. It has been more recently proposed that the switch to aerobic glycolysis serves as a mechanism to provide cancer cells with not only energy but also the building blocks for macromolecule synthesis, such as carbohydrates, proteins, and lipids.

In addition to the Warbug effect, alteration in lipid metabolism is increasingly being recognized as an important distinctive feature of cancer cells [71].

Cells acquire fatty acids to meet their metabolic demand from two major sources: (1) exogenously derived (dietary) fatty acids and (2) de novo endogenous synthesis. Most adult normal cells preferentially use exogenous fatty acids, except in the case of the lactating breast and the cycling endometrium, where biosynthesis of endogenous de novo fatty acids is catalyzed by the enzyme fatty acid synthase (FASN). Other human tissues acquire the majority of required fatty acids from the circulation, and accordingly de novo lipogenesis and expression of lipogenic enzymes, such as FASN, are low. In contrast, cancer cells preferentially endogenously synthesize fatty acids, even in an abundance of extracellular fatty acids; in fact, de novo synthesis may account for more than 93% of triacylglycerol fatty acids in tumor cells [72]. The relationship between this neoplastic lipogenesis and the tumor-associated "glycotic switch" remains poorly understood. The lipogenic phenotype almost certainly conveys a growth advantage, with dysregulated lipid metabolism and associated signaling pathways altered to meet the abnormal demand for cancer cell proliferation and survival.

Continuous de novo lipogenesis provides cancer cells with membrane-building blocks, signaling lipid molecules and energy to support rapid cell proliferation; however, the implications of increased fatty acid synthesis for tumor biology and cancer cell proliferation remain largely unknown. Increased fatty acid synthesis has been assumed to lead to the upregulation of phospholipid synthesis, which meets the increased need for membrane production in highly proliferative cancer cells. These phospholipids facilitate signal transduction, intracellular trafficking, and polarization, as well as providing essential structural lipids. It has been proposed that in epithelial cancer cells, FASN is primarily involved in production of phospholipids, which are incorporated into detergent-resistant membrane microdomains or rafts [73]. This was hypothesized because the main effect of FASN modulation in cancer cells was a reduction in the synthesis of major phosphatidyl phospholipids, primarily phosphatidylcholine followed by phosphatidylethanolamine and phosphatidylserine and phosphatidylinositol. In addition, choline kinase, the first enzyme in the Kennedy pathway responsible for the de novo synthesis of phosphatidylcholine, and phosphatidylethanolamine, the most abundant phospholipid in the cell membrane, appear to be markedly overexpressed in many cancers.

Human cells are unable to synthesize polyunsaturated and highly unsaturated fatty acids from the fully saturated precursors produced by FASN, and as such, the newly synthesized phospholipids are mainly saturated and mono-unsaturated (short-chain fatty acid are converted into mono-unsaturated fatty acids by the rate-limiting action of the enzyme SCD1 and the resulting products converted by a series of enzymes to generate mono-unsaturated fatty acids [59]). Diet-derived lipids do contain polyunsaturated fatty acids, and de novo lipogenesis in cancer cells may lead to not only quantitative changes in membrane lipids required for cell growth but also compositional change [59].

The alteration of lipid metabolism in cancer functions downstream from oncogenic signaling. The expression and upregulation of enzymes involved in fatty acid synthesis start at an early stage and, in addition to increased levels of saturated phosphatidylcholine, have been associated with aggressive tumor type, risk of recurrence, and poor prognosis in breast cancer [54] and reduced disease-free survival in CRC. As such FASN is a potential target both for prognostic stratification and therapeutic intervention, and alterations in structural membrane phospholipids may be an important feature of cancer cell metabolism.

4.7 "INTELLIGENT" SURGICAL DEVICES

More than 300,000 new cancer cases are diagnosed in the United Kingdom each year, and 1.8 million diagnostic, curative, or palliative surgical procedures are performed as part of treatment. Surgical excision remains the gold standard of care for the majority of solid tissue tumors. From an oncologic perspective, a surgical intervention is determined to be curative with the complete removal of tumor tissue along with an associated border of microscopically healthy tissue, often referred to as the tumor "margin with clearances." However, the functional or cosmetic success of surgery is also dependent on minimizing the excessive removal of noncancerous or normal tissue. Evidence suggests that current oncosurgical techniques are frequently inadequate; 20% of breast cancer patients treated with breast-conserving surgery (ie, lumpectomy) require further surgery to clear positive margins [74]. This is important because the surgical resection margin remains one of the most important prognostic factors in a large number of cancers [75,76]. Furthermore, re-excision is not always possible, as in the case of soft tissue tumors or colon cancer, necessitating further adjuvant therapies.

Therefore smarter surgical devices are required to provide biological feedback on critical tumor features such as the resection margin, which, in some tumors, exists in three dimensions. These technologies are not "intelligent" in the true sense of the term, but they do allow the surgeon to interpret complex clinical pathologies in a more precise manner. In situations where surgical resection margins may be compromised, the most commonly used method for tissue identification is the intraoperative frozen section, which is the current gold standard for intraoperative determination of resection margins, lymph node status, and tissue identification. The frozen section process involves sending the resected tissue to the pathology laboratory, where it is frozen in a cryostat, cut with a microtome, stained, and examined by a pathologist. The result is then relayed back to the surgical team, and the procedure continues accordingly. This process, however, has several limitations:

1. It is slow—it takes approximately 30 minutes for the tissue to be sent to the pathology laboratory, examined, and the results relayed to the operating theater team [1].

2. It places a burden on histopathology services because a trained pathologist must be on site to examine the prepared tissue section.

3. The process can be technically demanding (both preparation and interpretation), especially under pressure of time.

4. The quality of the slides produced by frozen section is morphologically inferior to standard formalin-fixed, paraffin-embedded tissue processing, making fixed tissue processing the preferred method for more accurate diagnosis. Low sensitivity and a high incidence of false negatives in all types of malignancy are the primary concerns, with overall sensitivity being 83% in the detection of positive margins in breast-conserving surgery [1].

5. The number of points that can reasonably be sampled during a single surgical resection is limited, so the frozen section must be targeted to particular areas of concern, which can be difficult to identify in, for example, a wide local excision for breast cancer.

6. The process is expensive, with an average cost of $300.

Intraoperative diagnostics have, thus far, been unable to provide information quickly enough for the rapid on-table surgical decision-making process. With recent advances in spectroscopic techniques, a multitude of spectroscopic methods for the identification of individual tissue morphologies have been developed, aimed at providing the clinician with histologic type, tissue-specific information, quickly and accurately within the operating theater environment. The ideal characteristics for an intraoperative diagnostic instrument are outlined in Table 4.3.

TABLE 4.3 Requirements for Real-Time Diagnostic Instruments in the Operating Room Environment, and Some Key Surgical Areas Where These Technologies Will Provide Significant Impact in Surgery

Major Requirement for Real-Time Margin Detection and Tissue Phenotyping Technology in Surgery	Important Clinical Areas of Application

<ul><li>Rapid (not significantly increase the duration of the surgery or interrupt the surgical workflow).</li><li>High throughput (ability to be used on many patients).</li><li>Objective and robust.</li><li>Greater accuracy than currently available methods.</li><li>Non-destructive.</li><li>Minimal tissue preprocessing.</li><li>Cost-effective.</li><li>High diagnostic accuracy (with high negative predictive value in cancer).</li><li>Safe (eg, no exposure to ionizing radiation).</li></ul>	Anatomical orientation and identification
	Cancer
	    Diagnostics, prognostics, and chemical staging
	    Real-time phenotyping for stratification of therapy and surgical therapy
	    Resection margin detection
	    Lymph node involvement and metastases identification
	    Detection (eg, within inflammation)
	Inflammation
	    Diagnosis and severity scoring
	    Resection margin (eg, Crohn disease)
	Ischemia
	    Diagnosis, detection, and prognosis
	Therapeutic efficacy measurement
	    Radiofrequency ablation
	    Laser ablation

The majority of significant advances in technologies for intraoperative margin determination have been made in the field of breast cancer, particularly since the introduction of breast-conserving surgery as the preferred surgical approach for the majority of patients (Table 4.4). These include techniques for examining the excised specimen or the surgical cavity with the aim of reducing positive margins and therefore the need for re-excision. Traditional pathologic methods such as frozen section and imprint cytology perform well in terms of sensitivity and specificity but significantly increase operating time for limited sampling points [77]. More novel methods such as digital specimen mammography, radiofrequency spectroscopy, and intraoperative ultrasonography were found to be significantly more time efficient in a recent systematic review but suffered from poor accuracy or had an incomplete evidence base [77]. A summary of technologies, both in use and in development, for intraoperative margin assessment in breast cancer are summarized in Table 4.4.

The MarginProbe, a new radiofrequency spectroscopy probe, performs well in terms of concordance with formal histopathology [77]. Spectroscopic techniques are attractive because they provide reproducible, quantitative biochemical information about the morphologic nature of tissues nondestructively, noninvasively, and without the use of ionizing radiation. The MarginProbe is a commercially available example of radiofrequency spectroscopy designed for intraoperative use. The device was initially developed for use in breast-conserving cancer resections and has been further developed for potential use in prostate cancer and other solid organ malignancies. The device measures the differing electromagnetic signature, which is reflected by both healthy and cancerous tissues, a result of the inherent structural differences between normal and malignant cells. The system comprises a single-use probe and a portable console, and the technique is rapid (<3 s) and nondestructive. Approved by the US Food and Drug Administration (FDA) in Jan. 2013, the device is currently commercially available in Europe and Israel. When used as an adjunctive tool to standard surgical care in 22 patients with impalpable ductal carcinoma in situ undergoing breast-conserving surgery, the use of the probe reduced re-excision rates from 38.8% to 18% [66]. A prospective randomized controlled trial with 300 recruited patients demonstrated an increase in the intraoperative identification and re-excision of all positive margins in the device group (60% vs 41%). Repeat lumpectomy rate was reduced in the device group compared with the control group (5.6% vs 12.7%) [67].

A number of alternative spectroscopic techniques are in development, including optical spectroscopy, Fourier transform infrared spectroscopy, and fluorescence spectroscopy. Optical spectroscopy, including diffuse reflectance spectroscopy, uses the interaction of light with tissue at the cellular and subcellular levels to provide a rapid, in vivo diagnosis based on differing absorption, reflectance, and scattering spectra produced by differences in cell morphology associated with malignant transformation [68]. A single optical probe, developed for use in skin lesions, has potential as a tool for the identification of

TABLE 4.4 Summary of Key Margin Detection Tools and Their Clinical Evidence Currently Being Deployed for Clinical Use

Technology	Study	Sensitivity	Specificity	Accuracy	Increased Operating Time
Intraoperative DSM	Kauffman et al. [78]	36%	71%	Not reported	−19 min
	Kim et al. [79]	Not reported	Not reported	Not reported	−2 min
Intraoperative ultrasonography	Moore et al. [80]	Not reported	Not reported	Not reported	−15 min
					−1 min
	Rahusen et al. [81]	Not reported	Not reported	Not reported	Not reported
	James et al. [82]	Not reported	Not reported	Not reported	Not reported
	Olsha et al. [83]	100%	74%	99.6%	Not reported
	Doyle et al. [84]	25%	95%	Not reported	Not reported
	Ramos et al. [85]	80%	86.6%	Not reported	Not reported
Radiofrequency spectroscopy probe	Karni et al. [86]	71%	68%	Not reported	+7.37 min
	Allweis et al. [87]	Not reported	Not reported	Not reported	Not reported
	Thill et al. [88]	Not reported	Not reported	73%	Not reported
	Rivera et al. [89]	Not reported	Not reported	Not reported	Not reported
Intraoperative frozen section	Olson et al. [90]	73%	99%	98%	Not reported
	Weber et al. [91]	80%	87.5%	83.8%	Not reported
	Dener et al. [92]	Not reported	Not reported	94%	+25 min
	Jorns et al. [93]	Not reported	Not reported	98.3%	+30 min
	Sabel et al. [93]	91.1%	100%	Not reported	+27 min

(Continued)

TABLE 4.4 (Continued)

Technology	Study	Sensitivity	Specificity	Accuracy	Increased Operating Time
Touch imprint cytology	Creager et al. [94]	80%	85%	85%	+20 min
	Martin et al. [95]	70%	97.1%	93.8%	Not reported
	Sumiyoshi et al. [113]	85%	100%	89.7%	+25 min
Optical coherence tomography	Nguyen et al. [96]	100%	82%	90%	Not reported
Gamma camera	Duarte et al. [97]	Not reported	Not reported	Not reported	Not reported
	Paredes et al. [98]	Not reported	Not reported	60%	+5 min
					Not reported
Gross tissue inspection/specimen radiography	Chagpar et al. [99]	Not reported	Not reported	Not reported	Not reported
	Fleming et al. [100]	73%	88%	Not reported	Not reported
	Cabioglu et al. [101]	92%	78%	87%	Not reported
					Not reported
Two-view specimen mammography	McCormick et al. [102]	55%	88%	Not reported	+15 min

malignant and benign tissues in vivo [69]. A similar single-fiber optical probe system has been described for the intraoperative detection of low-grade and high-grade brain tumors and the identification of positive tumor margins [70]. Diffuse reflectance spectroscopy using β-carotene/scattering ratios has been applied in the assessment of surgical margins during breast-conserving surgery. Positive margins were identified with sensitivity and specificity of 79% and 67%, respectively [71]. Fluorescence spectroscopy uses either native fluorescence from endogenous fluorophores such as flavins, elastin, and collagen or emission from exogenously administered fluorescent substances for in vivo quantitative tissue diagnostics. Advances in light delivery and collection systems (particularly fiberoptic systems) have facilitated the growth of fluorescence-based techniques as potentially translatable tools for the clinical environment. A near-infrared-capable laparoscopic tool has been optimally adapted to the fluorescence characteristics of indocyanine green, and neuro-endoscopes to allow for endoscopically assisted 5-ALA-fluorescence-guided resections during surgery for malignant brain tumors. The WavSTAT system by SpectraScience has, thus far, focused on the endoscopic detection of dysplastic and malignant colorectal tissues using laser-induced autofluorescence spectroscopy.

Despite significant interest and investment in the development of technologies for in vivo tissue diagnostics, none of these systems has, as yet, become part of routine clinical practice.

4.8   RAPID EVAPORATIVE IONIZATION–MASS SPECTROMETRY

To date, three forms of ambient MS have been deployed for use in the operating room: (1) Laser desorption ionization–mass spectrometry is an ambient ionization technique that can utilize surgical lasers in the ultraviolet and far-infrared wavelengths [103]. (2) Venturi easy ambient sonic-spray ionization [104] has been described as a means for directly linking the liquefied tissue extraction tubing to a Venturi air jet pump at the MS interface. (3) A third ambient ionization method that can be coupled to a surgical instrument for intraoperative measurements is rapid evaporative ionization–mass spectrometry (REIMS) [105]. In the surgical environment, electrosurgical devices are commonly employed for hemostasis, for improved accuracy of dissection, and for therapeutic purposes. Numerous energy devices are deployed on the basis of monopolar and bipolar electrocautery in most operating theaters globally (Fig. 4.3A). A major byproduct of their use is the production of "smoke" from the evaporating tissue as it is being resected. Historically, surgical smoke has been considered toxic and an irritant; therefore, it is often dispelled from the operative field. It is, in fact, a rich source of biological information, and the REIMS technique employs standard electrosurgical methods as a means of converting tissue components into gas-phase ionic species amenable to MS analysis [106]. Tissue specificity of the REIMS method has been shown to be similar to that of other DI-MS

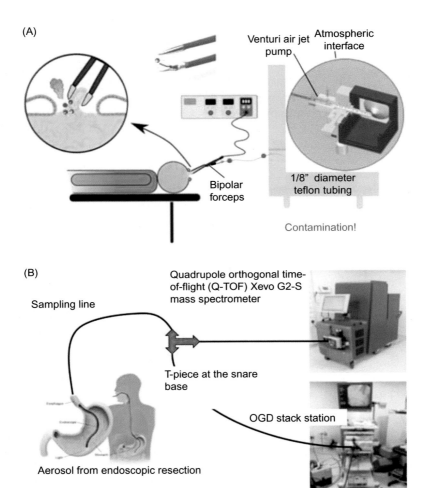

FIGURE 4.3   (A) Illustration of REIMS set up. (B) Adaption for use in endoscopic analysis and therapy of gastrointestinal luminal tumors, although this would also be applicable to other systems (eg, otolaryngology, urology, and the respiratory system).

methods [107]. Although the spatial resolution of REIMS technique is limited by the probe being a handheld one and the geometry, 100 μm is generally achievable with instantaneous (<0.9 s) feedback to the operator. Because the technique does not require the modification of electrosurgical tools, testing in the surgical environment involves only the modification of the mass spectrometer and coupling it to the surgical instruments (Fig. 4.3A). The resulting MS profiles are highly specific to the type of tissue analyzed, allowing for tissue identification as well as characterization on a level comparable with histopathologic analysis.

This analytical coupling creates new chemical information sets that describe the tissue and its associated pathology. The technology is dependent on the creation of an ex vivo database of histologically validated spectral data. Database and multivariate statistical tissue identification algorithms are then used for sample or spectral classification during surgery.

In contrast to MSI, Rapid Evaporative Ionization Mass Spectrometry (REIMS) was developed exclusively for in situ, even in vivo chemical characterization of tissues. The development of the REIMS technique was based on the discovery that all surgical instruments that utilize thermal evaporation approaches (electrosurgery, laser surgery, radiofrequency ablation, microwave ablation, etc.) ionize the molecular constituents of biological tissues. The subsequent hyphenation of surgical instruments with MS has yielded a novel approach capable of identifying tissues and their pathologic subtypes during surgical or diagnostic interventions [90]. The data generated from REIMS bears a striking similarity to other MS imaging techniques (MALDI-MSI or DESI-MSI) data [91]. Although REIMS is a more suitable technology for the surgical environment, the latter guarantees a high histologic specificity, with quantitative and potentially automated histopathologic analysis of tissue specimens. Despite this, low-molecular-weight metabolites, including amino acids, peptides, oligosaccharides, bile acids, phospholipids, lipopolysaccharides, and eicosanoids (among many other metabolically significant species) can readily be detected by using the REIMS technology.

The REIMS technology has now been successfully tested in the surgical environment [90], and to date, 2933 ex vivo specimens consisting of normal and cancerous tissues from 302 individual patients have been analyzed by using this technology. The samples contained stomach, colon, liver, lung, and breast tissues, as well as other tissue types. Spectra from these samples was detected in the mass range 600–900, representing predominantly lipids, and the data collected were correlated with histopathology before applying multivariant statistics (principal component analysis and linear discriminant analysis). Cross-validation revealed the sensitivities and specificities of the correct tissue classification, ranging from 69% for mixed adenosquamous cell carcinoma to 100% for healthy tissue from the lung and the liver [105]. Further validation was obtained by measuring in vivo data in the operating theater for 81 patients. The sensitivities and specificities from this analysis ranged between 95–100% and 92–100%, respectively. The measurement data were obtained within 2.5 s, resulting in near real time measurements during the operation. The rapid data acquisition and the ease of incorporation into the clinical environment are the main advantages of REIMS.

The technology has also been successfully tested in neurosurgical brain tumor excision of astrocytomas, meningiomas, metastatic brain tumors, and healthy brain tissue with similar sensitivity and specificity [93]. Thus, lipidomic/metabonomic MS provides real-time, descriptive in vivo data that are directly comparable with postinterventional histologic analysis. Since the technique

(A)

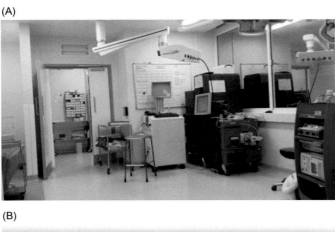

(B)

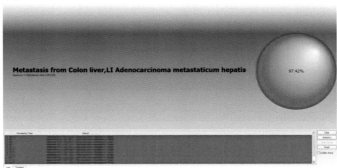

FIGURE 4.4 (A) REIMS instrument stored in an operating theater, ready for deployment. (B) Visualization of spectral data during surgery. The color signifies this is a malignant pathology (green is for benign) and provides data on the accuracy percentage and histologic subtype.

allows for the detection of genus-specific or serotype-specific markers of bacterial strains, the data also provide indirect information on the composition of microbial communities (Fig. 4.4).

The REIMS technology has now been developed for deployment via endoscopic and other minimally invasive routes. The applications are, therefore, numerous and vital for the further development of bedside diagnostic chemistry. A large number of colonoscopies are performed in Europe each year, and the number is increasing rapidly as REIMS becomes a prominent tool in colon cancer screening strategies. Electronic nose devices (also known as "artificial olfaction") based on volatile organocarbon sensing via gas chromatography (GC) and MS instruments have been used in studies of respiratory inflammatory states and in a single animal study of colitis, which demonstrated volatile patterns from the feces of patients with ulcerative colitis, and *Clostridium difficile* and *Campylobacter jejuni* infections. However, this technology has not yet been deployed via the endoscope, since it lacks sensitivity and is not suitable for detailed mucosal phenotyping. At least nine different scoring systems are used as outcome measures of disease severity in clinical trials. Endoscopy has an important role in most, and its contribution is index-specific. Consequently, interobserver variation in assessing endoscopic activity is important because any disagreement can alter the proportion of patients identified as being in remission and can influence regulatory decisions. White light endoscopy with 40–50 random biopsies has been promoted for surveillance but may miss a significant proportion of lesions. However, the yield of random biopsies to detect intraepithelial neoplasia is low, and random biopsies are expensive and labor intensive and distract attention from scrutiny of the colon. Recently, new endoscopic techniques have been proposed, and these distinguished as those that utilize a point measurement and still image and those that utilize real-time imaging approaches. Point measurement techniques include Raman spectroscopy, elastic (light) scattering spectroscopy, fluorescence spectroscopy, optical coherence tomography, and confocal laser microscopy. Chromoendoscopy significantly increases the sensitivity of detecting subtle dysplastic lesions and has emerged as a new standard of cancer surveillance in patients with inflammatory bowel disease. Pan-colonic chromoendoscopy has shown a significantly increased detection rate for diminutive adenoma [108]. Reports on chromoendoscopic findings in microscopic colitis are scarce, although a recent small-scale study showed some promise. Real-time technologies include narrow band imaging, fluorescence imaging, computer-generated endoscopic imaging, spectral videocolonoscopy, the application of photosensitizers (such as 5-aminolevulinic acid), immunoscopy (coupling of fluorescence dyes to tumor-related antigens), and endocytoscopy. However, these are costly and largely deployed as experimental adjuncts. Moreover, none of these approaches provides real-time biochemical or microbiological data to the clinician for diagnostic, prognostic, or therapeutic purposes. REIMS can be deployed to augment all of these approaches, rather than using it as an alternative. Metabolic phenotyping approaches have been extensively applied to the detection of inflammatory bowel conditions, and considerable evidence supports its capacity for distinguishing between inflammatory and noninflammatory states [109]. Therefore the deployment of chemically augmented endoscopic technologies heralds a significant advantage in the analysis and clinical management of this important condition.

Several other types of ambient MS have also been described in an attempt to overcome the destructive nature of REIMS (see Table 4.5). However, none of these has been deployed prospectively in human trials. Their advantages include short processing times of less than a minute and higher signal intensity in comparison with DESI-MS from the same tissue, and for paper spray–mass spectrometry (PS-MS) there is the potential to directly measure intracellular lipids without homogenization for probe electrospray ionization–mass spectrometry (PESI-MS) [110].

TABLE 4.5 Summary of Ambient Mass Spectrometry Techniques That Have Been Deployed in the Analysis of Surgical Tissues

	Name	In Vivo Human Experimental Data
DESI	Desorption electrospray–ionization	Yes
REIMS	Rapid Evaporative ionization–mass spectrometry	Yes
PS-MS [111] [2]	Paper spray–mass spectrometry	No
PESI-MS [112] [3]	Probe electrospray ionization–mass spectrometry	No
SPME [108]	Solid-phase microextraction	No

4.9 SUMMARY AND CONCLUSIONS

The future of precision surgery is dependent on the application of the "-omics" technologies in the operating room. Significant obstacles still remain; however, the rapid development of these technologies in recent years indicates that there is now a real opportunity to radically alter how surgical decisions are made in the operating room environment. These approaches may provide the surgeon with data on the oncological phenotype, the microbiology and even a patients likely drug response to adjuvant therapies. Moreover, technologies such as REIMS, DESI, and even MAS-NMR can be deployed in an integrated package to provide a deep analysis of common surgical conditions. They may also be used to augment current best practices and to work synergistically with complementary technologies, for example, in endoscopy. DESI has the potential to automate some components of current histologic practice with significant economic and translational impacts. Virtual histochemistry and analysis at the bedside is a revolutionary concept and one that has the potential to radically alter how cancer is treated. However, these technologies are still in their infancy, and the results of prospective randomized control trials are now awaited to determine if the age of precision surgery has truly arrived.

REFERENCES

[1] Kinross JM, Holmes E, Darzi AW, Nicholson JK. Metabolic phenotyping for monitoring surgical patients. Lancet 2011;377:1817–9.

[2] Mirnezami R, Kinross JM, Vorkas PA, Goldin R, Holmes E, Nicholson J, et al. Implementation of molecular phenotyping approaches in the personalized surgical patient journey. Ann Surg 2012;255:881–9.

[3] Nicholson JK, Holmes E, Kinross JM, Darzi AW, Takats Z, Lindon JC. Metabolic phenotyping in clinical and surgical environments. Nature 2012;491:384–92.

[4] Li JV, Ashrafian H, Bueter M, Kinross J, Sands C, le Roux CW, et al. Metabolic surgery profoundly influences gut microbial-host metabolic cross-talk. Gut 2011;60:1214–23.

[5] Mutch DM, Fuhrmann JC, Rein D, Wiemer JC, Bouillot JL, Poitou C, et al. Metabolite profiling identifies candidate markers reflecting the clinical adaptations associated with Roux-en-Y gastric bypass surgery. PLoS One 2009;4.

[6] Zhang H, DiBaise JK, Zuccolo A, Kudrna D, Braidotti M, Yu Y, et al. Human gut microbiota in obesity and after gastric bypass. Proc Natl Acad Sci USA 2009;106:2365–70.

[7] Digital first: clinical transformation through pathology innovation. Available at: <http://www.england.nhs.uk/wp-content/uploads/2014/02/pathol-dig-first.pdf>; 2014.

[8] Junttila MR, de Sauvage FJ. Influence of tumour micro-environment heterogeneity on therapeutic response. Nature 2013;501:346–54.

[9] Shah AA, Craig DM, Sebek JK, Haynes C, Stevens RC, Muehlbauer MJ, et al. Metabolic profiles predict adverse events after coronary artery bypass grafting. J Thorac Cardiovasc Surg 2012;143:873–8.

[10] Mao H, Wang H, Wang B, Liu X, Gao H, Xu M, et al. Systemic metabolic changes of traumatic critically ill patients revealed by an NMR-based metabonomic approach. J Proteome Res 2009;8:5423–30.

[11] Chen J, Wen H, Liu J, Yu C, Zhao X, Shi X, et al. Metabonomics study of the acute graft rejection in rat renal transplantation using reversed-phase liquid chromatography and hydrophilic interaction chromatography coupled with mass spectrometry. Mol BioSyst 2012;8:871.

[12] Kim C-D, Kim EY, Yoo H, Lee JW, Ryu do H, Noh DW, et al. Metabonomic analysis of serum metabolites in kidney transplant recipients with cyclosporine A- or tacrolimus-based immunosuppression. Transplantation 2010;90:748–56.

[13] Girlanda R, Cheema AK, Kaur P, Kwon Y, Li A, Guerra J, et al. Metabolomics of human intestinal transplant rejection. Am J Transplant 2012;12.

[14] GLOBAL STATUS REPORT on noncommunicable diseases. Available at: <http://apps.who.int/iris/bitstream/10665/148114/1/9789241564854_eng.pdf?ua=1>; 2014.

[15] Newton KF, Newman W, Hill J. Review of biomarkers in colorectal cancer. Colorectal Dis 2012;14:3–17.

[16] Van Cutsem E, Peeters M, Siena S, Humblet Y, Hendlisz A, Neyns B, et al. Open-label phase III trial of panitumumab plus best supportive care compared with best supportive care alone in patients with chemotherapy-refractory metastatic colorectal cancer. J Clin Oncol 2007;25:1658–64.

[17] Tol J, Nagtegaal ID, Punt CJA. BRAF mutation in metastatic colorectal cancer. N Engl J Med 2009;361:98–9.

[18] Pietsch EC, Humbey O, Murphy ME. Polymorphisms in the p53 pathway. Oncogene 2006;25:1602–11.

[19] Samowitz WS, Curtin K, Ma KN, Edwards S, Schaffer D, Lepper MF, et al. Prognostic significance of p53 mutations in colon cancer at the population level. Int J Cancer 2002;99:597–602.

[20] Abbas T, Dutta A. p21 in cancer: intricate networks and multiple activities. Nat Rev Cancer 2009;9:400–14.

[21] Sim SH, Kang MH, Kim YJ, Lee KE, Kim DW, Kang SB, et al. P21 and CD166 as predictive markers of poor response and outcome after fluorouracil-based chemoradiotherapy for the patients with rectal cancer. BMC Cancer 2014;14:241.

[22] Popat S, Hubner R, Houlston RS. Systematic review of microsatellite instability and colorectal cancer prognosis. J Clin Oncol 2005;23:609–18.

[23] Wang W, Li YF, Sun XW, Chen G, Zhan YQ, Huang CY, et al. Correlation analysis between loss of heterozygosity at chromosome 18q and prognosis in the stage-II colon cancer patients. Chin J Cancer 2010;29:761–7.

[24] Newton RC, Noonan DP, Vitiello V, Clark J, Payne CJ, Shang J, et al. Robot-assisted transvaginal peritoneoscopy using confocal endomicroscopy: a feasibility study in a porcine model. Surg Endosc 2012;26:2532–40.

[25] Yen JL, McLeod HL. Should DPD analysis be required prior to prescribing fluoropyrimidines? Eur J Cancer 2007;43:1011–6.

[26] Grützmann R, Molnar B, Pilarsky C, Habermann JK, Schlag PM, Saeger HD, et al. Sensitive detection of colorectal cancer in peripheral blood by septin 9 DNA methylation assay. PLoS One 2008;3.

[27] Ouyang DL, Chen JJ, Getzenberg RH, Schoen RE. Noninvasive testing for colorectal cancer: a review. Am J Gastroenterol 2005;100:1393–403.

[28] Wang J-Y, Wu CH, Lu CY, Hsieh JS, Wu DC, Huang SY, et al. Molecular detection of circulating tumor cells in the peripheral blood of patients with colorectal cancer using RT-PCR: significance of the prediction of postoperative metastasis. World J Surg 2006;30:1007–13.

[29] Opstad KS, Bell BA, Griffiths JR, Howe FA. Toward accurate quantification of metabolites, lipids, and macromolecules in HRMAS spectra of human brain tumor biopsies using LCModel. Magn Reson Med 2008;60:1237–42.

[30] Wright AJ, Fellows GA, Griffiths JR, Wilson M, Bell BA, Howe FA. Ex-vivo HRMAS of adult brain tumours: metabolite quantification and assignment of tumour biomarkers. Mol Cancer 2010;9:66.

[31] Wu C-L, Jordan KW, Ratai EM, Sheng J, Adkins CB, DeFeo EM, et al. Metabolomic imaging for human prostate cancer detection. Sci Transl Med 2010;2:16ra8.

[32] Bertilsson H, Tessem MB, Flatberg A, Viset T, Gribbestad I, Angelsen A, et al. Changes in gene transcription underlying the aberrant citrate and choline metabolism in human prostate cancer samples. Clin Cancer Res 2012;18:3261–9.

[33] Dill AL, Ifa DR, Manicke NE, Ouyang Z, Cooks RG. Mass spectrometric imaging of lipids using desorption electrospray ionization. J Chromatogr B Analyt Technol Biomed Life Sci 2009;877:2883–9.

[34] Cheng LL, Chang IW, Smith BL, Gonzalez RG. Evaluating human breast ductal carcinomas with high-resolution magic-angle spinning proton magnetic resonance spectroscopy. J Magn Reson 1998;135:194–202.

[35] Beckonert O, Monnerjahn J, Bonk U, Leibfritz D. Visualizing metabolic changes in breast-cancer tissue using 1H-NMR spectroscopy and self-organizing maps. NMR Biomed 2003;16:1–11.

[36] Sitter B, Lundgren S, Bathen TF, Halgunset J, Fjøsne HE, Gribbestad IS. Comparison of HR MAS MR spectroscopic profiles of breast cancer tissue with clinical parameters. NMR Biomed 2006;19:30–40.

[37] Bathen TF, Jensen LR, Sitter B, Fjøsne HE, Halgunset J, Axelson DE, et al. MR-determined metabolic phenotype of breast cancer in prediction of lymphatic spread, grade, and hormone status. Breast Cancer Res Treat 2007;104:181–9.

[38] Sitter B, Bathen TF, Singstad TE, Fjøsne HE, Lundgren S, Halgunset J, et al. Quantification of metabolites in breast cancer patients with different clinical prognosis using HR MAS MR spectroscopy. NMR Biomed 2010;23:424–31.

[39] Li M, Song Y, Cho N, Chang JM, Koo HR, Yi A, et al. An HR-MAS MR metabolomics study on breast tissues obtained with core needle biopsy. PLoS One 2011;6:e25563.

[40] Cao MD, Giskeødegård GF, Bathen TF, Sitter B, Bofin A, Lønning PE, et al. Prognostic value of metabolic response in breast cancer patients receiving neoadjuvant chemotherapy. BMC Cancer 2012;12:39.

[41] Giskeødegård GF, Lundgren S, Sitter B, Fjøsne HE, Postma G, Buydens LM, et al. Lactate and glycine-potential MR biomarkers of prognosis in estrogen receptor-positive breast cancers. NMR Biomed 2012;25:1271–9.

[42] Bathen TF, Geurts B, Sitter B, Fjøsne HE, Lundgren S, Buydens LM, et al. Feasibility of MR metabolomics for immediate analysis of resection margins during breast cancer surgery. PLoS One 2013;8:e61578.

[43] Jimenez B, Mirnezami R, Kinross J, Cloarec O, Keun HC, Holmes E, et al. 1H HR-MAS NMR spectroscopy of tumor-induced local metabolic 'field-effects' enables colorectal cancer staging and prognostication. J Proteome Res 2013;12:959–68.

[44] Mirnezami R, Jimenez B, Li JC, Kinross JM, Veselkov K, Goldin RD, et al. Rapid diagnosis and staging of colorectal cancer via high-resolution magic angle spinning nuclear magnetic resonance (HR-MAS NMR) spectroscopy of intact tissue biopsies. Ann Surg 2014;259:1138–49.

[45] Jackson SN, Wang HYJ, Woods AS. Direct profiling of lipid distribution in brain tissue using MALDI-TOFMS. Anal Chem 2005;77:4523–7.

[46] Eberlin LS, Ferreira CR, Dill AL, Ifa DR, Cooks RG. Desorption electrospray ionization mass spectrometry for lipid characterization and biological tissue imaging. Biochim Biophys Acta 2011;1811:946–60.

[47] Eberlin LS, Norton I, Dill AL, Golby AJ, Ligon KL, Santagata S, et al. Classifying human brain tumors by lipid imaging with mass spectrometry. Cancer Res 2012;72:645–54.

[48] Caprioli RM, Farmer TB, Gile J. Molecular imaging of biological samples: localization of peptides and proteins using MALDI-TOF MS. Anal Chem 1997;69:4751–60.

[49] Yalcin EB, de la Monte SM. Review of matrix-assisted laser desorption ionization-imaging mass spectrometry for lipid biochemical histopathology. J Histochem Cytochem 2015.

[50] Caldwell RL, Gonzalez A, Oppenheimer SR, Schwartz HS, Caprioli RM. Molecular assessment of the tumor protein microenvironment using imaging mass spectrometry. Cancer Genomics Proteomics 2006;3:279–87.

[51] Zhang MH, Xu XH, Wang Y, Ling QX, Bi YT, Miao XJ, et al. A prognostic biomarker for gastric cancer with lymph node metastases. Anat Rec 2013;296:590–4.

[52] Gámez-Pozo A, Sanchez-Navarro I, Nistal M, Calvo E, Madero R, Diaz E, et al. MALDI profiling of human lung cancer subtypes. PLoS One 2009;4.

[53] Schwamborn K, Krieg RC, Grosse J, Reulen N, Weiskirchen R, Knuechel R, et al. Serum proteomic profiling in patients with bladder cancer. Eur Urol 2009;56:989–96.

[54] Chughtai K, Jiang L, Greenwood TR, Glunde K, Heeren RM. Mass spectrometry images acyl-carnitines, phosphatidylcholines, and sphingomyelin in MDA-MB-231 breast tumor models. J Lipid Res 2013;54:333–44.

[55] Flatley B, Malone P, Cramer R. MALDI mass spectrometry in prostate cancer biomarker discovery. Biochim Biophys Acta 2014;1844:940–9.

[56] Kazmierczak M, Luczak M, Lewandowski K, Handschuh L, Czyz A, Jarmuz M, et al. Esterase D and gamma 1 actin level might predict results of induction therapy in patients with acute myeloid leukemia without and with maturation. Med Oncol 2013;30:725.

[57] Gustafsson JOR, Oehler MK, Ruszkiewicz A, McColl SR, Hoffmann P. MALDI imaging mass spectrometry (MALDI-IMS)-application of spatial proteomics for ovarian cancer classification and diagnosis. Int J Mol Sci 2011;12:773–94.

[58] Mirnezami R, Spagou K, Vorkas PA, Lewis MR, Kinross J, Want E, et al. Chemical mapping of the colorectal cancer microenvironment via MALDI imaging mass spectrometry (MALDI-MSI) reveals novel cancer-associated field effects. Mol Oncol 2014;8:39–49.

[59] Guo S, Wang Y, Zhou D, Li Z. Significantly increased monounsaturated lipids relative to polyunsaturated lipids in six types of cancer microenvironment are observed by mass spectrometry imaging. Sci Rep 2014;4:5959.

[60] Bregeon F, Brioude G, De Dominicis F, Atieh T, D'Journo XB, Flaudrops C, et al. MALDI-ToF mass spectrometry for the rapid diagnosis of cancerous lung nodules. PLoS One 2014;9: e97511.

[61] Seeley EH, Caprioli RM. Molecular imaging of proteins in tissues by mass spectrometry. Proc Natl Acad Sci USA 2008;105:18126–31.

[62] Takáts Z, Wiseman JM, Cooks RG. Ambient mass spectrometry using desorption electrospray ionization (DESI): instrumentation, mechanisms and applications in forensics, chemistry, and biology. J Mass Spectrom 2005;40:1261–75.

[63] Wiseman JM, Puolitaival SM, Takáts Z, Cooks RG, Caprioli RM. Mass spectrometric profiling of intact biological tissue by using desorption electrospray ionization. Angew Chemie 2005;117:7256–9.

[64] Santagata S, Eberlin LS, Norton I, Calligaris D, Feldman DR, Ide JL, et al. Intraoperative mass spectrometry mapping of an onco-metabolite to guide brain tumor surgery. Proc Natl Acad Sci USA 2014;111:11121–6.

[65] Calligaris D, Caragacianu D, Liu X, Norton I, Thompson CJ, Richardson AL, et al. Application of desorption electrospray ionization mass spectrometry imaging in breast cancer margin analysis. Proc Natl Acad Sci USA 2014;111:15184–9.

[66] Eberlin LS, Tibshirani RJ, Zhang J, Longacre TA, Berry GJ, Bingham DB, et al. Molecular assessment of surgical-resection margins of gastric cancer by mass-spectrometric imaging. Proc Natl Acad Sci USA 2014;111:2436–41.

[67] Ward PS, Patel J, Wise DR, Abdel-Wahab O, Bennett BD, Coller HA, et al. The common feature of leukemia-associated IDH1 and IDH2 mutations is a neomorphic enzyme activity converting alpha-ketoglutarate to 2-hydroxyglutarate. Cancer Cell 2010;17: 225–34.

[68] Choi C, Ganji SK, DeBerardinis RJ, Hatanpaa KJ, Rakheja D, Kovacs Z, et al. 2-hydroxyglutarate detection by magnetic resonance spectroscopy in IDH-mutated patients with gliomas. Nat Med 2012;18:624–9.

[69] Chen CH, Lin Z, Garimella SVB, Zheng L, Shi R, Cooks RG. Development of a mass spectrometry sampling probe for chemical analysis in surgical and endoscopic procedures. Anal Chem 2013;85:11843–50.

[70] Veselkov KA, Mirnezami R, Strittmatter N, Goldin RD, Kinross J, Speller AVM, et al. Chemo-informatic strategy for imaging mass spectrometry-based hyperspectral profiling of lipid signatures in colorectal cancer. Proc Natl Acad Sci USA 2014;111:1216–21.

[71] Kuhajda FP. Fatty acid synthase and cancer: new application of an old pathway. Cancer Res 2006;66:5977–80.

[72] Menendez JA, Lupu R. Fatty acid synthase and the lipogenic phenotype in cancer pathogenesis. Nat Rev Cancer 2007;7:763–77.

[73] Li J, Dong L, Wei D, Wang X, Zhang S, Li H. Fatty acid synthase mediates the epithelial-mesenchymal transition of breast cancer cells. Int J Biol Sci 2014;10:171–80.

[74] Jeevan R, Cromwell DA, Trivella M, Lawrence G, Kearins O, Pereira J, et al. Reoperation rates after breast conserving surgery for breast cancer among women in England: retrospective study of hospital episode statistics. BMJ 2012;345:e4505.

[75] Westgaard A, Tafjord S, Farstad IN, Cvancarova M, Eide TJ, Mathisen O, et al. Resectable adenocarcinomas in the pancreatic head: the retroperitoneal resection margin is an independent prognostic factor. BMC Cancer 2008;8:5.

[76] Wibe A, Rendedal PR, Svensson E, Norstein J, Eide TJ, Myrvold HE, et al. Prognostic significance of the circumferential resection margin following total mesorectal excision for rectal cancer. Br J Surg 2002;89:327–34.

[77] Butler-Henderson K, Lee AH, Price RI, Waring K. Intraoperative assessment of margins in breast conserving therapy: a systematic review. Breast 2014;23:112–9.

[78] Kaufman CS, Jacobson L, Bachman BA, Kaufman LB, Mahon C, Gambrell LJ, et al. Intraoperative digital specimen mammography: rapid, accurate results expedite surgery. Ann Surg Oncol 2007;14:1478–85.

[79] Kim SHH, Cornacchi SD, Heller B, Farrokhyar F, Babra M, Lovrics PJ. An evaluation of intraoperative digital specimen mammography versus conventional specimen radiography for the excision of nonpalpable breast lesions. Am J Surg 2013;205:703–10.

[80] Moore MM, Whitney LA, Cerilli L, Imbrie JZ, Bunch M, Simpson VB, et al. Intraoperative ultrasound is associated with clear lumpectomy margins for palpable infiltrating ductal breast cancer. Ann Surg 2001;233(6):761–8.

[81] Rahusen FD, Bremers AJ, Fabry HF, van Amerongen AH, Boom RP, Meijer S. Ultrasound-guided lumpectomy of nonpalpable breast cancer versus wire-guided resection: a randomized clinical trial. Ann Surg Oncol 2002;9(10):994–8.

[82] James TA, Harlow S, Sheehey-Jones J, Hart M, Gaspari C, Stanley M, et al. Intraoperative ultrasound versus mammographic needle localization for ductal carcinoma in situ. Ann Surg Oncol 2009;16:1164–9.

[83] Olsha O, Shemesh D, Carmon M, Sibirsky O, Abu Dalo R, Rivkin L, et al. Resection margins in ultrasound-guided breast-conserving surgery. Ann Surg Oncol 2011;18:447–52.

[84] Doyle TE, Factor RE, Ellefson CL, Sorensen KM, Ambrose BJ, Goodrich JB, et al. High-frequency ultrasound for intraoperative margin assessments in breast conservation surgery: a feasibility study. BMC Cancer 2011;11:444.

[85] Ramos M, Diaz JC, Ramos T, Ruano R, Aparicio M, Sancho M, et al. Ultrasound-guided excision combined with intraoperative assessment of gross macroscopic margins decreases the rate of reoperations for non-palpable invasive breast cancer. Breast 2013;22:520–4.

[86] Karni T, Pappo I, Sandbank J, Lavon O, Kent V, Spector R, et al. A device for real-time, intraoperative margin assessment in breast-conservation surgery. Am J Surg 2007;194:467–73.

[87] Allweis TM, Kaufman Z, Lelcuk S, Pappo I, Karni T, Schneebaum S, et al. A prospective, randomized, controlled, multicenter study of a real-time, intraoperative probe for positive margin detection in breast-conserving surgery. Am J Surg 2008;196:483–9.

[88] Thill M, Roder K, Diedrich K, Dittmer C. Intraoperative assessment of surgical margins during breast conserving surgery of ductal carcinoma in situ by use of radiofrequency spectroscopy. Breast 2011;20:579–80.

[89] Rivera RJ, Holmes DR, Tafra L. Analysis of the impact of intraoperative margin assessment with adjunctive use of marginprobe versus standard of care on tissue volume removed. Int J Surg Oncol 2012;2012:868623.

[90] Galili O, Versari D, Sattler KJ, Olson ML, Mannheim D, McConnell JP, et al. Early experimental obesity is associated with coronary endothelial dysfunction and oxidative stress. Am J Physiol Heart Circ Physiol 2007;292:H904–11.

[91] Weber WP, Engelberger S, Viehl CT, Zanetti-Dallenbach R, Kuster S, Dirnhofer S, et al. Accuracy of frozen section analysis versus specimen radiography during breast-conserving surgery for nonpalpable lesions. World J Surg 2008;32:2599–606.

[92] Dener C, Inan A, Sen M, Demirci S. Interoperative frozen section for margin assessment in breast conserving energy. Scand J Surg 2009;98:34–40.

[93] Jorns JM, Visscher D, Sabel M, Breslin T, Healy P, Daignaut S, et al. Intraoperative frozen section analysis of margins in breast conserving surgery significantly decreases reoperative rates: one-year experience at an ambulatory surgical center. Am J Clin Pathol 2012;138:657–69.

[94] Creager AJ, Shaw JA, Young PR, Geisinger KR. Intraoperative evaluation of lumpectomy margins by imprint cytology with histologic correlation: a community hospital experience. Arch Pathol Lab Med 2002;126:846–8.

[95] Martin DT, Sandoval S, Ta CN, Ruidiaz ME, Cortes-Mateos MJ, Messmer D, et al. Quantitative automated image analysis system with automated debris filtering for the detection of breast carcinoma cells. Acta Cytol 2011;55:271–80.

[96] Nguyen FT, Zysk AM, Chaney EJ, Adie SG, Kotynek JG, Oliphant UJ, et al. Optical coherence tomography: the intraoperative assessment of lymph nodes in breast cancer. IEEE Eng Med Biol Mag 2010;29:63–70.

[97] Duarte GM, Cabello C, Torresan RZ, Alvarenga M, Telles GH, Bianchessi ST, et al. Radioguided Intraoperative Margins Evaluation (RIME): preliminary results of a new technique to aid breast cancer resection. Eur J Surg Oncol 2007;33:1150–7.

[98] Paredes P, Vidal-Sicart S, Zanon G, Roe N, Rubi S, Lafuente S, et al. Radioguided occult lesion localisation in breast cancer using an intraoperative portable gamma camera: first results. Eur J Nucl Med Mol Imaging 2008;35:230–5.

[99] Chagpar A, Yen T, Sahin A, Hunt KK, Whitman GJ, Ames FC, et al. Intraoperative margin assessment reduces reexcision rates in patients with ductal carcinoma in situ treated with breast-conserving surgery. Am J Surg 2003;186:371–7.

[100] Fleming FJ, Hill AD, McDermott EW, O'Doherty A, O'Higgins NJ, Quinn CM. Intraoperative margin assessment and re-excision rate in breast conserving surgery. Eur J Surg Oncol 2004;30:233–7.

[101] Cabioglu N, Hunt KK, Sahin AA, Kuerer HM, Babiera GV, Singletary SE, et al. Role for intraoperative margin assessment in patients undergoing breast-conserving surgery. Ann Surg Oncol 2007;14:1458–71.

[102] McCormick JT, Keleher AJ, Tikhomirov VB, Budway RJ, Caushaj PF. Analysis of the use of specimen mammography in breast conservation therapy. Am J Surg 2004;188:433–6.

[103] Sachfer KC, Szaniszlo T, Schultz S, Balog J, Denes J, Keseru M. In situ, real-time identification of biological tissues by ultraviolet and infrared laser desorption ionization mass spectrometry. Anal Chem 2011;83:1632–40.

[104] Schafer KC, Balog J, Szaniszlo T, Szalay D, Mezey G, Denes J, et al. Real time analysis of brain tissue by direct combination of ultrasonic surgical aspiration and sonic spray mass spectrometry. Anal Chem 2011;83:7729–35.

[105] Balog J, Sasi-Szabo L, Kinross J, Lewis MR, Muirhead LJ, Veselkov K, et al. Intraoperative tissue identification using rapid evaporative ionization mass spectrometry. Sci Transl Med 2013;5:194ra93.

[106] Schafer KC, Denes J, Albrecht K, Szaniszlo T, Balog J, Skoumal R, et al. In vivo, in situ tissue analysis using rapid evaporative ionization mass spectrometry. Angew Chem Int Ed Engl 2009;48:8240–2.

[107] Balog J, Szaniszlo T, Schaefer KC, Denes J, Lopata A, Godorhazy L, et al. Identification of biological tissues by rapid evaporative ionization mass spectrometry. Anal Chem 2010;82:7343–50.

[108] Kobayashi Y, Hayashino Y, Jackson JL, Takagaki N, Hinotsu S, Kawakami K. Diagnostic performance of chromoendoscopy and narrow band imaging for colonic neoplasms: a meta-analysis. Colorectal Dis 2012;14:18–28.

[109] Williams HRT, Walker DG, North BV, Taylor-Robinson SD, Orchard TR. Metabonomics in ulcerative colitis. J Proteome Res 2010;9:2794–5.

[110] Yoshimura K, Chen LC, Mandal MK, Nakazawa T, Yu Z, Uchiyama T, et al. Analysis of renal cell carcinoma as a first step for developing mass spectrometry-based diagnostics. J Am Soc Mass Spectrom 2012;23:1741–9.

[111] Wang H, Manicke NE, Yang Q, Zheng L, Shi R, Cooks RG, et al. Direct analysis of biological tissue by paper spray mass spectrometry. Anal Chem 2011;83:1197–201.

[112] Hiraoka K, Nishidate K, Mori K, Asakawa D, Suzuki S. Development of probe electrospray using a solid needle. Rapid Commun Mass Spectrom 2007;21:3139–44.

[113] Sumiyoshi K, Nohara T, Iwamoto M, Tanaka S, Kimura K, Takahashi Y, et al. Usefulness of intraoperative touch smear cytology in breast-conserving surgery. Experimental and Therapeutic Medicine 2010;1(4):641–5.

Chapter 5

High-Throughput Metabolic Screening

Anthony C. Dona

MRC-NIHR National Phenome Centre, Department of Surgery & Cancer, Imperial College London, London, UK; Kolling Institute of Medical Research, Northern Clinical School, University of Sydney, St Leonards, NSW, Australia

Chapter Outline

5.1 INTRODUCTION

Metabonomics (or metabolomics or metabolic profiling) is a growing field of research, which, unlike genomics, measures the combined effects of the genome and the exposome providing an overall picture of a subject's health at a given point in time. Metabolic profiling has come a long way from its humble beginnings of analytically deciphering the small molecule content of single biofluid samples [1]. The last couple of decades have provided methodologies and technologies that have allowed for accurate high-throughput screening of many

E. Holmes, J.K. Nicholson, A.W. Darzi & J.C. Lindon (Eds): Metabolic Phenotyping in Personalized and Public Healthcare. DOI: http://dx.doi.org/10.1016/B978-0-12-800344-2.00005-7

types of biofluids. The fastest moving and most commonly used technologies for metabolic studies include nuclear magnetic resonance (NMR) spectroscopy [2–4], gas chromatography (GC), liquid chromatography (LC), and mass spectrometry (MS), along with various couplings of the techniques. Each analytical technique has its own advantages and disadvantages in the context of small molecule analysis in complex biofluids, commonly exploited in conjunction with one another to enable global coverage, reproducibility, and accurate quantification of potential biomarkers.

Urine and either blood plasma or serum are considered the most useful biofluids in both clinical and epidemiologic science research. However, methods have been introduced to analyze fecal water, cerebrospinal fluid (CSF) and many other biofluids, tissue extracts or whole tissue (muscle, heart, brain, kidney, liver, etc.), tears, breath condensate, and cell constituents (media, extracts, and supernatants). Assay development and optimization for a given biofluid by each analytical platform have been a focus of research over recent years. Accurate and reproducible analysis across various large cohorts of biofluid samples has been made possible by multiple analytical platforms. Validation experiments aimed at testing the interinstrument and interlaboratory reproducibility [2,5] have helped establish quality control criteria and displayed possibilities of integrating data obtained at different sites. Consequently, current research has enabled large-scale coverage of global metabolic profiling with both current and longitudinal or historical physiologic data, which is a prerequisite for personalized health care.

Metabolic profiling has demonstrated enormous potential in furthering the understanding of disease processes, prognosis and diagnosis, toxicologic mechanisms, dietary impact on health, and biomarker discovery. Metabolic project design is often based on the understanding of development of a single aspect of disease or population dynamics; however, inadvertently, each hereditary, dietary, or environmental factor is inherently a confounding variable, and all measurable health factors must be considered when trying to reach biochemical conclusions.

There is immense potential for prognosis, diagnosis, therapy monitoring and intervention, and biochemical understanding of pathogenesis of many diseases. With the ability to detect an ever-increasing group of metabolites, time is now spent developing methods to monitor altered biochemistry with an ever-decreasing amount of time and cost. It is thought that the human body contains upward of 5000 metabolites, and information on many of them has been stored in a database [6]. Unlike other "-omics" areas of research, all of which deal with much larger estimated numbers of constituents, each metabolite exists across a broad range of continuous concentrations rather than exhibiting a presence or absence. Changes in biological status or disease diagnosis are usually based on the detection of perturbations in the concentrations of metabolites related to disease (undetected metabolites are normally observed only in the specialized area of inborn errors of metabolism). There has also been much interest in the analysis of samples stored in biobanks provided by groups that have follow-up medical data. By statistically powering such metabolic epidemiologic studies, prognostic metabolites for various health problems developed by populations can be assessed. A simple example would be collecting biofluid samples during a patient's journey while the patient is undergoing hospital treatment, leading to prediction of the outcome of the particular treatment. The collection of a cohort of patient journey samples generally takes much less time compared with the average epidemiologic study. However, such patients are usually administered a wide variety of drugs, and this can lead to a plethora of detected drug metabolites that can complicate subsequent analyses. The complexity of clinical data sets is also enhanced by comorbidities developed by patients during hospital treatment. Nevertheless, the development of metabolic profiling of large cohorts with minimal batch effect has further advanced both epidemiologic and patient journey metabolic research.

Toxicology and drug metabolism studies not only can elucidate the efficacy of drug therapies but also can help understand pathophysiologic conditions. Toxicologic mechanisms are closely related to metabolic studies, providing knowledge of disturbed biochemical pathways. Metabolic profiling is the most rapid method of identifying potential targets of a hazardous compound in animal studies. Perturbations in metabolic profiles due to toxic effects are generally more easily discovered than initial changes in the genome, transcriptome, and proteome [7,8]. The other major advantage of metabonomics approaches to toxicologic studies is the information that is available from noninvasive or minimally invasive sampling, which would only otherwise be available by histopathologic means [9,10].

Metabolites measured in patients' biofluids, although very useful for disease and toxicology research, vary from individual to individual due to differences in nutritional intake. Although nutritional biomarkers are often overlooked in disease research, important developments concerning predictive health, the exposome, and complex pathobiology are set to accentuate the role of diet and nutrition in metabolic models of health and disease. Much like patient care or the following of a patient journey, the goal is to expand knowledge based on increasing understanding of genomics, transcriptomics, and proteomics to allow dietary interventions at a personalized level to improve human health and wellbeing. Until recently, nutritional sciences have been, in terms of simple cause-and-effect relationships (focused on one nutrient at a time), quite similar to clinical diagnosis, which often measures one biomarker at a time. Single nutritional deficiencies or extremes are being used to predict responses in homogeneous groups of individuals. Emerging techniques and experimental procedures permit the investigation of all complexities involved in nutritional interactions within a single human of unique genome, exposures, and dietary history.

All metabolic profile studies are most often geared toward finding metabolite concentrations that are statistically correlated to a disease, toxicology, or dietary group when compared with a control group. Unfortunately, spectral analysis does not elucidate a set of metabolites but rather a set of spectral features of interest. The interpretation of metabolite identification from spectral features and placing them in a biologically relevant context are the most time consuming. One-dimensional proton NMR data are generally used to distinguish the spectral features of disease groups, but the technique, by itself, is almost never enough to make a solid assignment of metabolites. Tentative assignments can be confirmed by NMR spectroscopy by either measuring further popular homonuclear or heteronuclear experiments, including J-resolved (JRES) [11], correlation spectroscopy (COSY), total correlation spectroscopy (TOCSY), heteronuclear single quantum correlation spectroscopy (HSQC), and heteronuclear multiple-bond correlation spectroscopy (HMBC). These sequences provide extra structural information, aiding the confirmation of the identity of molecules of interest and overcoming problems arising from lack of spectroscopic peak resolution in biofluid analysis. Another approach to metabolite verification is to compare standard spectra (both one-dimensional and multidimensional) of possible metabolite matches with biofluid spectra. Additionally, "spiking" is a technique in which a small amount of a standard substance is added to a biofluid containing the metabolite of interest and running basic one-dimensional or two-dimensional NMR experiments between each "spiking." Spiking is a feasible option for metabolite confirmation if the standard is available along with enough volume of a biofluid sample containing the metabolite. Similar to NMR analysis, annotation of spectral features by LC–MS (liquid chromatography–mass spectrometry) or GC–MS (gas chromatography–mass spectrometry) generally is not achieved from spectral features alone, as measured in a basic assay. Often, spiking a standard into a biofluid sample is also necessary here. Otherwise, comparison of features from a standard spectrum (including retention time, detected mass, fragmentation pattern, and isotopic ratio) with features of interest from a cohort of biofluid samples improves the validity of an assignment. Quantification of metabolites of interest is also possible by spiking a labeled standard of known concentration, which is detected with a mass almost exactly 1 Da heavier (1.00616 for deuterium or 1.00335 for ^{13}C labeled) than the feature of interest. This signal from the labeled standard can be used to create a calibration curve to reference the concentration of the particular metabolite targeted.

Advances in technologies across scientific fields closely related to metabolic screening have driven a requirement for large-scale metabolic phenotyping. In the performance of chemical syntheses, thousands of compounds are developed in a month in certain laboratories; sequencing of DNA is constantly improving in terms of cost and time; and analytical technologies have shown better resolution and dynamic range more than ever. With these advances, metabolic profiling laboratories are moving toward developing analytical assays and databases that allow for comparison of metabolic data internationally. Currently, metabolic studies are limited to particular diseases within populations; however, the proposed harmonization of phenotyping would benefit disease research by encompassing entire populations.

5.2   SAMPLE PROCEDURES

5.2.1   Collection Protocols

Sample manipulation from the point of collection can lead to physical or chemical changes in the metabolic content of such biofluids [12]. Therefore, it is important to adhere to a strict protocol for collection so that the spectral results for samples can be compared directly with one another. In fact, this principle should be applied for the entirety of the experimental process because any increased variation at any stage can lead to complications in the interpretation of results.

In general, and if possible, patients should be asked to fast, avoid alcohol or medications, and avoid excessive exercise before collection of any biofluid. Within an animal population, the diversity of a biofluid is able to be kept more controlled compared with human populations, as major confounding factors such as dietary and drug intake can be easily controlled. Animal studies therefore allow for the exploration of interesting disease-metabolite correlations with the use of much smaller cohort sizes. Greater control during the design of metabolic studies in animals allows for exploration of interesting disease correlations with the use of much smaller cohort sizes. As this chapter is concerned with human patient and population samples, biofluid sample collection and studies in animals will not be discussed here.

5.2.1.1   Human Urine Collection and Storage

Research projects targeted at optimizing the first step of biofluid analysis, namely the sample collection, is ongoing. Whether urine samples should be collected as spot collections or 24 h collections, whether they should be collected into a bacteriostatic agent or protease inhibitors, what type of receptacle should be used, and how to aliquot samples for long-term storage are considered on a per-study basis. However, to harmonize phenotyping studies, researchers should follow a single optimized collection protocol.

There is great interest in replacing inconvenient 24 h collections with simpler collection methods to assess dietary metabolites. Whether alternative collection methods (overnight, timed, or spot) are reliable remains uncertain. Research based on the systematic review of salt excretion measured with 24 h collection compared with spot urine collection has revealed a range of correlation coefficients ($r = 0.17–0.94$) of urinary salt concentrations [13]. The best alternative to 24 h collection, as well as the biological basis for the variability remains unclear; thus, 24 h collection is generally recommended.

Azide is a common bacteriostat used to halt bacterial fermentation during collection of urine. Bacteriostats are particularly important in metabolic research but are often overlooked in collection of urine for protein biomarker discovery, where protease inhibitors are added to terminate protein degradation upon collection [14,15]. As always, consistency in following collection protocols would be ideal for interdisciplinary (metabonomic and proteomic) comparison in disease research. In this case, however, protease inhibitors are essential for protein research, although they could affect the analytical result (or even compromise the quality of a high-throughput run) when measuring small molecules by LC–MS or NMR. Multi-omics analysis will, therefore, need to be a compromise.

Long-term storage of urine samples are generally conducted in one of three ways: (1) in a −80°C freezer, (2) in a −20°C freezer, or (3) in liquid nitrogen cryogenic storage. How long a sample can be refrigerated (or left at room temperature) before long-term storage without causing chemical or physical changes in their metabolic content is still being debated. Similarly, the length of time a sample can remain in these conditions between thawing, preparation, and analytical analysis is also controversial. Recent research has concluded that urine samples should be preserved in azide and stored in a freezer (at −80°C), freeze thaw cycles between collection and analysis are to be avoided, and filtration is generally not advised (although in some circumstances such as fluids containing infectious agents, it is recommended) because it removes large molecules and can also "wash" small molecules into the biofluid of interest [16].

5.2.1.2 Human Plasma or Serum Collection

Human plasma or serum collection for metabolic profiling is considered more complicated than urine collection. Human blood is composed, as all human biofluids are, of a complex mixture of small molecules and a range of different cell types, including red blood cells (erythrocytes) and white blood cells. Serum and plasma are prepared from whole blood by allowing clotting to occur or by centrifugation, respectively. However, both serum and plasma consist of molecules ranging in size from small molecule metabolites to proteins (albumin being the most abundant). The complexity of biofluids is increased by biochemical assemblies of lipids and proteins, called *lipoproteins*. These molecules have multiple functions but are often used by the body to transport water-insoluble molecules such as triglycerides and cholesterol among cells, tissues, and various organs. A rigorous collection and preparation procedure needs to be strictly followed to ensure that the complex construction of plasma or serum samples remain intact at the molecular level.

Human blood should be collected into an evacuated tube system with interchangeable plastic tubes that have color-coded stoppers. It is often desirable to collect both plasma and serum. The several kinds of anticoagulant used to store plasma can affect metabolic profiling assays in various ways, so it is necessary to use caution to avoid problems with certain laboratory applications. Ethylenediaminetetraacetic acid (EDTA), for example, chelates with metals (calcium [Ca^{2+}] and magnesium [Mg^{2+}]), which makes it well suited for DNA-based assays, but in the case of NMR spectroscopic metabolic profiling, large peaks overlapping with regions of interest obscure the ^1H NMR spectrum. Heparin collection results in a spectrum that, although not obviously altered, contains broad heparin proton resonances, which are not easily identifiable.

These signals complicate the analysis of other resonances, including charac-terization of broad lipoprotein peaks into subclasses [17]. Lesser used antico-agulants such as citrate heparin or acid citrate dextrose generally produce more interpretable RNA or DNA analyses; however, much like EDTA collection, they produce sharp spectral lines that interfere with global metabolic profiling by NMR spectroscopy [18–20].

As with urine collection, the recommended amount of time allowed to elapse between collection and storage of blood should be minimal, as should the time between preparation and analysis; as well, freeze thaw cycles should be avoided. Variations in the protocol would cause the complex chemistry in plasma and serum samples to change the composition. Physically, plasma and serum samples form a concentration gradient under the force of gravity, which affects most forms of small molecule analyses performed during metabolic pro-filing studies. Therefore, the amount of time a sample is left at room tempera-ture should be kept to an absolute minimum.

5.2.2   Sample Preparation

Sample preparation (much like collection) is ideally performed in such a way that little variation in the metabolic content of samples is allowed during the process. In many cases, however, preparative and analytical variations can-not be avoided. Inconsistencies in quantification of metabolites can arise from many sources during sampling, sample storage, sample extraction, derivatiza-tion, analysis, and/or detection. During the sampling and preparation of sam-ples, undesired changes in the metabolic content of samples may occur due to enzyme activity, high reactivity of metabolites, or breakdown or degradation of metabolites. Due to the nature of metabonomics analyses, a very inert analytical preparation is required to minimize absorption and degradation of metabolites, particularly the relatively polar compounds. Also, the degree of adsorption and degradation can vary among different samples with different biomass concen-trations and different sample matrices. Consequently, such matrix effects should be evaluated.

Preparation of quality control samples (both internal and external to the project design) helps determine the amount of preparative and analytical vari-ance introduced during a study [21]. Three sample types are useful for these measures: (1) a composite quality control (QC), (2) a stable quality control, and (3) an external quality control. A composite study reference (SR) is sim-ply a reference solution made from equal amounts of each sample within the study. A composite QC is made from the SR in the same way each individual sample is prepared. A stable QC is prepared in a much larger scale with the necessary reagents or buffers and then transferred into multiple tubes or vials for analysis. An external QC is simply produced with samples of the biofluid of interest, obtained completely independently of the study, and prepared simi-larly to each individual sample. Buffer blanks can also be prepared by simply replacing a sample with pure water. These samples are important to ensure that the robot used for sample transfer to the analytical instrument is performing as expected and that the prepared buffer has not been contaminated in the preparation process.

The following section will concentrate on details of sample preparation for NMR spectroscopy, LC–MS, and GC–MS, which are generally considered the most useful analytical platforms for high-throughput metabolic global profiling.

5.2.2.1 Sample Preparation for NMR Spectroscopy

The major advantage of NMR analysis of complex solutions is the reproducibility of spectral results if the chemical environment is well controlled, specifically the temperature, pH and ionic strength of solution. Detailed protocols have been reported in the literature [3,4] and if they are followed carefully, they should produce comparable data sets across instrumentation and laboratories. In theory, NMR spectroscopy is the only technique that currently exists to facilitate true population screening and offers the ability to build an internationally useful metabolic database. Standard spectra of metabolically interesting small molecules (as measured by NMR spectroscopy) already exist in publicly available databases [6,22].

With the current protocols, there exists the possibility of the generation of complex solution (urine, plasma) databases linked with clinical/dietary information of the patient (see Section 4.5). Physiologic features (age, gender, ethnicity, body mass index [BMI], etc.) that confound metabolic disease information are often overlooked in studies aiming to increase the understanding of altered metabolism due to disease. Generating a biofluid spectral database (even of samples from patients considered healthy) would allow for the comprehensive understanding of healthy metabolism, sanctioning rapid advances into disease prognosis and diagnosis.

Preparation of biofluid samples can be either manual (pipette preparation) or completed using a syringe-mounted robot. Manual preparation, if performed well, is not necessarily the more inconsistent of the two preparative methods. However, when manually preparing a large cohort of samples, certain factors such as timing, pipette calibration, and personnel should be considered. Ideally, samples are prepared immediately before analysis, using the same pipettes, and by a single person for consistency, although, unfortunately, in large-scale projects, these criteria become unfeasible.

Robotic preparation workflows are becoming more common as formats for automated sample analysis are more compatible with liquid-handling robots. The reverse is also true in that robotic solutions are becoming more compatible with classic 5 mm/7 in. NMR tubes. Robotic preparation reduces the need for personnel time, although there is a common misconception that robotic preparation is more precise and accurate than manual preparation. Exact dispensed amounts of biofluid and reagents, when measured by weight, can vary from the amounts programmed. Intrinsically, robots that use push solvents introduce, through capillary interaction, an amount of push solvent into samples during the aspiration and dispensing of solutions. For similar reasons, capillary interactions regularly lead to "carry-over" during high-throughput sample preparation. Understanding of the amounts of each reagent, sample, and push solvent introduced during preparation and the variations in each of these amounts is important in understanding robotic accuracy and precision during the preparation procedure.

5.2.2.2   Sample Preparation for LC–MS

Sample preparation for LC–MS analysis is a critical step. Much like other techniques, the subsequent data quality is dependent on the quality of the sample preparation. Additionally, the type and amount of metabolite detected are also heavily dependent on the sample preparation process. The chromatographic element of LC–MS analysis is particularly sensitive to variation in the molecular properties of a cohort of prepared samples. During a high-throughput run, factors such as changes in chemical properties of chromatographic surfaces and variations in the solvent flow rate due to localized molecular blockages in the column compromise the accuracy of the measured retention time of a small molecule. Mistreatment of a single sample during the preparation process can subsequently compromise the integrity of the analysis of an entire cohort. The complexity of plasma and serum samples includes polar molecules such as sugars and amino acids as well as nonpolar molecules such as lipids. The type of chromatographic column (commonly hydrophilic interaction chromatography [HILIC] or reversed phase chromatography [RPC]) will determine which type of molecule will bind strongly to the analytical column. Lipids, in particular, tend to bind strongly to RPC columns, ultimately modifying the stationary phase, which changes the dynamics of chemical interactions and thus the molecular retention time and the lifetime of a column. QC samples are often used in real time to monitor instrument stability of chromatographic retention time and mass spectral detector sensitivity.

In many cases, the technologies do exist; however, the necessary methods and protocols for the preparation of an assay developed to target a defined group of molecules has not yet been tested and validated rigorously. Significant efforts are being made to further develop the preparation and validation of targeted LC–MS assays.

5.3   ANALYTICAL PLATFORMS

5.3.1   NMR Spectroscopy

NMR spectroscopy fundamentally measures the frequency of spin resonances of many different kinds of atomic nuclei. NMR spectroscopy is commonly used for many applications outside of global metabolite profiling; however, it is rapidly becoming more popular in the area of profiling complex solutions. NMR spectroscopy is applied widely in material science, chemical analysis, solution structure, molecular dynamics, protein folding, and drug screening, and more recently, metabolic phenotyping has become a major area of application.

Similar to drug screening experiments, metabolic profiling studies often have large numbers of samples, generally requiring full automation [23]. Most high-throughput laboratories will develop a workflow that determines tube type, automation type, spectrometer and probe type, which experiments are to be run, whether samples will be barcoded and managed by a Laboratory Information Management System (LIMS), and eventual storage. The automation process often starts with the submission of a request, followed by preparation, transfer to the spectrometer, automated probe tuning, shimming, and pulse length calibration, followed by data acquisition and processing and, finally, data distribution and archiving.

The type of automation used for high-throughput NMR restricts which type of sample tube can be used for analysis. The tube type is, however, normally specified on the basis of the volume of sample available. In the case where an ample sample volume is accessible, 5 mm NMR tubes are routinely used, since there are currently multiple types of automated sample handlers compatible with standard 5 mm tubes (SampleCase, SampleXpress, SampleJet). The most advanced of the automated systems (SampleJet) is also compatible with Match NMR tubes, which provide 5, 3, or 1.7 mm microtubes in a 96-well-plate format that enables more compatible and compact preparation and eventual storage of samples. Each sample can be barcoded with a small two-dimensional barcode on the sample cap and tracked throughout the preparative and analytical process.

The many facets of the analytical acquisition process have to be automated in a robust fashion to fully automate high-throughput NMR spectroscopy. It is important that data quality is monitored periodically on each instrument. Regular calibrations of the temperature within the probe, probe performance, and quantification references help prevent instrument inconsistencies. These checks also allow for the calibration of instruments and control for small variations in the performance of the NMR spectrometer.

During a run, samples must be inserted into the magnet, and then the probe should automatically tune and match and find a solvent lock. Modern methods will then wait for the temperature of the sample to equilibrate. Furthermore, by measuring the temperature of the sample at very small intervals, the probe can confirm the stability of the sample temperature. Once the temperature has been stabilized, spectrometers will run the latest shim routines, ensuring homogeneity of the magnetic field in three dimensions across the sample detection area. Routine experiments can then proceed after an optimal pulse length calibration. A subset of one or more of four possible experiment types is routinely run on a sample cohort from the following:

1. **One-dimensional ^{1}H nuclear Overhauser effect spectroscopy (NOESY) presaturation**: The ^{1}H NMR spectra are measured by using this specific water suppression pulse sequence, which employs the first increment of

a NOESY pulse sequence with continuous wave irradiation at the water resonance frequency during the relaxation delay and also during the mixing time. Application of the gradients ensures that dispersive residual water signals are filtered out and do not contribute to the final spectrum. All other protons in solution register a signal and so are measured (Fig. 5.1)

2. **One-dimensional ^{1}H Carr–Purcell–Meiboom–Gill (CPMG) spin-echo with presaturation**: A spin-echo spectrum using the CPMG sequence is

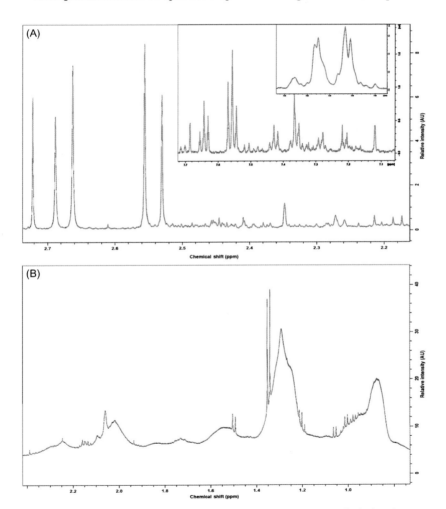

FIGURE 5.1 Representative portions of a (A) human urine NMR spectrum displaying sharp resonances of varying intensities and shapes. *Inset:* Representative portions of the aromatic region of the NMR spectrum of a human urine sample. (B) Representative region of a plasma sample illustrating broad signals from large lipoprotein complexes, overlapping with sharp metabolite signals from fast-diffusing molecules. *Image from Ref. [23].*

generally acquired for each plasma/serum sample (or other biofluids with large molecules present). CPMG sequences attenuate peaks from molecules with slow rotational correlation times such as the lipids and proteins commonly found in plasma or serum, which otherwise give rise to many broad, interfering resonances. Therefore, only the fast-diffusing, sharp, small molecule resonances are measured.

3. **One-dimensional ^{1}H diffusion-edited sequence**: A diffusion-edited pulse sequence is often run on plasma or serum samples because, in contrast to a CPMG sequence, it attenuates peaks from the fast-diffusing (ie, generally low-molecular-weight) molecules and only acquires signals from protons attached to slower-moving proteins and larger lipids. The application of this technique has been successful in the past, linking lipoprotein subfractions at a metabolic level with genetic associations [24,25].

4. **Two-dimensional ^{1}H J-resolved sequence**: a J-resolved experiment separates the chemical shift from the coupling constant information of each resonance by flipping each multiplet into a second dimension. This experiment is particularly useful in complex mixture profiling because spectral information often lost due to a lack of resolution can be determined.

The resulting free induction decay (FID) for each experiment of each sample is taken through a validated set of processing steps to produce the resulting spectra. The FID is often multiplied by an exponential window to increase the signal-to-noise ratio; however, this operation does broaden the resonance and has been known as *line broadening*. Each intensity–frequency spectral window must be phased, baselined, and calibrated to an internal reference. During high-throughput studies, a quantification reference is commonly used to synthesize an electronic signal, which is utilized to determine metabolite concentrations [26]. An electronic reference to access *in vivo* concentrations (ERETIC) can be applied for absolute quantification of various signals [27]. Both commercial and academic groups are creating methods to enable automated deconvolution and quantification of metabolite signals to feed detailed small molecule data back to even the least experienced NMR user. Completely automated methods for quantifying metabolites during analysis of biofluids such as plasma/serum [17,28] or urine [29] are still in their infancy, whereas less complex solutions such as fruit juice or wine are already being comprehensively screened quantitatively by NMR spectroscopy [29].

5.3.2 Liquid Chromatography

High-performance liquid chromatography (HPLC) is coupled efficiently with mass spectroscopy to provide a multidimensional separation technique. HPLC relies on a pressurized aqueous liquid biofluid sample running through a column filled with solid adsorbent material. Each molecule type interacts slightly differently with the adsorbent material, causing different flow rates of metabolites that are detected either upon elution by an ultraviolet (UV) visible detector or simply on the basis of the intensity of molecules measured by an online mass spectrometer. More often than not, modern laboratories use ultra-high performance liquid chromatography (UPLC). UPLC is a specific type of HPLC column that contains smaller particle sizes (and thus larger surface areas) packed into columns, allowing the pressure in the column to be raised, enabling efficient, better-resolved methods essential for the application of high-throughput screening.

The most popular varieties of chromatography used for biofluid analysis are hydrophilic interaction liquid chromatography (HILIC) and reversed phase liquid chromatography (RP-LC). RP-LC is the obvious choice for biofluid analysis because of the aqueous nature of the mobile phase. However, human urine contains a large number of highly polar metabolites such as amino acids, organic acids, sulfate and glucuronic acid conjugates, and sugars. The polar metabolites generally elute together close to the dead time, and thus retention time makes no contribution to identification [30,31]. HILIC applications are complementary to those of RP-LC and similar to those of normal-phase liquid chromatography (NP-LC). HILIC methods replace the nonaqueous mobile phase used in NP-LC with an eluent containing a high content of water-miscible organic solvent (typically acetonitrile). As the method promotes hydrophilic interactions between the analyte and a water-enriched hydrophilic stationary phase, it is well suited for online coupling with electrospray ionization–mass spectrometry (ESI–MS). Acetonitrile gradients (50–95%) are typically chosen as the organic phase because compared with methanol, the retention of analytes is better and the resolution of the method is improved due to a lower viscosity [32–35].

5.3.3 Gas Chromatography

Much like HPLC, GC is normally coupled with MS to separate biofluid metabolites, further increasing the resolution of the technique. GC is, however, far from ideal for metabonomics analysis, specifically high-throughput analysis, because during the sample treatment stages, all the nonvolatile compounds must be carefully removed before analysis. GC is limited to compounds that are either volatile or can be made volatile through a chemical derivatization process. However, clearly no one analytical platform is capable of detecting the whole set of metabolites in any given biological sample type.

Volatile, low-molecular-weight metabolites can be sampled and analyzed directly; however, many metabolites contain functional groups and are thermally labile at the temperatures required for their separation or are not volatile at all. Therefore, derivatization with an oximation reagent followed by silylation or solely silylation prior to GC analysis is needed to extend the application range of GC-based methods. As silylation reagents are the most versatile and universally applicable derivatization reagents, these are more suitable for comprehensive GC analysis of biofluids.

5.3.4 Mass Spectrometry

MS can be either run by direct infusion or coupled with GC or LC techniques. For metabonomics, MS is increasingly being used as a biomarker discovery tool in epidemiology and stratified medicine, for example, to identify subgroups of patients with a distinct mechanism for disease or disease recovery. Such investigations require the analytical platform to be robust to large-scale study designs and large cohort sizes in order to appropriately power statistical analysis [36]. Large cohorts generate a dataset that requires a multiple-batched experimental design, increasing the impact of analytical or technical variations arising particularly from MS or more evidently from chromatographic separation techniques. Improvements in data processing algorithms aimed at "stitching" batches of data together or accurately aligning and extracting metabolic features across a large cohort of biofluid samples for comparison comprise a highly active area of metabolic research.

Coupling the technique with either LC or MS tends to produce a data set for each sample that is far more difficult to manage from a preprocessing perspective and with a considerably larger storage size. Coupling analytical techniques does, however, reduce the sample complexity by spreading it over two dimensions of separation, alleviating matrix effects during ionization. Further strength is given to an experimental design by the use of any one of the multiple forms of quality control sample discussed earlier (see Section 2.2). Intrabatch or interbatch variations occurring as a result of changes in the chemical properties of chromatographic surfaces over time and variations in the solvent flow rates arising from localized molecular blockages, column changes, and mass spectroscopic instability are essentially corrected by using algorithms. Many forms of algorithms used to map linear and nonlinear temporal variation in the "identical" QC samples aim to re-align analytical variances. An example is quality control–robust spline correction (QC-RSC), which aims to reduce analytical variations and increase the proportion of variations [37].

Direct infusion mass spectrometry (DIMS) is becoming increasingly more prevalent as an analytical platform for metabonomics and lipidomics. Advantages of the use of the technique include relatively the small volumes of biofluid required, very rapid analysis times, high technique reproducibility [38], and spectral results as information-rich as LC–MS data [39]. Disadvantages of the technique include ion suppression, co-elution of metabolites, and complex spectra that yield an accurate mass only of small molecule features, inherently limiting metabolite identification.

5.3.4.1 Liquid Chromatography–Mass Spectrometry

LC–MS is potentially the most promising technique in global metabolic profiling and is destined to become a major source of metabonomic data output (Fig. 5.2). Many challenges are associated with LC–MS–based techniques before the analytical output can be confidently interpreted to provide quantified metabolite

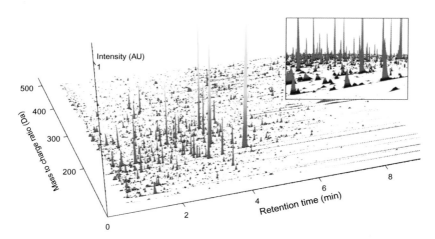

FIGURE 5.2 UPLC–MS data from a human urine sample displaying the two axes of separation. *Inset:* An enlarged portion of the same human urine data file highlighting the sensitivity and detail in the small molecule features collected.

information. Particularly, MS coupled with RPC has proven specifically useful for the profiling of body fluids with minimal preparative stages. However, polar and polar ionic metabolites such as amino acids and sugars are not retained on reverse phase columns. After investigations into other chromatographic methods, including HILIC, C18, amino, and phenyl-hexyl columns, it is evident that the current selection of chromatographic techniques does not represent a complete solution for metabonomics. Given the array of samples types to be analyzed, it may never be possible to simply recommend a single standard LC–MS method for metabolite profiling.

In the interests of high-throughput profiling, LC–MS techniques are scrutinized for accuracy and consistency using instruments of the same model and manufacturer. If metabonomics results are to be meaningful, there needs to be confidence in the community that factors such as drift are controlled in both the chromatographic and accurate mass measurement.

5.3.4.2 Gas Chromatography–Mass Spectrometry

The major limitation of GC–MS is, again, largely due to the capability of the analytical platform only being able to analyze volatile compounds or those that can be made volatile by derivatization. The technique otherwise is used to an alternative to LC–MS and is most commonly used in plant metabonomics studies. To compensate for variation in retention times across the long life of a column or across instruments, retention indices rather than retention times are used. To formulate a retention index, a homologous series of chemicals are spiked into each sample to compensate for retention time drift over time. The retention index of a metabolite bracketed by two retention index standards will remain consistent even if the retention times of all three analytes change. Large-scale metabolic studies have the advantage of less variation in the retention index measured across a study compared with the retention time measured during an LC–MS study.

5.3.5 Differences Between Urine and Plasma or Serum Metabolic Profiles

Different biofluids provide different degrees of information on the human metabolic status. Plasma or serum samples, for instance, provide a snapshot of a whole organism's metabolic status. The homeostatic mechanism of organisms commonly means that perturbations in diet, environment, or disease state take a relatively long time to be reflected in the metabolic composition of blood products. Urine, in contrast, is a "pool" of a number of hours of sampling (or even 24 h given a full day collection) but will be constantly disturbed by internal and external influences on the metabolome. Anticipating the sources of variations in a data set is important for understanding the confounding effects but must not hinder the exploratory global profiling approach.

5.4 BIOINFORMATICS PROCESSES FOR HIGH-THROUGHPUT ANALYSIS

High-throughput laboratories demand a bioinformatics expertise that plays a role during each process along the workflow, from sample storage, tracking, acquisition, processing, and interpretation. A good bioinformatics team manages each stage of the laboratory's processes in a way that is user friendly for scientists either inputting samples or data into the system or teams looking to pull data, information, or results from the system.

5.4.1 Laboratory Information Management Systems

Sometimes referred to as a Laboratory Information System (LIS) or a Laboratory Management System (LMS), a LIMS system supports the operation of a modern laboratory, particularly high-throughput laboratories, where sample numbers are large and processes require automation. Besides the key functions of sample management, instrument application, and data transfer, LIMS are also able to provide basic laboratory functions such as document management, calibration, maintenance requirements, and data entry.

There are many open-source LIMS customized toward specific types of biomaterials or research data. More generic or general LIMS have come about as a necessity to catalogue large-scale biobanks of human specimens and manage and track their process through a cascade of complex analytical procedures [40].

5.4.2   Quality Assurance/Control Workflows

There are many components to a quality assurance (QA) program. Starting from the top, it is important to have a detailed management structure in place for the laboratory. This effectively allows for each role in the laboratory to be well defined and dedicated to a particular person within the laboratory. Stemming from a management structure, it is important that each member of the laboratory is adequately trained to perform the areas and activities delegated to them. The level of training required or criteria for correct levels of training for a particular task should be outlined in the standard operating procedures (SOPs) for tasks.

SOPs provide the core of the day-to-day running of a laboratory and the QA program. They are the laboratory's internal reference manuals for all procedures, detailing every relevant step in each procedure, allowing anyone with the relevant training to perform the procedure. Some scientists feel that they do not need SOPs because of their technical experience and the availability of manuals or published literature. However, in practice, SOPs present the procedure in a way that avoids differences in interpretation and thus circumvents subtle differences in the way the procedures are performed or equipment is used.

QA and QC compliance can be checked either manually or by building automated routines. There are multiple stages to QA, and samples can be dropped from a study for many different reasons. First, it may not be possible to analyze a biofluid sample successfully by using a particular analytical platform. An example would be a biofluid that contains particulate matter or paramagnetic molecules, which would not allow successful shimming in an NMR magnet. Alternatively, samples with very high concentrations of protein would block a UPLC column designed for small molecule separation. Although it generally only becomes obvious after the analysis, these samples should be removed from the sample set before the analysis because they can compromise portions of a run or even an entire high-throughput analysis. If these samples pertain to a particular disease group in a study, it may be likely that the analytical platform of choice is not the best for exploring the physiologic condition at hand.

Furthermore, analytical data should be scrutinized and only kept for interpretation if they are to meet selection criteria confirming that the analysis has been a success. If they fail this test, the analytical data should be omitted from further analysis. By taking an analytically sound data set forward, samples of patients who are clearly not compliant with the study restrictions can be accessed. For example, subjects with diabetes or those who have not fasted when the study design specifies inclusion of patients without diabetes and fasted patients can be identified from their metabolic profiles. Patients with diabetes can be clearly identified in metabolic profiling studies by determining the amount of glucose excreted; generally, metabolic profiling studies identify a number of undiagnosed patients with diabetes. Also, patients who have not complied with the instruction to fast before sample collection are identified from the concentration of particular metabolites in plasma and urine samples. Although the analytical data may be excellent, the data are not worth taking forward to interpretation because these patients have been noncompliant with the study criteria.

5.4.3 Feature Extraction

The process chosen for converting spectral features into identified and quantified metabolites is dependent on the type of spectral data, the biofluid, and even the number of samples analyzed. However, the desired results of all known analytical techniques when applied to metabolic phenotyping are similar. Projects often require complete analysis of all the metabolites contained within the given biofluid along with an accurate sample-wise quantification of each metabolite. In the case that a metabolite can be identified unambiguously and quantified, statistical processes are able to elucidate biomarkers of pathologic interest. If the biomarkers are assigned to particular molecules, hypotheses regarding the biochemical pathways involved in the disease processes can be generated. This is the ultimate goal of hypothesis-generating metabolic phenotyping.

NMR spectroscopic analysis has the advantage of being completely quantitative in principle, accurately measuring the amount of protons, under given conditions. Spectral data can thus be directly compared and spectral features of interest elucidated. Databases of standards are generally easy to generate, and a lot of information is available to assign NMR peaks. Unfortunately, NMR spectroscopy suffers from comparably poor resolution and also has issues relating to molecules of similar structure producing similar magnetic resonance signatures, making them difficult to confidently assign without further work. Current research is aimed at producing reliable tools that deconvolve NMR signals from complex biofluid spectra and automatically provide an accurate measure of the number of metabolites in large profiling studies. Issues of variations in the chemical shifts among samples, along with overlapping signals of varying shape, cause difficulties in creating automated packages to measure metabolites and their concentrations from spectral data and form a topic of ongoing research.

Another successful example for feature extraction from one-dimensional NMR is statistical correlation spectroscopy (STOCSY) [41]. Currently, there are many forms of STOCSY across different types of spectral data, and they utilize various statistical methods [42], all stemming from STOCSY, along with various other statistical methods, take advantage of the multicollinearity of the intensities of signal across a cohort of spectra. It generates a pseudo–two-dimensional spectrum, which displays the correlation among the intensities of various peaks across the sample cohort. The method is able to register similar intramolecular connectivities, similar to standard two-dimensional NMR spectroscopic methods. Moreover, multiple molecules involved in the same biochemical pathway can present high intermolecular correlations or anticorrelations (Fig. 5.3).

LC–MS feature extraction has goals similar to that of other types of data in that all signals caused by true ions are to be detected as well as attempting to

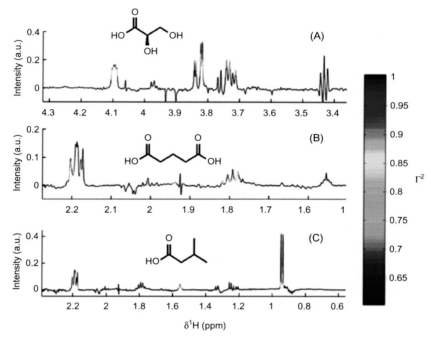

FIGURE 5.3   Representative examples of 1D STOCSY analysis across a cohort of samples run by NMR for three resonances demonstrating their correlation direction and strength (*color bar*). (A) δ 3.818 ppm, glycerate; (B) δ 1.792 ppm, isovalerate; (C) δ 0.947 ppm, glutarate. *Image from Ref. [41].*

provide quantitative information on ion concentrations. In practice, this is rarely performed perfectly and so is an ongoing area of development. The first step to most LC–MS data extraction techniques is to convert the ion counts into peaks (so-called centroiding). This process has the advantage of simplifying the spectral data, ideally producing a data set where a single data point is a single ion. It also cuts the data set storage down by a large factor, which becomes significant when storing data from large cohort studies and making the raw data publicly available. The continuous data are binned into mass–charge ratio values, summing the intensities of particular ion intensities.

At this stage, noise reduction methods are aimed at removing random noise from the measured signal. Unlike noise from impurities in buffers, random noise is generally attributed to the detector. Random noise reduction techniques are typically implemented by filtering in the chromatographic direction and fitting a polynomial in the mass–charge direction of measurement. Finally, the non-zero baseline of the spectral data is removed from the data by first estimating the baseline shape and then subtracting this shape from the spectrum. The baseline shape distortion is affected by both random noise distortions and chemical noise.

Feature detection and extraction can finally be performed on the preprocessed LC–MS data set. There are currently three main strategies to detecting features:

1. The first strategy independently detects peaks in both the mass–charge ratio direction and the retention time direction. Peak detection methods search for data points, with intensities above a particular threshold defined as peaks [43]. Similarly, feature detection can be performed in two directions, with additional constraints on allowed peak shapes in either direction.
2. The second strategy slices the data set into extracted ion chromatograms, producing a large number of one-dimensional spectra covering a narrow mass–charge ratio range. These chromatograms can be processed independently in the time domain by using either a Gaussian filter to find peak inflection points or by calculating a threshold from the mean or median of chromatographic levels [44].
3. The final strategy for feature extraction is to fit a model to the original raw spectral data set. For example, a generic isotope pattern is fitted and subtracted from the original spectrum (all metabolites contain around 1.1% of naturally occurring ^{13}C, for example). The process begins with the largest signal and is iteratively repeated until the highest remaining peak is of the order of magnitude as the background noise [45,46].

To validate these strategies, direct comparison of the raw data has been performed, or visualization of statistical testing of differences between multiple data sets is used on replicate samples. Further work has evolved by grouping peaks from ions, ion adducts, fragments, and different charge states. Deconvolution techniques are used to assign different fragments from the same molecule, based on the fact that ions will have the same retention time and their profiles across multiple samples are highly correlated. Complications occur in metabonomics experiments because biological matrices tend to have a large number of overlapping peaks, since the resolution in the chromatography direction usually means that several metabolites co-elute. Also, there are problems of ion suppression in complex systems and the concentrations of many biologically active metabolites correlate with one another across a cohort of samples due to similarities in the regulatory biological systems, leading to difficulties in deconvolution.

5.4.4   Data Analysis

5.4.4.1   Cross-Platform Phenotyping

In the area of metabolic phenotyping, platforms with many corresponding methods have been developed to gain a broad coverage of metabolites in a number of biofluids. Unfortunately, there is no one method that covers all possible biomarker candidates. Within this research area, there is a need to expand beyond single analytical platforms and develop workflows that utilize multiple platforms. Also, two distinct philosophies are applied to metabolic phenotyping studies, depending on whether an open investigation is ongoing (global, nontargeted profiling) or an existing hypothesis is being tested (targeted screening). The study designs and analytical methods used often produce data that are difficult to compare in a comprehensive fashion, especially on the scale that many modern high-throughput studies are analyzed.

Metabonomics is unique among the "-omics" technologies because it makes it possible to explore environmental influences and the exposome on human health. However, proteomic and genomic phenotyping are still considered the most direct ways of measuring genotype relationships with disease. Many analytical technologies exist in genotyping areas of phenotyping, and the final result is in a format very different from metabolite measurements. A major challenge in the integration of "-omics" data is how best to compare and correlate these large data sets to provide the most meaningful biological context. This challenge is being addressed at the data level, using statistical and chemometric methods, and at the biology level, using pathway analysis. In either case, distinguishing or deconvoluting disease effects that may be relevant to a particular disease can be particularly challenging.

5.4.4.2   Data Fusion

Fusion of metabonomics data leads to a comprehensive view on the metabolome of an organism or biological system. Such data are becoming more and more abundant, and eventually proper tools for fusing these types of data sets will be required. The first decision to be made is the level of fusion that generally depends on the type of data to be fused and whether it is feasible to fuse at a preprocessed level. Therefore, the preprocessing stage is important for each data set, as is the subsequent multivariate analysis, as both can have a dramatic effect on the end result. Finally, the criteria for variable selection based on a robust correlation between the metabolic and disease class variables must be determined. The variable selection is particularly difficult in situations with a high variable–sample ratio (as is usually found with metabolic data sets) because classic selection methods (eg, forward selection and stepwise regression) have failed as a result of the large number of measured variables. Several alternatives to these methods have been suggested in the literature, mainly in the area of partial least squares regression analysis [47,48].

5.4.5   Data Archiving

High-throughput laboratories with standardized protocols are producing spectral data that are harmonized across instruments and long periods of time. Real power comes from these methods when data are deposited in a central database and archived such that future analysis can be designed directly from stored data. The main problem is data archiving because of the large file sizes of analytical data (particularly two-dimensional or three-dimensional data sets such as those produced by LC–MS) and the large sample numbers, which necessitate storage with many petabytes of disk space.

Publicly available metabolic phenotype data repositories are beginning to be available to achieve harmonization across laboratories. Depositing raw analytical data into public databases benefits all laboratories, promoting better practice and enabling new studies. The results deposited in the databases cover a variety of techniques, and a wide range of metabolic species believed to be linked to various physiologic factors have already been identified. MetaboLights, for instance, is a general-purpose, open-access repository for metabonomics study results, including raw data and associated metadata, and is maintained by one of the major open-access data providers in molecular biology [49]. Similarly, MetabolomicsWorkbench is a repository in the United States. MetabolomeXchange is a search engine that allows the user to search all known online repositories for spectral data. Particularly for human clinical trial data, often many hurdles need to be overcome, on a case-by-case basis, with regard to permission to upload because of the requirements and limitations of ethical approvals and data ownership.

5.4.6 Data Visualization

Metabolic phenotyping data sets, along with other phenomic data, generate complex data tables, which can be difficult to summarize and visualize. For many years, continuous spectral data, such as those produced by an NMR spectrometer, has been statistically correlated with disease groups. Orthogonal partial least squares discriminant analysis (OPLS-DA) is an important and much used technique for the visualization and identification of spectral regions associated with a discrete or continuous variable of physiologic interest, for example, disease class, or a continuous clinical variable such as blood pressure. Displaying highly correlated areas of the complex spectrum can even help identify the molecules responsible for the given signals or metabolic variation. Moreover, multivariate analysis techniques that discriminate based on a combination of analytical techniques provide a more extensive coverage of the metabolome [50].

The complexity of three-dimensional data sets that arise from LC–MS or GC–MS data pose additional problems [51]. The standard use of chemometrics tools such as principal component analysis (PCA), partial least squares regression (PLS), and OPLS-DA have been robust methods for modeling such information-rich biological data sets (Fig. 5.4). These approaches to produce validated statistical models of disease class, for example, have led to techniques to improve visualization of results. For instance, the so-called S-plot visualizes both the covariance and correlation between metabolites and the modeled class designation. Metabolites identified as most statistically and potentially biochemically significant lie farther from the origin of the plot.

If the selection of interesting metabolites was based solely on the correlation with the disease type, a great number of biochemical compounds would be produced as hits. To counteract this and to avoid false-positive classification, it is necessary to remove the variation in the data set that could be caused by

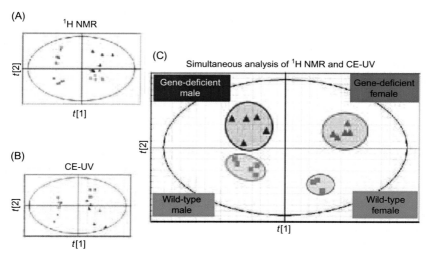

FIGURE 5.4   Principal components analysis of urine samples taken from mice of both sex and both wild type and knock-out mice using (A) NMR data only; (B) capillary electrophoresis with ultraviolet detection (CE-UV) data only; (C) NMR and CE-UV combined dataset. Combining the spectral data sets provides clearer metabolic differentiation between groups as more of the metabolome is covered. *Image from Ref. [50].*

confounding conditions. For example, if the interest is in identifying biomarkers of a particular cardiovascular disease, then the variations in the metabolic profiles, resulting from dietary differences, blood lipid variation, statin use, ethnicity, age and gender, have to be removed.

Moreover, as metabolic phenotyping data have a large dynamic range (this is one of the issues when choosing an analytical technique to globally profile biofluids), covariate analysis will highlight metabolites of high concentration and ignore the hundreds of small molecules generally found in concentrations that are many factors less. Analysis solely using the covariate approach often leads to a small number of high-concentration metabolites being continually highlighted. Selection of compounds that are biochemically relevant relies on a visualization that depends on both covariation and correlation.

As mentioned above, large metabonomic data sets always contain variations that relate to associated endpoints or further physiologic measures of interest that potentially confound models. A shared and unique structure plot (SUS-plot) is a useful technique for comparing the results of multiple models, since it will highlight the metabolites that are shared or unique between a set of different models. The visualization plots the correlation of the predictive component of each model against one another. Depending on which model they are uniquely associated with, unique biomarkers are found to plot along the *x*-axis or *y*-axis.

REFERENCES

[1] Piotto M, Saudek V, Sklenar V. Gradient-tailored excitation for single-quantum NMR spectroscopy of aqueous solutions. J Biomol NMR 1992;2:661–5.

[2] Choi Y, Vincent L, Lee A, Dobke M, Engler A. Mechanical derivation of functional myotubes from adipose-derived stem cells. Biomaterials 2012;33:2482–91.

[3] Arnold J, Dharmatti S, Packard M. Chemical effects on nuclear induction signals from organic compounds. J Chem Phys 1951;19:507.

[4] Craig Richardson J, Bowtell R, Karsten M, Melia C. Pharmaceutical applications of magnetic resonance imaging (MRI). Adv Drug Deliv Rev 2005;57:1191–209.

[5] Swann J, Spagou K, Lewis MR, Nicholson JK, Glei D, Seeman T, et al. Microbial– mammalian cometabolites dominate the age-associated urinary metabolic phenotype in Taiwanese and American populations. J Proteome Res 2013;12:3166–80.

[6] Wishart DS, Tzur T, Knox C, Eisner R, Guo AC, Young N, et al. HMDB: the human metabolome database. Nucl Acids Res 2007;35:D521–526.

[7] Strauss V, Mellert W, Wiemer J, Leibold E, Kamp H, Walk T, et al. Increased toxicity when fibrates and statins are administered in combination—a metabolomics approach with rats. Toxicol Lett 2012;211:187–200.

[8] Van Ravenzwaay B, Cunha GC, Leibold E, Looser R, Mellert W, Prokoudine A, et al. The use of metabolomics for the discovery of new biomarkers of effect. Toxicol Lett 2007;172:21–8.

[9] Ebbels TMD, Keun HC, Beckonert O, Bollard ME, Lindon JC, Holmes E, Nicholson JK. Prediction and classification of drug toxicity using probabilistic modeling of temporal metabolic data: the consortium on metabonomic toxicology screening approach. 2007;6:11.

[10] Want E, Masson P, Michopoulos F, Wilson ID, Theodoridis G, Plumb R, et al. Global metabolic profiling of animal and human tissues via UPLC–MS. Nat Protoc 2013;8:17–32.

[11] Ernst R, Anderson W. Application of fourier transform spectroscopy to magnetic resonance. Rev Sci Instrum 1966;37:93.

[12] Rist MJ, Muhle-Goll C, Görling B, Bub A, Heissler S, Bernhard W, et al. Influence of freezing and storage procedure on human urine samples in NMR-based metabolomics. Metabolites 2013;3:243–58.

[13] Ji C, Sykes L, Paul C, Dary O, Legetic B, Campbell N, et al. Systematic review of studies comparing 24-hour and spot urine collections for estimating population salt intake. Pan Am J Public Health 2012;32:307–15.

[14] Eric Thomas C, Sexton W, Benson K, Sutphen R, Koomen J. Urine collection and processing for protein biomarker discovery and quantification. Cancer Epidemiol Biomarkers Prev 2010;19:953–9.

[15] Zhou H, Yuen PST, Pisitkun T, Gonzales PA, Yasuda H, Dear JW, et al. Collection, storage, preservation, and normalisation of human urinary exosomes for biomarker discovery. Kidney Int 2006;69:1471–6.

[16] Saude EJ, Sykes BD. Urine stability for metabolomic studies: effects of preparation and storage. Metabolomics 2007;3:19–27.

[17] Ala-Korpela M. Critical evaluation of ^{1}H NMR metabonomics of serum as a methodology for disease risk assessment and diagnostics. Clin Chem Lab Med 2008;46:27–42.

[18] Vaught JB. Blood collection, shipment, processing and storage. Cancer Epidemiol Biomarkers Prev 2006;15:1582.

[19] Holland NT, Smith MT, Eskenazi B, Bastaki M. Biological sample collection and processing for molecular epidemiological studies. Mutat Res 2003;543:217–34.

[20] Landi M, Caporaso N. Sample collection, processing, and storage In: Applications of biomarkers in cancer epidemiology. IARC Sci Publ 1997;142:223–36.

[21] Kamleh MA, Ebbels TMD, Spagou K, Masson P, Want EJ. Optimizing the use of quality control samples for signal drift correction in large-scale urine metabolic profiling studies. Anal Chem 2012;84:2670–7.

[22] Albers M, Butler T, Rahwa I, Bao N, Keshari K, Swanson M, et al. Evaluation of the ERETIC method as an improved quantitative reference for ^{1}H HR-MAS spectroscopy of prostate tissue. Magnet Reson Med 2010;61:525–32.

[23] Dona AC, Jimenez B, Schafer H, Humpfer E, Spraul M, Lewis MR, et al. Precision high-throughput proton NMR spectroscopy of human urine, serum and plasma for large-scale metabolomic phenotyping. Anal Chem 2014;86:9887–94.

[24] Petersen A, Stark K, Musameh M, Nelson C, Römisch-Margl W, Kremer W, et al. Genetic associations with lipoprotein subfractions provide information on their biological nature. Hum Mol Genet 2012;21:1433–43.

[25] Kaess B, Jóźwiak J, Nelson C, Lukas W, Mastej M, Windak A, et al. The relation of rapid changes in obesity measures to lipid profile—insights from a nationwide metabolic health survey in 444 Polish cities. Public Library Sci One 2014;9:e86837.

[26] Da Silva L, Godejohann M, Martin F, Collino S, Bürkle A, Moreno-Villanueva M, et al. High-resolution quantitative metabolome analysis of urine by automated flow injection NMR. Anal Chem 2013;85:5801–9.

[27] Albers A, Butler T, Rahwa I, Bao N, Keshari K, Swanson M, et al. Evaluation of the ERETIC method as an improved quantitative reference for ^{1}H HR-MAS spectroscopy of prostate tissue. Magnet Reson Med 2009;61:525–32.

[28] Ala-Korpela M, Lankinen N, Salminen A, Suna T, Soininen P, Laatikainen R, et al. The inherent accuracy of ^{1}H NMR spectroscopy to quantify plasma lipoproteins is subclass dependent. Atherosclerosis 2007;190:352–8.

[29] Link M, Spraul M, Schaefer H, Fang F, Schuetz B. Novel NMR-technology to access food quality and safety. Int J Biol Veterinary Agric Food Eng 2013;7:621–4.

[30] Hodson M, Dear G, Griffin J, Haselden J. An approach for the development and selection of chromatographic methods for high throughput metabolomic screening of urine by ultra pressure LC–ESI–ToF–MS. Metabolomics 2009;5:166–82.

[31] Guy P, Tavazzi I Bruce S, Ramadan Z, Kochhar S. Global metabolic profiling analysis on human urine by UPLC–TOFMS: issues and method validation in nutritional metabolomics. J Chromatogr 2008;871:253–60.

[32] Alpert A. Hydrophilic-interaction chromatography for the separation of peptides, nucleic acids and other polar compounds. J Chromatogr 1990;499:177–96.

[33] Alpert A, Shukla M, Shukla A, Zieske L, Yuen S, Ferguson M, et al. Hydrophilic-interaction chromatography of complex carbohydrates. J Chromatogr 1994;676:191–202.

[34] Bajad S, Lu W, Kimball E, Yaun J, Peterson C, Rabinowitz J. Separation and quantitation of water soluble cellular metabolites by hydrophilic interaction chromatography–tandem mass spectrometry. J Chromatogr 2006;1125:76–88.

[35] Schlichtherle-Cerny H, Affolter M, Cerny C. Hydrophilic interaction liquid chromatography coupled to electrospray mass spectrometry of small polar compounds in food analysis. Anal Chem 2003;75:2349–54.

[36] Blaise B. Data-Driven Sample Size Determination for Metabolic Phenotyping Studies. Anal Chem 2013;85:8943–50.

[37] Kirwan J, Broadhurst D, Davidson R, Viant MR. Characterising and correcting batch variation in an automated direct infusion mass spectrometry (DIMS) metabolomics workflow. Anal Bioanal Chem 2013;405:5147–57.

[38] Han J, Danell R, Patel J, Gumerov D, Scarlett C, Paul Speir J, et al. Towards high-throughput metabolomics using ultrahigh-field Fourier transform ion cyclotron resonance mass spectroscopy. Metabolomics 2008;4:128–40.

[39] Lin L, Yu Q, Yan X, Hang W, Zheng J, Xing J, et al. Direct infusion mass spectrometry or liquid chromatography mass spectrometry for human metabolomics? A serum metabonomic study of kidney cancer. Analyst 2010;135:2970–8.

[40] Voegele C, Alteyrec L, Caboux E, Smans M, Lesueur F, Le Calvez-Kelm F, et al. A sample storage management system for biobanks. Bioinformatics 2010;26:2798–800.

[41] Cloarec O, Dumas M, Craig A, Barton R, Trygg J, Hudson J, et al. Statistical total correlation spectroscopy: an exploratory approach for latent biomarker identification from metabolic ^{1}H NMR data sets. Anal Chem 2005;77:1282–9.

[42] Robinette S, Lindon J, Nicholson JK. Statistical spectroscopic tools for biomarker discovery and systems medicine. Anal Chem 2013;85:5297–303.

[43] Hastings C, Norton S, Roy S. New algorithms for processing and peak detection in liquid chromatography/mass spectrometry data. Rapid Commun Mass Spectrom 2002;16:462–7.

[44] Radulovic D, Jelveh S, Ryu S, Hamilton T, Foss E, Mao Y, et al. Informatics platform for global proteomic profiling and biomarker discovery using liquid chromatography–tandem mass spectrometry. Mol Cell Proteomics 2004;3:984–97.

[45] Hermansson M, Uphoff A, Kakela R, Somerharju P. Automated quantitative analysis of complex lipidomes by liquid chromatography/mass spectrometry. Anal Chem 2005;77:2166–75.

[46] Leptos K, Sarracino D, Jaffe J, Krastins B, Church G. MapQuant: open-source software for large-scale protein quantification. Proteomics 2006;6:1770–82.

[47] Wold S, Sjostrom M, Eriksson L. PLS-regression: a basic tool of chemometrics. Chemometr Intell Lab Syst 2001;58:109–30.

[48] Perez-Enciso M, Tenenhaus M. Prediction of clinical outcome with microarray data: a partial least squares discriminant analysis (PLS-DA) approach. Hum Genet 2003;112:581–92.

[49] Hang K, Salek R, Conesa P, Hastings J, de Matos P, Rijmbeek M, et al. MetaboLights—an open-access general-purpose repository for metabolomics studies and associated meta-data. Nucl Acids Res 2013;41:781–6.

[50] Garcia-Perez I, Villaseñor A, Wijeyesekera A, Posma J, Jiang Z, Stamler J, et al. Urinary metabolic phenotyping the slc26a6 (chloride-oxalate exchanger) null mouse model. J Proteome Res 2012;11:4425–35.

[51] Wiklund S, Johansson E, Sjostrom L, Mellerowicz E, Edlund U, Shocker J, et al. Visualization of GC/TOF–MS-based metabolomics data for identification of biochemically interesting compounds using OPLS class models. Analy Chem 2008;80:115–22.

Chapter 6

Pharmacometabonomics and Predictive Metabonomics: New Tools for Personalized Medicine

Jeremy R. Everett[1], John C. Lindon[2] and Jeremy K. Nicholson[2]

[1]*Medway Metabonomics Research Group, University of Greenwich, Kent, UK*
[2]*Department of Surgery and Cancer, Imperial College London, London, UK*

Chapter Outline

E. Holmes, J.K. Nicholson, A.W. Darzi & J.C. Lindon (Eds): Metabolic Phenotyping in Personalized and Public Healthcare. DOI: http://dx.doi.org/10.1016/B978-0-12-800344-2.00006-9

6.1 INTRODUCTION

6.1.1 Introduction to Metabonomics and Metabolite Profiling

Metabolic profiling of biological fluids such as urine, plasma, and cerebrospinal fluid has a history going back several decades [1,2]. These experiments are typically conducted using technologies such as nuclear magnetic resonance (NMR) spectroscopy, or mass spectrometry (MS) coupled with a separation technology such as high-performance liquid chromatography (HPLC–MS), ultra-performance liquid chromatography (UPLC–MS), or gas chromatography (GC–MS). In the NMR spectroscopy community, this methodology was originally known as biofluid NMR, and it was used to study a number of different areas such as drug metabolism for high-dose drugs [3–7], the occurrence and origin of toxicity of drugs in terms of target organs and the biochemical mechanisms [8–12], the changes associated with aging [13] and with diseases and their treatments [14,15] (particularly inborn errors of metabolism [16–18]), and, in animals, changes caused by specific genetic mutations [19,20].

The choice of methodology, ie, NMR versus MS, to detect metabolites and quantify them depends on the type of metabolites of interest and the nature of the experiments. Both technologies are capable of detecting hundreds of metabolites in a biofluid such as human urine, and they are at their most powerful when used together. The characteristics of the two techniques are summarized in Table 6.1.

In 1999, Jeremy Everett and Jeremy Nicholson defined *metabonomics* as "the study of the metabolic response of organisms to disease, environmental change or genetic modification" [21]. This concept was introduced to provide a coherent framework for a series of experiments conducted jointly by Pfizer Global Research and Development (R&D) and Imperial College London, integrating genomics, proteomics, metabolic profiling, and clinical biochemistry data for the early prediction of drug safety. The definition of metabonomics is interventional in nature; that is, the focus is on *changes* in metabolite profiles occasioned by an intervention such as drug administration or the onset of disease. The definition in the literature became a little confusing with the later definition of *metabolomics* by Fiehn as a "comprehensive analysis in which all the metabolites of a biological system are identified and quantified" [22]. In contrast to the interventional definition of *metabonomics*, the definition of *metabolomics* is observational in nature. In addition, given that a biofluid such as urine contains an immense number of metabolites, with a concentration range of many orders of magnitude, this definition can only be aspirational. Although there was originally some discrimination in the use of the two terms, they are now used interchangeably in the literature. More recently, the term *metabolic phenotyping* has come into common use [23]. In this review, we will use the original term *metabonomics* throughout.

All well-designed metabolic profiling or metabonomics experiments follow a certain sequence of events: (1) definition of experimental aims, for example,

TABLE 6.1 A Comparison of the Characteristics of NMR Spectroscopy and Mass Spectrometry as Metabolite Detectors in Metabonomics Experiments

NMR Spectroscopy	Mass Spectrometry
• Powerful molecular structure elucidation capability	• Powerful molecular structure elucidation capability, generally hyphenated with a separation technology such as gas chromatography (GC) or ultra-performance liquid chromatography (UPLC) for optimal performance
• The ^{1}H NMR spectrum of a metabolite contains a host of important information on the molecular structure and the connectivities between the hydrogen atoms in that metabolite, but the information is usually short range, through two to three bonds	• Provides information on the molecular weight and fragments of molecules and, when operating at high resolution, molecular formulae as well • Spectra may be complicated by the presence of multiple species such as from multiple derivatizations in GC–MS, and multiple adductions and multiple charge species in LC–MS • May be difficult to discriminate isomers
• Inherently quantitative	• Ionization suppression or enhancement can lead to complications with quantitation without careful use of reference standards
• Low sensitivity compared with MS, but improved considerably since the introduction of cryo-cooled and low-volume probes	• High sensitivity
• Stable technology: No contact between sample and spectrometer	• Less stable: Direct introduction of the sample into the spectrometer, which can affect sensitivity and stability
• Can be completely automated, and arrays of sophisticated experiments can be run on batches of samples in tube or flow mode	• Automated injection of samples is possible
• Minimal or no sample preparation required	• Studies by GC–MS generally require extraction of metabolites and their derivatization in order to make them volatile enough for analysis
• Open window detector; will report on the presence of any metabolite above the detection threshold	• May be metabolite class specific, depending on the detection techniques used

to determine if patients treated with a drug exhibit changes in their metabolite profiles consistent with recovery from the disease state; (2) gaining of ethical approval for the experiments; (3) collection of samples; (4) storage of samples; (5) preparation of samples for analysis; (6) data acquisition; (7) quality control of the data; (8) spectroscopic data preprocessing steps, which for NMR spectroscopy would include Fourier transformation of the time-domain data into the frequency domain spectrum, baseline correction, and so on; (9) statistical data preprocessing steps such as peak alignment, scaling, and normalization; (10) statistical analysis of data, for example, to determine if there are significant changes in the metabolite profiles of the biofluids of patients as a result of drug treatment; (11) identification of the metabolites responsible for any statistically significant differences; and (12) rationalization of the metabolite changes found on the basis of physiology and biochemistry.

These 12 steps have been subject to considerable scrutiny over the past decade or more, and detailed protocols are available for most of the steps for both NMR-based [24–30] and MS-based metabonomics approaches [31–38].

A major bottleneck in metabonomics experiments, whether conducted with MS-based or NMR-based detection, is metabolite identification. Most of the metabolites that will be detected will be known compounds. However, categorically determining the identities of these metabolites can be challenging. For instance, the differences between 2-hydroxyisobutyric acid and 3-hydroxyisobutyric acid may not be immediately obvious by LC–MS. Conversely, the differences between caproylglycine and capryloylglycine will not be easy to determine with NMR spectroscopy. For this reason, the Metabolomics Standards Initiative established a Chemical Analysis Working Group (CAWG), which defined a four-level system for known metabolite identification in 2007 [39]. Unfortunately, this system has not been used much or further developed since that time [40]. However, new metabolite identification efficiency proposals based on molecular spectroscopic information analysis have recently been published [41], and it is hoped that this will help research in the area to move forward.

Another factor that plagued early preclinical metabonomics experiments focused on drug safety assessment was heterogeneity in response to drug administration, and it was in investigating the origin of this problem that pharmacometabonomics was discovered.

6.1.2 Introduction to Personalized Medicine

Personalized Medicine [42] is sometimes also interchangeably referred to as *precision medicine* or *stratified medicine*. It has various definitions, one of which is "application of genomic and molecular data to better target the delivery of health care, facilitate the discovery and clinical testing of new products, and help determine a person's predisposition to a particular disease or condition" [43].

The concept is an old one and also one that many doctors adhered to well before the advent of genomics: specifically, that all patients should be treated as individuals with differing needs, rather than being treated as having some uniform condition warranting uniform treatment. The definition above is rather complex. With respect to the effective delivery of medicines to patients, Personalized Medicine can be more simply defined as the use of genomic, molecular, and clinical information to select medicines that are more likely to be both effective and safe for that patient.

The need for improvements in the personalized delivery of medicines to patients is striking. Estimates by Pomeranz et al. showed that in US hospitals in 1994 alone, around 2,216,000 patients had serious adverse drug reactions (ADRs), with serious ADR defined as requiring hospitalization, being permanently disabling, or resulting in death [44]. It was estimated that around 100,000 patients suffered fatal ADRs. The personal toll of these fatalities is incalculable; the financial impact of serious ADRs to the US economy is estimated at between $30 billion and $100 billion per year [45]. Finding a way to predict which groups of patients would or would not suffer ADRs is, thus, a very important goal. It is also well known that many medicines only work in subsets of patients for reasons that are often obscure. It is therefore an imperative for 21st century medicine that improvements are made to the personalized delivery of drugs.

Until the discovery of pharmacometabonomics (see Section 6.2) nearly all studies of Personalized Medicine used human genomic data, in particular the analysis of single nucleotide polymorphisms (SNPs) to generate predictions of efficacy or safety—which is pharmacogenomics [46–48]. Although pharmacogenomics works well in certain circumstances, it is a limited approach for three main reasons: (1) the physiologic and biochemical status of the human body is not simply the sum of the influences of our genes; it is now well known that the human microbiome, particularly the gut microbiome, exerts a strong and complex influence over metabolism, physiologic status, and disease in the human body [49–51]; (2) analysis of the genome of a person will give information on what *might* happen to that person at some point in the future (dependent on the expression of the genes of interest and the availability of substrates for those gene products to work on, etc.), whereas analysis of metabolic profiles gives information on what *is* happening in a person's body at the time of sampling [52]; and (3) the problem of phenoconversion, induced by co-administration of drugs, which can lead to the conversion of *genotypical* extensive metabolizers into *phenotypical* poor metabolizers and thus confound a pharmacogenomics analysis [53]; these are important considerations. Recent reviews of the value of pharmacogenomics in randomized clinical trials in a variety of disease areas, including cardiovascular disease (CVD) [54], diabetes [55], and depression [56], have failed to show clear evidence of the value of the technology in clinical practice. Thus, the potential of pharmacogenomics is clear, and several examples of its utility exist, particularly in oncology, but broader clinical utility still appears elusive. In this situation, it may be helpful for a complementary technology to emerge to help deliver the promise of Personalized Medicine: pharmacometabonomics.

6.2 THE DISCOVERY OF PHARMACOMETABONOMICS

6.2.1 Preclinical Pharmacometabonomics

At the end of the 1990s, Imperial College, London, and Pfizer Global R&D embarked on a collaboration to detect early safety signals from drugs, using an integrated combination of metabonomics, proteomics, genomics, and conventional clinical biochemistry measurements. The ultimate aim of the collaboration was to develop metabonomics methodologies that could reduce safety attrition in Pfizer drug discovery projects with the early detection of toxicity signals. A confounding factor in these experiments was the variability in the results obtained. In rats of the same weight, age, gender, and strain, all of which were treated identically, highly variable results could be obtained. In fact, this variability in response of animals to drug dosing had been observed in collaborations between Beecham Pharmaceuticals and Birkbeck College, University of London, in the previous decade. The differences in response could be so extreme that, on occasion, it was doubted whether all the animals had been uniformly dosed. These differential results were conventionally ascribed to the so-called biological variability or to technical incompetence. At a Pfizer–Imperial College collaboration review meeting on October 18, 2000, in Amboise, France, further inconsistencies in results were observed, with some rats within a study group exhibiting results completely different from others. A radical new idea was proposed at this meeting, that is, the differences in animal responses were real and important and related to the physiologic and biochemical status of the animal *prior to dosing*. A series of experiments was designed to test this hypothesis, which culminated in a study of the metabolism and safety of paracetamol (acetaminophen) in Sprague-Dawley rats. The hypothesis was that the nature of the metabolism of paracetamol and its degree of toxicity in individual animals should be predictable based on an analysis of their predose biofluids. To cut a long story short, the hypothesis was proven [57]. Using an unsupervised (unbiased) principal components analysis (PCA) of the data, the degree of liver histopathology in the rats at the borderline toxic dose chosen was found to be weakly but significantly correlated with the predose urine metabolite profiles of the rats (Fig. 6.1). Analysis of the PCA loadings plot showed that the significant *predose* metabolites giving rise to a differential toxicity response to paracetamol were taurine together with trimethylamine-*N*-oxide (TMAO) and betaine. Higher levels of taurine in the predose urine were associated with lower liver histopathology after dosing. This was an interesting result, since taurine is known to be hepato-protective for paracetamol dosing, although other metabolic rationales are possible [57]. Fig. 6.1 shows the results of the unsupervised

PCA of the preintervention urine of the rats subsequently dosed orally with 600 mg/kg of paracetamol.

In addition to this prediction of toxicity, it was also found that a projection to latent structure (PLS) model [58] of the *predose urine* was able to model the *postdose* ratio of the drug metabolite paracetamol glucuronide (G) to parent paracetamol (P). This model was cross-validated and showed a positive correlation between the G/P ratio and the integrals of the bins for the region 5.06–5.14 ppm in the ^{1}H NMR spectrum of the predose urine. This result can be rationalized on the basis that this region of the NMR spectrum contains the signals of the H1 protons of endogenous ether glucuronides. It is therefore logical that an increased propensity to form ether glucuronides before dosing should

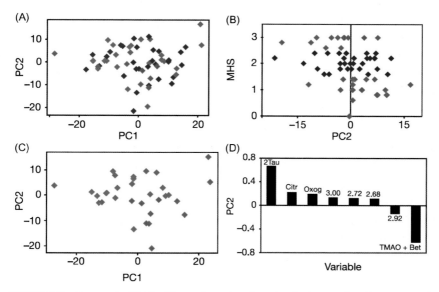

FIGURE 6.1   Unsupervised PCA of the *predose* urine of rats dosed subsequently with paracetamol (oral, 600 mg/kg). A) PCA scores plot of the multivariate analysis of the binned 600 MHz ^{1}H NMR spectra of the predose rat urine. Each diamond represents a different animal color-coded according to postdose liver histopathology outcome: Class 1, no or minimal liver necrosis, green; Class 2, mild necrosis, blue; and Class 3, moderate necrosis, red. A partial separation is observed across principal component 2 (PC2) between Class 1 and Class 3. B) A plot of the mean liver histopathology score (MHS) for each animal plotted against PC2: a weak but significant correlation is observed. C) PCA Scores plot of the multivariate analysis of the binned 600 MHz ^{1}H NMR spectra of the predose rat urine, with the same color-coding as in (A) for Classes 1 and 3 only: the partial separation across PC2 is more readily observed. D) The PCA loadings plot showing the bins of the ^{1}H NMR spectrum of the predose urine that are responsible for the separations across PC2 and the direction of influence. Abbreviations: *Tau*, taurine; *Citr*, citrate; *Oxog*, 2-ketoglutarate; *TMAO+ Bet*, trimethylamine-N-oxide (TMAO) and betaine. Where numbers are given, they refer to the ^{1}H NMR chemical shift in parts per million (ppm) at the center of the bin responsible for the separation. *Reproduced from Nature Publishing Group [57].*

translate into an increased formation of paracetamol glucuronide after dosing with paracetamol [57].

This study proved for the first time that the quantitative analysis of the metabolites present in *predose* biological fluids could be used to predict *postdose* responses such as drug metabolism and pathology and represents the discovery of pharmacometabonomics, which was defined as "the prediction of the outcome (for example, efficacy or toxicity) of a drug or xenobiotic intervention in an individual based on a mathematical model of preintervention metabolite signatures" [57]. Thus, pharmacometabonomics is the metabolic profiling equivalent of pharmacogenomics, where genetic profiles are used to aid the prediction of drug responses. Pharmacometabonomics is seen to be a complementary technology to pharmacogenomics, and, as discussed below, it can be beneficial to use the two technologies in concert.

It is worth repeating here that pharmacometabonomics does have some advantages relative to pharmacogenomics. First, the predictions made in pharmacometabonomics are based on measurements of the levels of one or more metabolites, which reflect the actual physiologic and biochemical status of the subject, in contrast to pharmacogenomics, which measures gene polymorphisms that may or may not translate into differences in phenotype at some future time point. Second, and most important, pharmacometabonomics experiments can provide information on metabolites produced by the host genome, the microbiome, and those co-produced by both the host genome and the microbiome. Human genomics is blind to events mediated by the microbiome. Finally, since pharmacometabonomics measures the actual metabolic phenotype, it is not subject to errors due to phenoconversion [53], which can be a problem for pharmacogenomics.

6.2.2   Definition of Pharmacometabonomics

As stated above, pharmacometabonomics is defined as "the prediction of the outcome (eg, efficacy or toxicity) of a drug or xenobiotic intervention in an individual based on a mathematical model of preintervention metabolite signatures" [57]. Pharmacometabonomics is thus distinguished from metabonomics as follows: (1) metabonomics is concerned with the discovery and understanding of *metabolite profile changes* that occur as a result of an *intervention*, whereas (2) pharmacometabonomics is concerned with the *prediction* of the effects of the specific intervention of *drug dosing* based on the analysis of predose metabolite profiles. As for metabonomics, the term *pharmacometabonomics* is also used interchangeably with the later term *pharmacometabolomics*; there should be no difference in the definition of the two terms. Unfortunately, in the recent literature, there is a tendency to degrade the definition of the term *pharmacometabolomics* such that it applies to any study that merely observes the effect of drug administration on metabolite profiles after dosing. These experiments are straightforward metabonomics experiments with no element of outcome prediction. The same phenomenon has occurred with the use of the term *pharmacoproteomics*. This is to be avoided; otherwise the literature will become confusing.

6.3   THE USE OF PHARMACOMETABONOMICS FOR THE PREDICTION OF PHARMACOKINETICS AND DRUG METABOLISM

6.3.1   Demonstration of Pharmacometabonomics in Humans and Prediction of the Metabolism of Paracetamol

Soon after the discovery of pharmacometabonomics [57], the Pfizer–Imperial College team set out to find out if the initial findings made in rats could have application in humans. This was an exciting prospect, since the ability to predict drug effects before dosing is a key aim of Personalized Medicine, and, as seen above, there is a significant need to improve the delivery of Personalized Medicine [45,46]. The aim of the human trial that was devised was to test if the analysis of predose biological fluids could provide a prediction of the metabolism of the important analgesic drug paracetamol (acetaminophen, see Fig. 6.2) [59].

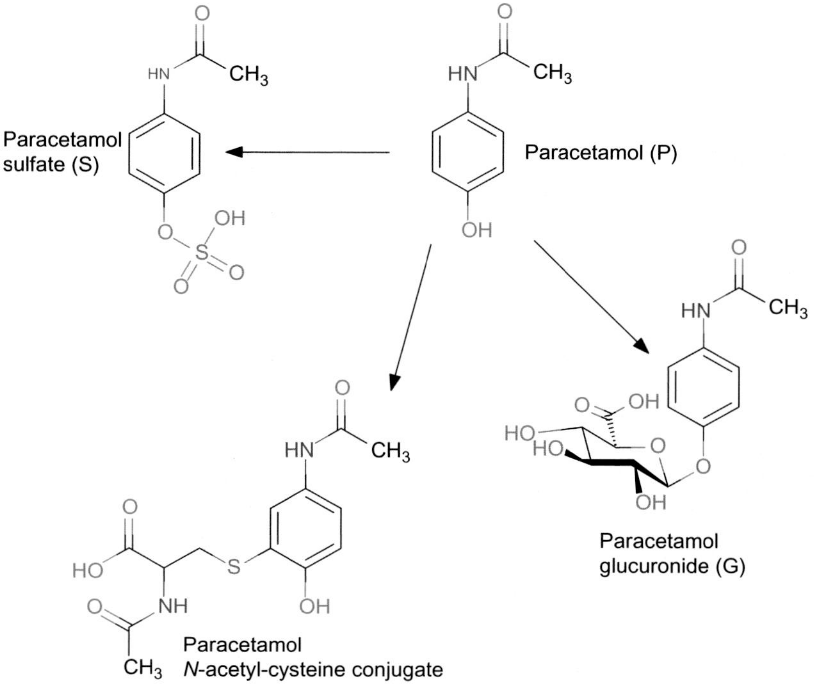

FIGURE 6.2   The molecular structures of paracetamol and its major metabolites in humans.

Paracetamol has two main pathways of metabolism in humans: sulfation and glucuronidation, to increase hydrophilicity prior to excretion. The drug dosing was conducted between March and April 2003, after ethical approval, in a group of 100 fit and healthy, nonsmoking, male volunteers, aged 18–64 years, who provided urine samples both prior to and 0–3 h and 3–6 h after ingestion of a standard dose (2 × 500 mg tablets) of paracetamol with water. No diet was imposed. However, each volunteer was asked not to take drugs, herbal remedies, dietary supplements, or alcohol prior to the trial. The predose and postdose 600 MHz ^{1}H NMR spectra from two different volunteers on the trial are shown in Fig. 6.3. Comparing the predose and 0–3 h postdose ^{1}H NMR spectra of volunteer 1, the main differences are due to the excretion of paracetamol sulfate (S, peaks shown as 7 in Fig. 6.3B) and paracetamol glucuronide (G, 8 in Fig. 6.3B) into the urine. There is significantly less S than G excreted into the postdose urine (ratio of peaks S/G <1). For volunteer 2 (Fig. 6.3C and D), a similar pattern was observed, but here the ratio S/G was >1. Note that the N-acetyl CH$_3$ signals of paracetamol and its metabolites resonate as sharp singlets in the region c. 2.15–2.18 ppm, whereas the phenyl ring aromatic protons resonate as pseudo-doublets at c. 7.1–7.5 ppm.

The signals of a large number of endogenous urine metabolites are also observed in these spectra. The analysis of the remaining volunteers' urine spectra showed similar patterns with variability in the ratio of S/G excreted in the postdose urine. A key observation was that the low ratio of S/G in volunteer 1's postdose urine was associated with high levels of a signal due to an unknown endogenous metabolite labeled "4" (singlet at c. 2.35 and pseudo-doublets between 7.2 and 7.3 ppm in Fig. 6.3A) in the person's predose urine. The opposite occurred for volunteer 2, where no signals for metabolite 4 were clearly observed in the predose urine (Fig. 6.3C) and the S/G ratio was relatively high (Fig. 6.3D). The critical finding was that the pattern observed in Fig. 6.3 for volunteers 1 and 2 was repeated in the remaining volunteers; high levels of metabolite 4 in the predose urine were associated with low ratios of S/G in the corresponding postdose urine spectra (Fig. 6.4).

Further analysis gave a clearer picture of the relationship between the predose levels of metabolite 4 and the postdose S/G ratio. Fig. 6.5 shows a plot of the postdose ratio of S/G against the predose ratio of metabolite 4 normalized to creatinine. Fig. 6.5 clearly shows that when the predose level of metabolite 4, normalized to creatinine, is above 0.06, then the 0–3 h postdose ratio of S/G is always low and below 0.8. A Mann–Whitney U test in conjunction with a Bonferroni correction of 100 (to correct for multiple hypothesis testing) showed that this finding was statistically significant ($P = 1.0 \times 10^{-4}$, with cutoff for 95% confidence $= 5 \times 10^{-2}/100 = 5 \times 10^{-4}$). However, if the predose level of metabolite 4, normalized to creatinine, is below the cutoff of 0.06, then the postdose ratio of S/G can take a very wide range of values, and no prediction of this metabolite ratio is possible.

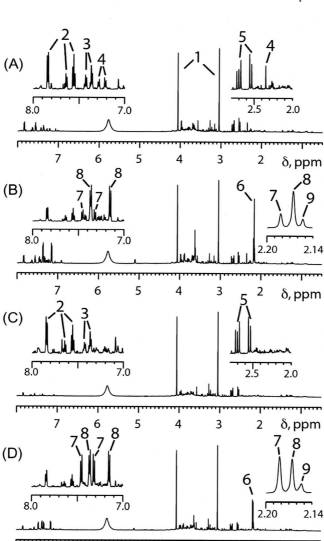

FIGURE 6.3   The 600 MHz ^{1}H NMR spectra of urine from two human volunteers on the paraceta-mol trial in spring 2003. (A) Predose spectrum of volunteer 1 from c. 0.5 to 8.0 ppm, together with expansions of the regions from c. 2.0 to 2.8 and from c. 7.0 to 8.0 ppm, where the signals of a number of endogenous metabolites can be observed. (B) Corresponding 0–3 h postdose urine spectrum of volunteer 1 with expansions from 2.14 to 2.20 and from 7.0 to 8.0 ppm. (C and D) Corresponding predose and postdose spectra, respectively, of volunteer 2. Key to metabolite signals: 1, creatinine; 2, hippurate; 3, phenylacetylglutamine; 4, metabolite 4 (unknown at the time: see text); 5, citrate; 6, cluster of N-acetyl groups from paracetamol-related compounds; 7, paracetamol sulfate; 8, paracet-amol glucuronide; 9, other paracetamol-related compounds. All spectra are referenced to deuterated trimethylsilylpropionate (TSP-d$_4$) at 0 ppm. *Reproduced from PNAS [59].*

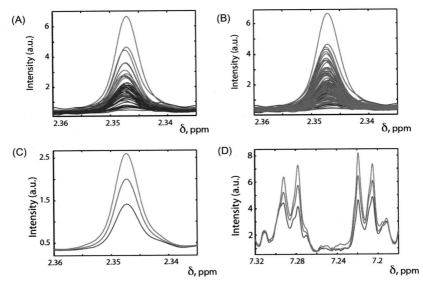

FIGURE 6.4 A series of expansions of the ¹H NMR spectra of the *predose* urine from the human volunteers on the paracetamol metabolism trial. (A) The spectra of the singlet CH₃ peak of metabolite 4 at ca 2.35 ppm, color-coded for the 25 volunteers with the highest (blue), and the 25 with the lowest (red) *postdose* ratio of S/G. (B) Corresponding expansions with the superimposition of the predose spectra of the 49 volunteers with intermediate *postdose* ratios of S/G (green). (C) The information in (B) processed to a single, synthetic, average spectrum for each group. (D) The corresponding aromatic region of the *predose* spectrum of metabolite 4. (C) and (D) have the same color-coding as (A) and (B). All spectra referenced to TSP-d₄ at 0 ppm. *Reproduced from PNAS [59].*

Fig. 6.6 shows that exactly the same effect was observed for the later 3–6 h postdose time point; high predose levels of metabolite 4, normalized to creatinine, were significantly associated with low postdose ratios of S/G ($P=1.2 \times 10^{-4}$).

It thus became imperative to identify the unknown endogenous metabolite 4 in the predose urine of some of the human volunteers. The predose ¹H NMR spectra showed that metabolite 4 had a singlet signal at c. 2.35 ppm, indicating that it was a methyl group attached to an sp2 carbon. Given that two additional, mutually coupled, pseudo-doublet, aromatic signals for 4 were observed between 7.2 and 7.3 ppm, it was hypothesized that the methyl group was one of the substituents on a *para*-disubstituted phenyl ring. Confirmation that the other substituent was an -O-sulfate group came from the finding that treatment of urine with sulfatase enzyme resulted in the transformation of metabolite 4 to another molecule, identified as *para*-cresol by comparison with an authentic reference standard. Final confirmation that metabolite 4 was *para*-cresol sulfate came after the material was synthesized and spiked into a representative predose human urine [59].

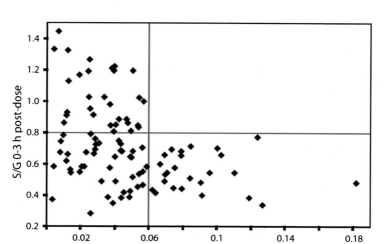

FIGURE 6.5   The ratio of paracetamol sulfate (S) to paracetamol glucuronide (G) in the 0–3 h *postdose* urine plotted against the ratio of metabolite 4, normalized to creatinine, in the corresponding *predose* urine for each volunteer on the human paracetamol trial. Each black diamond represents the results for one human volunteer. *Reproduced from PNAS [59].*

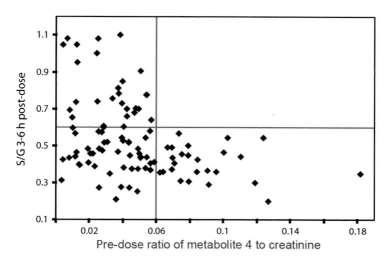

FIGURE 6.6   The ratio of paracetamol sulfate (S) to paracetamol glucuronide (G) in the 3–6 h *postdose* urine plotted against the ratio of metabolite 4, normalized to creatinine, in the corresponding *predose* urine for each volunteer on the human paracetamol trial. Each black diamond represents the results for one human volunteer. *Reproduced from PNAS [59].*

The identification of metabolite 4 in the predose human urine as *para*-cresol sulfate came as a shock, as this molecule is not of human origin. Most *para*-cresol sulfate in the body arises from the sulfation of *para*-cresol excreted by bacteria in the gut microbiome, particularly the *Clostridium* species. Thus, it appeared that the biomarker indicating the metabolism pathway for paracetamol for some of the human volunteers was not of human origin [59].

The rationale for these findings is as follows: the sulfation capacity of the human body is not high [60]. If a person is exposed to a significant level of a toxic metabolite like *para*-cresol from the gut microbiome, then it will be metabolized and excreted as rapidly as possible. Unlike rodents, which metabolize *para*-cresol to both sulfate and glucuronide metabolites, humans principally sulfate *para*-cresol [61,62]. A high predose level of *para*-cresol sulfate indicates that a person has been subject to a significant sulfation challenge. If that person then has a subsequent dose of paracetamol, it is possible that the sulfation capacity of the person is depleted and the metabolism of paracetamol will then switch to glucuronidation, as opposed to sulfation.

Not only are *para*-cresol and paracetamol sulfated by the same sulfatase enzymes (such as SULT1A1), they compete for the same co-factor 3'-phospho-adenosine 5'-phosphosulfate (PAPS, see Fig. 6.7).

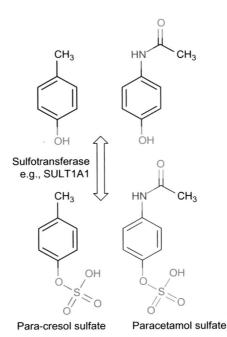

Para-cresol sulfate   Paracetamol sulfate

FIGURE 6.7   The molecular structures of *para*-cresol (*left, top*) and paracetamol (*right, top*) and their transformation by sulfotransferases into the corresponding, and related sulfate metabolites.

This study represents a series of significant findings: (1) the very first demonstration of pharmacometabonomics in humans; (2) the demonstration of the influence of microbial metabolites on human drug metabolism; and, because of this, (3) the limitations of human genomics to contribute to the prediction of drug effects in the presence of significant microbiome influences [59].

6.3.2   Other Pharmacometabonomics Studies Focused on the Prediction of Drug Metabolism and Pharmacokinetics

Yoon and Hwang [63] used an LC–MS-based approach and partial least squares data analysis to show that predose urine levels of cortisol, 1-methylguanosine, acetylarginine, and phosphoethanolamine could be used to classify 29 human volunteers into low, medium, and high drug area under the curve (AUC) cohorts after oral dosing with the immunosuppressant tacrolimus. In order to control the experiment in this cohort, the healthy volunteers were hospitalized for the duration of the study, were on a strict diet, and were taking no other medications. Blood concentrations of tacrolimus were also measured using LC–MS. This work is potentially important because tacrolimus has a narrow therapeutic index, but further studies in patients will be needed to test its more general utility in a clinical setting.

Turning to the prediction of metabolism again, Kemsley et al. used a ^{1}H NMR-based approach to predict cytochrome P450 3A4 (CYP3A4) induction in a cohort of 301 female twin volunteers aged 45–84 years [64]. The volunteers took a standardized extract of St John's Wort three times daily for 2 weeks in order to induce CYP3A4. On the evening of day 14 of this treatment, each volunteer was also administered a 300 mg tablet of quinine sulfate. CYP3A4 metabolizes quinine (Q) to 3-hydroxyquinine (3-OHQ). UPLC–MS was used to measure the ratio of 3-OHQ:Q in the day 15, posti-ntervention urine as a measure of CYP3A4 activity. Multiple linear regression methods were used to determine that the optimal model for predicting the postintervention 3-OHQ:Q ratio comprised a number of predose NMR bins and some covariates, including the body mass index of the volunteers and the batch codes for the UPLC–MS. The NMR bins of significance included signals for the following metabolites: proline, betaine, glycine, and *scyllo*-inositol. However, no metabolic link to CYP3A4 induction was found and the correlation of 3-OHQ:Q with UPLC–MS batch code was unexplained.

Cho et al. took a different approach to the prediction of CYP3A4 activity [65]. LC–MS was used to determine the clearance of midazolam by measurement of the parent drug and the 1'-hydroxymidazolam and 4-hydroxymidazolam metabolites in postdose blood plasma from a total of 24 healthy male volunteers split into three cohorts: (1) midazolam alone (CYP3A4 control group), (2) midazolam 4 days after pretreatment with 400 mg ketoconazole daily (CYP3A4 inhibited cohort), and (3) midazolam after 10 days pretreatment with 600 mg rifampicin daily (CYP3A4 induced cohort). Pre- and post-midazolam urine samples were taken, extracted and derivatized for GC–MS analysis. Postdose midazolam blood clearance was best predicted with an equation that included (1) the predose urinary ratio of 7β-hydroxydehydroepiandrosterone (7β-hydroxy-DHEA): DHEA; (2) the predose urinary ratio of 6β-hydroxycortisone: cortisone; and (3) the CYP3A5 genotype. Thus, this interesting prediction combines both metabolic and genomics features.

6.4 THE USE OF PHARMACOMETABONOMICS FOR THE PREDICTION OF DRUG SAFETY AND DRUG EFFICACY

6.4.1 The Use of Pharmacometabonomics for the Prediction of Drug Safety

In the first pharmacometabonomics study on patients, as opposed to volunteers, Keun et al. used ^{1}H NMR spectroscopy to show that baseline serum lactate, alanine, and body fat were all prognostic for weight gain in a cohort of 21 postmenopausal women undergoing 5-fluorouracil-, cyclophosphamide-, or epirubicin-based intravenous chemotherapy every 3 weeks [66]. Weight gain in these patients is a common problem and is associated with poor quality of life and poor outcomes. It is hoped that a predictive model such as the one discovered could target effective therapy to those women most at risk of a poor outcome.

Keun et al. also used ^{1}H NMR spectroscopy to establish that higher predose serum levels of low-density lipoprotein-derived lipids, including polyunsaturated fatty acids and choline phospholipids, were associated with high levels of toxicity in a cohort of 54 patients being treated with capecitabine for inoperable colorectal cancer [67]. The levels of toxicity were graded using version 2.0 of the National Cancer Institute Common Toxicity Criteria. Again, it is hoped that the ability to predict patient outcomes in advance of chemotherapy could help determine the optimal treatments for these patients.

Winnike et al. used ^{1}H NMR spectroscopic analysis of predose urine to try to predict mild liver injury in a group of 71 male and female volunteers aged 18–55 years, who were taking a high oral dose of 4 g of paracetamol (acetaminophen) daily for 7 days [68]. Liver toxicity was judged by elevations in the postdose serum levels of alanine aminotransferase (ALT) and the volunteers were thereby classified as responders or nonresponders. Although no statistically significant prediction of postdose liver toxicity could be obtained from analysis of the predose urine spectra, it was possible to use the NMR data derived shortly after dosing, and before conventional measures of liver injury gave any signals, to separate the responders from the nonresponders. This was termed *early-onset pharmacometabonomics*.

6.4.2 The Use of Pharmacometabonomics for the Prediction of Drug Efficacy

This has been an area of significant recent activity with many studies published. A series of papers have been published by Kaddurah-Daouk et al. on prediction of the efficacy of the lipid-lowering drug simvastatin. In the first paper, a targeted lipidomic analysis of plasma from a total of 36 (12 + 24) responders and non-responders from the 944 participants on the Cholesterol and Pharmacogenetics (CAP) study was conducted [69]. The methodology involved solvent extraction from the plasma, thin layer chromatography (TLC), and GC analysis of derivatized samples removed from TLC. Response to treatment was assessed by changes in both low-density lipoprotein cholesterol (LDL-C) and C-reactive protein (CRP) levels after treatment. The predose plasma lipids, whose levels were most predictive of postdose LDL-C reduction, were a number of phosphatidylcholine (PC), cholesterol ester (CE), and free fatty acid (FA) metabolites, including PC18:2n6 and related PC lipid families, CE18:1n7, and FA18:3n3 and the related FAn3. There was no overlap between the predose metabolites predictive of response as judged by LDL-C levels and that judged by CRP levels. Predose plasma levels of phosphatidyl ethanolamine (PE) plasmalogens and PC plasmalogens were positively and negatively correlated (respectively), with response as measured by greater reductions in postdose CRP. A number of other predose plasma PC and CE metabolites were associated with CRP response to treatment. A second paper from the same group used targeted GC–MS and LC–MS methods to measure the levels of 12 sterols and 14 bile acids in the predose plasmas from two groups of patients from the CAP study [70]. In a group of 100 patients exhibiting a full range of responses to simvastatin, *lower* baseline levels of the following bile acids correlated with improved efficacy, as measured by LDL-C reduction postdose: taurocholic acid (TCA), glycocholic acid (GCA), taurochenodeoxycholic acid (TCDCA), glycochenodeoxycholic acid, (GCDCA), and glycoursodeoxycholic acid (GUDCA). Table 2 of their paper [70] shows that baseline levels of taurodeoxycholic acid (TDCA) were also correlated with LDL-C reduction, but this was not commented on. In a separate analysis of 24 good responders and 24 poor responders, it was found that *higher* baseline plasma levels of lithocholic acid (LCA), taurolithocholic acid (TLCA), and glycolithocholic acid (GLCA), as well as coprostanol (COPR), were correlated with greater postdose percent LDL-C reduction. These are all secondary bile acids produced by gut bacteria. Interestingly, there is no overlap between the predictive metabolites in the two cohorts of this study in either metabolite identity or the direction of change in metabolite level. It was concluded that these differences may signify "differing specific relationships among extreme responders compared with the remainder of the population," but this is difficult to understand. Genetic evidence implicates the organic anion transporter SLCO1B1 with the hepatic uptake of both simvastatin and some bile acids, and it was hypothesized that competition between simvastatin and bile acids may affect both drug pharmacokinetics and efficacy. In a third publication on patients from the same CAP study, GC–MS was used to show that lower levels of the predose plasma metabolites xanthine, 2-hydroxyvaleric acid, succinic acid, and stearic acid and the higher predose levels of galactaric acid were correlated with increased response to simvastatin treatment, as measured by reductions in LDL-C levels after treatment [71].

TABLE 6.2 Currently Published Human Pharmacometabonomics Experiments[a]

Class of Experiment	Study	Authors and References
Prediction of pharmacokinetics (PK)	Prediction of tacrolimus PK	Phapale et al. [63]
Prediction of drug metabolism	Prediction of metabolism of paracetamol/acetaminophen	Clayton et al. [59]
	Prediction of CYP3A4 induction	Rahmioglu et al. [64]
	Prediction of CYP3A activity	Shin et al. [65]
Prediction of drug efficacy	Prediction of simvastatin efficacy in patients on the cholesterol and pharmacogenomics study	Kaddurah-Daouk et al. [69,70], Trupp et al. [71]
	Prediction of citalopram/escitalopram response in patients with major depressive disorder (MDD)	Ji et al. [72], Abo et al. [73]
	Prediction of sertraline and placebo responses in patients with MDD	Kaddurah-Daouk et al. [74,75], Zhu et al. [76]
	Prediction of efficacy of anti-psychotics	Condray et al. [77]
	Prediction of response to aspirin	Lewis et al. [78], Ellero-Simatos et al. [79]
	Prediction of efficacy with anti-tumor necrosis factor (TNF) therapies in rheumatoid arthritis	Kapoor et al. [80]
	Prediction of efficacy of L-carnitine therapy for sepsis	Puskarich et al. [81]
Prediction of adverse events	Prediction of weight gain in patients with breast cancer undergoing chemotherapy	Keun et al. [66]
	Prediction of toxicity in patients with inoperable colorectal cancer treated with capecitabine	Backshall et al. [67]
	Prediction of toxicity of paracetamol/acetaminophen ("early-onset pharmacometabonomics")	Winnike et al. [68]

[a]All publications found by searching for the terms "pharmacometabonomics," "pharmacometabolomics," "pharmaco-metabonomics," and "pharmaco-metabolomics" in February 2015.

Weinshilboum and Kaddurah-Daouk used a GC–MS approach to analyze baseline plasma samples from 40 patients with major depressive disorder (MDD) in the Mayo Clinic–NIH Pharmacogenetics Research Network (PGRN) Citalopram/Escitalopram Pharmacogenomics (Mayo-PGRN SSRI) study [72]. These patients received 10 or 20 mg of the selective serotonin reuptake inhibitor (SSRI) escitalopram daily for 8 weeks. Twenty of the patients had a Quick Inventory of Depressive Symptomatology–Clinician Rated (QIDS-C) score of ≤5 after therapy and were classified as "remitters." The other 20 were "nonremitters," with a QIDS-C score of >5 after 8 weeks of therapy. An analysis of responders (QIDS-C score reduced by ≥50% vs nonresponders) was also made. Significantly higher levels of pretreatment plasma glycine were observed in nonresponders versus responders ($P = 0.005$) but the effect was not significant for nonremitters versus remitters ($P = 0.059$, in Fig. 2 of the paper and 0.058 in text [72]). Lower pretreatment levels of hydroxylamine were also associated with improved escitalopram response, but this was not discussed further. In a new development, the pharmacometabonomics results were used to inform a pharmacogenomics analysis of SNPs in some genes whose products are responsible for glycine synthesis and degradation. The rs10975641 SNP in the glycine dehydrogenase gene *GLDC* was found to be significantly associated with remission status ($P = 0.008$). A validation of this study was conducted by genotyping for the rs10975641 SNP in 1245 patients with MDD from the Sequenced Treatment Alternatives to Relieve Depression (STAR*D) study, and it was found to be significantly associated with response in white, non-Hispanic subjects ($P = 0.016$). However, due to differences between the STAR*D participants who contributed DNA and those who did not, the authors recommended further studies to replicate the finding. Weinshilboum et al. then continued this study by using "1000 Genomes" SNP imputation to replicate the finding in the Mayo-PGRN SSRI trial patients that the rs10975641 SNP was associated with treatment outcome, among other results [73]. More recently, Weinshilboum et al. used a genome-wide association study to show that SNPs in or near the *CYP2C19* and the *CYP2D6* genes were significantly associated with plasma escitalopram and *S*-didesmethlycitalopram levels in 435 patients with MDD in the Mayo Clinic NIH-Pharmacogenomics Research Network-Antidepressant Medication Pharmacogenomic Study [82].

Kaddurah-Daouk et al. used a liquid chromatography electrochemical array (LC-ECA) approach to determine that pretreatment levels of plasma dihydroxyphenylacetic acid (DOPAC), 4-hydroxyphenyllactic acid (4-HPLA), serotonin (5-HT), and gamma tocopherol contributed to partial least square-discriminant analysis (PLS-DA) models that discriminated between responders and nonresponders in 43 outpatients with MDD treated with sertraline (between 50 and 150 mg per day) [74]. After cross-validation, the overall correct classification rate was 81%. Hypoxanthine, xanthine, uric acid, 5-methoxytryptophol (5-MTPOL), 5-HT, 3-hydroxykynurenine (3-OHKY), and 5-hydroxyindoleacetic acid (5-HIAA), DOPAC, cysteine, and several tocopherols contributed to a

PLS-DA model, which separated responders from nonresponders on placebo treatment. Thus, there was a partial overlap of the metabolites that contributed to the prediction of sertraline versus placebo responses. It was noted that 5-MTPOL is produced in the body from 5-HT and that 4-HPLA is in the phenylalanine pathway and interconnected with the tyrosine and catecholamine pathways. It was not noted that 5-HIAA is a metabolite of 5-HT [83]. A second publication on a largely overlapping group of 75 outpatients with MDD used LC-ECA and GC–MS analysis of plasma to determine that a good response to sertraline ($n=35$) was associated with higher pretreatment levels of 5-methoxytryptamine (5-MTPM), whereas in the placebo group ($n=40$), lower pretreatment levels of 5-MTPOL were associated with better outcomes, as assessed by 4-week changes in the 17-item Hamilton Rating Scale for Depression (HAMD17) [76]. The pretreatment values of several plasma metabolite ratios (melatonin to 5-HT, melatonin to 5-MTPM, N-acetyl 5-HT to 5-MTPM, and 5-HIAA to 5-MTPM) were also reported to be associated with sertraline treatment outcomes, and the ratio of 5-MTPOL to tryptophan was associated with placebo treatment outcome. However, the overlap of discriminating pretreatment metabolites between the two studies was limited. A third publication titled "Pharmacometabolomic mapping of early biochemical changes induced by sertraline and placebo" [75] was found to contain only metabonomics data and no pharmacometabonomics findings.

Yao et al. used LC-ECA analysis of pretreatment plasma metabolites from 25 first-episode, neuroleptic-naive patients with schizophrenia subsequently treated with five different drugs to show that low pretreatment levels of 3-hydroxykynurenine (3-OHKY) were associated with the greatest improvement in symptoms after 4 weeks of treatment [77]. Interestingly, high baseline levels of 3-OHKY were associated with lower baseline total clinical scores. It is not yet clear why this is the case, but several lines of analysis were given.

Kaddurah-Daouk et al. used a pharmacometabonomics-informed pharmacogenomics approach to study resistance to aspirin therapy [84]. Resistance or nonresistance was measured using *ex vivo* blood platelet aggregation after 2 weeks of aspirin therapy (81 mg per day). Participants were members of the Old Order Amish population from Lancaster County, PA, taking part in the Heredity and Phenotype Intervention Heart Study (HAPI). The Results section indicates that untargeted GC–MS was used to analyze pretreatment and posttreatment serum samples from 76 volunteers from HAPI, who underwent a 2-week aspirin intervention. However, the picture is confused as the Methods section states that in the metabolomics study, samples *originated* from the 76 HAPI participants but that "non-first-degree relatives of the first (good responders) and the fourth (poor responders) sex-specific drug–response quartiles were selected for metabolic profiling." In any event, no pharmacometabonomics results were reported from the first element of this study. In a second replication element, the Results section indicated that GC–MS analysis of the serum from 37 additional subjects from the HAPI study, who were either good or poor responders, showed that poor responders had higher pretreatment inosine levels than good responders ($P = 0.05$). Surprisingly, the Supplementary Information in this paper shows that poor responders also had significantly higher hypoxanthine levels before treatment ($P<0.05$) but this was not commented on. Confusingly, again, the Methods section stated that the replication element was conducted using 49 HAPI participants, of whom 19 were good responders and 20 were poor responders, which makes 39 in total rather than 37 or 49. Because of the involvement of purine metabolites in the prediction of response as well as the response to aspirin treatment, a "pharmacometabonomics-informed pharmacogenomics study" was conducted. Fifty-one SNPs in the adenosine kinase gene were found to be associated with change in platelet aggregation during aspirin therapy ($P<0.0005$) within a false discovery cutoff of 0.01. The SNP most strongly associated with response was the intronic variant rs16931294. The less common G allele was associated with higher platelet aggregation and higher posttreatment inosine and guanosine levels than the more common A allele. Exploratory metabolic profiling studies showed that the G allele was significantly associated with higher pretreatment levels of adenosine monophosphate, xanthine, and hypoxanthine compared with the A allele. Given that inosine is known to inhibit platelet aggregation, the association of the G allele of rs16931294 with both higher posttreatment aggregation and higher posttreatment inosine levels was perplexing, and more work was recommended. Surprisingly, a subset of the authors of the original aspirin paper published in *Clinical Pharmacology & Therapeutics* in October 2013 [84], reviewed this very same work in the November 2013 issue of the same journal [78]. A third publication from the same group used targeted UPLC–MS to study amine-containing metabolites in serum from healthy subjects on the same HAPI study [79]. Higher 5-HT (serotonin) before and after treatment with aspirin correlated with higher post-aspirin collagen-induced platelet aggregation in both the discovery ($n=80$) and replication ($n = 125$) cohorts.

Young et al. used ^{1}H NMR spectroscopy to analyze urine from 16 patients with rheumatoid arthritis (RA) and 20 patients with psoriatic arthritis (PsA), who were randomized to 12-month treatments with the anti-tumor necrosis factor (anti-TNF) agents infliximab and etanercept [80]. For the patients with RA, baseline urine levels of histamine, glutamine, xanthurenic acid, ethanolamine, and others were correlated with changes in the Disease Activity Score in 28 joints (DAS28), used as a measure of clinical efficacy. The four named metabolites were all found through three different statistical analysis methods, and all were thought to derive from amino acid degradation, although histamine could be derived from mast cells in synovial infiltrates.

Stringer et al. used ^{1}H NMR spectroscopic analysis of serum samples to study the effects of therapy on 31 patients with septic shock, who were randomized to either saline placebo ($n = 15$) or L-carnitine ($n = 16$) [81]. The ketone bodies 3-hydroxybutyrate and acetoacetate and the metabolite 3-hydroxyisovalerate were significantly higher, both before treatment and at 24h after treatment, in

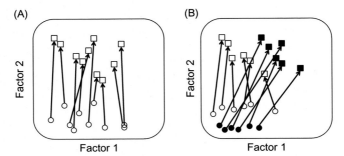

FIGURE 6.8 A conceptual visualization of the difference between metabonomics and pharmacometabonomics experiments. In each case, the outcome of the experiments is represented by a multivariate analysis, such as a PCA, where factor 1 and factor 2 would be the first and second principal components of a PCA scores plot. (A) In a metabonomics experiment, the change in the metabolic state of the subjects due to an intervention of some kind is monitored (arrows show the metabolic trajectory across factor 2 = PC2) and interpreted in terms of the changes caused to the physiology and biochemistry of the subjects. (B) By contrast, in a pharmacometabonomics experiment, the predose metabolic state of the subjects is analyzed, and a model is used to predict which subjects will behave in a certain manner, for example, responders versus nonresponders to drug treatment. In the example shown here, the subjects with the filled circle metabolic starting points are distinguished from the subjects with the open circles on the basis of their lower factor 2 scores. After intervention, the two subgroups of subjects undergo different metabolic trajectories to open- or closed-square final metabolic positions, dictated by their different starting positions.

the serum of nonsurvivors. Using the pretreatment median 3-hydroxybutyrate concentration of 153 μM, patients were categorized as having either low or high ketone levels. Using mortality endpoints, a clear benefit of L-carnitine therapy was seen in both patient groups. However, a marked trend of greater patient survival in L-carnitine-treated patients with low pretreatment ketones ($P=0.007$) was observed. It was concluded that pharmacometabonomics has the potential to provide better information to guide L-carnitine therapy decisions compared with standard Sequential Organ Failure Assessment (SOFA) scores.

Finally, several papers have been published recently with "pharmacometabonomics" or "pharmacometabonomics" in the titles, abstracts, or keywords, but, as far as it could be determined, contained no pharmacometabonomics data at all [75,85–87]. Thus some confusion has been created in the literature between what constitutes a metabonomics experiment and what a pharmacometabonomics experiment is. As is clear in Fig. 6.8, the two concepts are entirely different [21,57], and it is hoped that this regrettable blurring of definitions will cease.

6.5 PREDICTIVE METABONOMICS

Many examples of human pharmacometabonomics have been demonstrated in recent years, as illustrated above. In fact, pharmacometabonomics is just one example of a broader class of experiments termed *predictive metabonomics*.

This can be defined as follows: "the prediction of the outcome of an intervention in an individual based on a mathematical model of preintervention metabolite signatures" [88]. When originally defined [57], it was envisaged that the pre-intervention metabolic profiling conducted for pharmacometabonomics would also be able to predict the outcome of interventions other than drug administration. These interventions could include changes in diet, physical exercise, medical interventions, microbiologic changes, or merely the passage of time. These predictions, made in 2006 [57], have been borne out by the publication of a number of significant predictive metabonomics studies recently, and these studies are summarized here. The studies used preintervention metabolite analysis to predict a future event. In all cases, the "future event" was disease or death and the "intervention" was the passage of time.

Wang et al. used LC–MS to analyze baseline plasma samples from 189 individuals (from 2422 eligible subjects without diabetes in the Framingham Offspring Study) who went on to develop new-onset diabetes in a 12-year follow-up period [89]. Baseline plasma levels were compared with those from 189 propensity-matched controls who did not develop diabetes. Higher baseline concentrations of five plasma amino acids—leucine, isoleucine, valine, phenylalanine, and tyrosine—were found for the subjects who went on to develop diabetes ($P = 0.001$ or lower). The five amino acids were re-analyzed in the Malmo Diet and Cancer Study as a replication analysis. All except isoleucine were again found to be significantly associated with incident diabetes (P values from 0.009 to 0.04).

Wang-Sattler et al. used a prospective population-based approach to determine that lower baseline levels of glycine and lysophosphatidylcholine (LPC) (18:2) were predictive of future (baseline plus 7 years) impaired glucose tolerance and the development of type 2 diabetes in a subset of the Cooperative Health Research in the Region of Augsburg (KORA) cohort [90].

Shah et al. used targeted MS profiling of baseline plasma metabolites to identify an array of dicarboxylacylcarnitines, medium-chain acylcarnitines, and fatty acids that were predictive of future events, including death/myocardial infarction in a cohort of 2033 patients on the Measurement to Understand the Reclassification of Disease of Cabarrus and Kannapolis Cardiovascular Study (MURDOCK CV) [91]. In a separate study of 3903 patients undergoing elective, nonurgent coronary angiography at the Cleveland Clinic, Hazens et al. demonstrated that increased plasma levels of choline and betaine were associated with increased risk of major adverse cardiac events (MACEs) when levels of microbiome-derived TMAO were also high [92].

Federici et al. also used targeted MS methods to discover that two baseline serum "metabolite factors," 1 and 7, were independently associated with MACEs in 67 very old Italian participants with a high rate of previous CVD [93]. Factor 1 comprised a mixture of medium- to long-chain acylcarnitines, whereas factor 7 was just alanine. Only factor 1 significantly increased the prediction accuracy of the Framingham Recurring Coronary Heart Disease Score.

Finally, Purroy et al. used LC–MS analysis of the plasma samples from 131 patients with transient ischemic attack (TIA) to show that stroke recurrence was significantly associated with low concentrations of the lysophosphatidylcholine LysoPC[16:0] [94].

6.6 FUTURE DEVELOPMENTS

The examples above have shown that metabolic profiling ahead of drug administration, ie, pharmacometabonomics, although a subject area in its infancy, is able to provide a degree of prediction of patient outcome in some situations. This is a significant development for Personalized Medicine because, in contrast to pharmacogenomics and its focus on gene mutations, metabolic profiling and pharmacometabonomics are able to sample both genetic and environmental factors that can, and will, influence the transport, pharmacokinetics, metabolism, efficacy, and toxicity of a drug in a patient. We envisage that the combined use of pharmacometabonomics and pharmacogenomics will lead to the best possible prediction of patient outcomes. The advent of the broader field of predictive metabonomics is also in its infancy but promises to enable intervention in the management of patients who are predicted to be at higher risk of disease or death.

The future of Personalized Medicine based on metabolic profiling is predicted to be very bright. Already, new developments such as longitudinal pharmacometabonomics, where repeated sampling of a patients' metabolite profile is used to inform treatment options throughout the "patient journey," have emerged [95]. Another area that is predicted to develop and have a strong impact on Personalized Medicine is the linking of genetic information with metabolite profiles in genome-wide association studies (GWAS) [96–99]. The outcome of these studies could also significantly inform decision making in personalized health care.

We stand at the dawn of a new era of Personalized Medicine, where, instead of a single focus on genetic information, patient care decisions will be made by a combination of clinical, genetic, and metabolic factors that integrates important information from the patient's genome, the microbiome, and the broader environment. We look forward to this emerging science in the next decade or so and the benefits that it will bring to us all.

REFERENCES

[1] Lindon J, Nicholson J, Everett J. NMR spectroscopy of biofluids. Ann Rep NMR Spectrosc 1999;38:1–88.

[2] Gates SC, Sweeley CC. Quantitative metabolic profiling based on gas-chromatography. Clin Chem 1978;24(10):1663–73.

[3] Everett J, Jennings K, Woodnutt G, Buckingham M. Spin-Echo H-1-NMR spectroscopy—a new method for studying penicillin metabolism. J Chem Soc Chem Commun 1984;14:894–5.

[4] Everett J, Jennings K, Woodnutt G. F-19 NMR-spectroscopy study of the metabolites of flucloxacillin in rat urine. J Pharm Pharmacol 1985;37(12):869–73.

[5] Everett J, Tyler J, Woodnutt G. A study of flucloxacillin metabolites in rat urine by two-dimensional H-1, F-19 COSY NMR. J Pharm Biomed Anal 1989;7(3):397–403.

[6] Everett JR. High-resolution NMR spectroscopy of biofluids and applications in drug metabolism. Anal Proc 1991;28(6):181–3.

[7] Connor S, Everett J, Jennings K, Nicholson J, Woodnutt G. High-resolution H-1-NMR spectroscopic studies of the metabolism and excretion of ampicillin in rats and amoxicillin in rats and man. J Pharm Pharmacol 1994;46(2):128–34.

[8] Robosky LC, Robertson DG, Baker JD, Rane S, Reily MD. *In vivo* toxicity screening programs using metabonomics. Comb Chem High Throughput Scr 2002;5(8):651–62.

[9] Lindon JC, Keun HC, Ebbels TM, Pearce JM, Holmes E, Nicholson JK. The Consortium for Metabonomic Toxicology (COMET): aims, activities and achievements. Pharmacogenomics 2005;6(7):691–9.

[10] Coen M, Holmes E, Lindon JC, Nicholson JK. NMR-based metabolic profiling and metabonomic approaches to problems in molecular toxicology. Chem Res Toxicol 2008;21(1):9–27.

[11] Robertson DG, Watkins PB, Reily MD. Metabolomics in toxicology: preclinical and clinical applications. Toxicol Sci 2011;120:S146–70.

[12] Nicholson JK, Connelly J, Lindon JC, Holmes E. Metabonomics: a platform for studying drug toxicity and gene function. Nat Rev Drug Discov 2002;1(2):153–61.

[13] Mishur RJ, Rea SL. Applications of mass spectrometry to metabolomics and metabonomics: detection of biomarkers of aging and of age-related diseases. Mass Spectrom Rev 2012;31(1):70–95.

[14] Ellis DI, Dunn WB, Griffin JL, Allwood JW, Goodacre R. Metabolic fingerprinting as a diagnostic tool. Pharmacogenomics 2007;8(9):1243–66.

[15] Ala-Korpela M. Potential role of body fluid ^{1}H NMR metabonomics as a prognostic and diagnostic tool. Expert Rev Mol Diagn 2007;7(6):761–73.

[16] Iles RA, Hind AJ, Chalmers RA. Use of proton nuclear magnetic-resonance spectroscopy in detection and study of organic acidurias. Clin Chem 1985;31(11):1795–801.

[17] Moolenaar SH, Engelke UFH, Wevers RA. Proton nuclear magnetic resonance spectroscopy of body fluids in the field of inborn errors of metabolism. Ann Clin Biochem 2003;40:16–24.

[18] Iles RA. Nuclear magnetic resonance spectroscopy and genetic disorders. Curr Med Chem 2008;15(1):15–36.

[19] Murdoch TB, Fu H, MacFarlane S, Sydora BC, Fedorak RN, Slupsky CM. Urinary metabolic profiles of inflammatory bowel disease in interleukin-10 gene-deficient mice. Anal Chem 2008;80(14):5524–31.

[20] Saadat N, IglayReger HB, Myers MG, Bodary P, Gupta SV. Differences in metabolomic profiles of male db/db and s/s, leptin receptor mutant mice. Physiol Genomics 2012;44(6):374–81.

[21] Lindon J, Nicholson J, Holmes E, Everett J. Metabonomics: metabolic processes studied by NMR spectroscopy of biofluids. Concepts Magn Reson 2000;12(5):289–320.

[22] Fiehn O. Metabolomics—the link between genotypes and phenotypes. Plant Mol Biol 2002;48(1-2):155–71.

[23] Dona AC, Jimenez B, Schaefer H, Humpfer E, Spraul M, Lewis MR, et al. Precision high-throughput proton NMR spectroscopy of human urine, serum, and plasma for large-scale metabolic phenotyping. Anal Chem 2014;86(19):9887–94.

[24] Bylesjo M, Rantalainen M, Cloarec O, Nicholson JK, Holmes E, Trygg J. OPLS discriminant analysis: combining the strengths of PLS-DA and SIMCA classification. J Chemom 2006;20(8-10):341–51.

[25] Craig A, Cloarec O, Holmes E, Nicholson JK, Lindon JC. Scaling and normalization effects in NMR spectroscopic metabonomic data sets. Anal Chem 2006;78(7):2262–7.

[26] Beckonert O, Keun HC, Ebbels TMD, Bundy JG, Holmes E, Lindon JC, et al. Metabolic profiling, metabolomic and metabonomic procedures for NMR spectroscopy of urine, plasma, serum and tissue extracts. Nat Protoc 2007;2(11):2692–703.

[27] Claridge T. High-resolution NMR techniques in organic chemistry, 2nd ed. Oxford, UK: Elsevier; 2009.

[28] Posma JM, Garcia-Perez I, De Iorio M, Lindon JC, Elliott P, Holmes E, et al. Subset Optimization by Reference Matching (STORM): an optimized statistical approach for recovery of metabolic biomarker structural information from H-1 NMR spectra of biofluids. Anal Chem 2012;84(24):10694–701.

[29] Bouatra S, Aziat F, Mandal R, Guo AC, Wilson MR, Knox C, et al. The human urine metabolome. PLoS One 2013;8:9.

[30] Emwas A-HM, Salek RM, Griffin JL, Merzaban J. NMR-based metabolomics in human disease diagnosis: applications, limitations, and recommendations. Metabolomics 2013;9(5):1048–72.

[31] Fancy S-A, Beckonert O, Darbon G, Yabsley W, Walley R, Baker D, et al. Gas chromatography/flame ionisation detection mass spectrometry for the detection of endogenous urine metabolites for metabonomic studies and its, use as a complementary tool to nuclear magnetic resonance spectroscopy. Rapid Commun Mass Spectrom 2006;20:15.

[32] Crockford DJ, Holmes E, Lindon JC, Plumb RS, Zirah S, Bruce SJ, et al. Statistical heterospectroscopy, an approach to the integrated analysis of NMR and UPLC–MS data sets: application in metabonomic toxicology studies. Anal Chem 2006;78(2):363–71.

[33] Wu Z, Huang Z, Lehmann R, Zhao C, Xu G. The application of Chromatography–Mass spectrometry: methods to metabonomics. Chromatographia 2009;69:23–32.

[34] Want EJ, Wilson ID, Gika H, Theodoridis G, Plumb RS, Shockcor J, et al. Global metabolic profiling procedures for urine using UPLC–MS. Nat Protoc 2010;5(6):1005–18.

[35] Spagou K, Wilson ID, Masson P, Theodoridis G, Raikos N, Coen M, et al. HILIC–UPLC–MS for exploratory urinary metabolic profiling in toxicological studies. Anal Chem 2011;83(1):382–90.

[36] Theodoridis G, Gika HG, Wilson ID. Mass spectrometry-based holistic analytical approaches for metabolite profiling in systems biology studies. Mass Spectrom Rev 2011;30(5):884–906.

[37] Theodoridis GA, Gika HG, Want EJ, Wilson ID. Liquid chromatography–mass spectrometry based global metabolite profiling: a review. Anal Chim Acta 2012;711:7–16.

[38] Gika H, Theodoridis G. Sample preparation prior to the LC–MS-based metabolomics/metabonomics of blood-derived samples. Bioanalysis 2011;3(14):1647–61.

[39] Sumner LW, Amberg A, Barrett D, Beale MH, Beger R, Daykin CA, et al. Proposed minimum reporting standards for chemical analysis. Metabolomics 2007;3(3):211–21.

[40] Salek RM, Steinbeck C, Viant MR, Goodacre R, Dunn WB. The role of reporting standards for metabolite annotation and identification in metabolomic studies. GigaScience 2013;2(1) 13-.

[41] Everett JR. A new paradigm for known metabolite identification in metabonomics/metabolomics: metabolite identification efficiency. Comput Struct Biotechnol J 2015;13(0):131–44.

[42] Pokorska-Bocci A, Stewart A, Sagoo GS, Hall A, Kroese M, Burton H. 'Personalized medicine': what's in a name? Per Med 2014;11(2):197–210.

[43] Abrahams E, Ginsburg GS, Silver M. The personalized medicine coalition—goals and strategies. Am J Pharmacogenom 2005;5(6):345–55.

[44] Lazarou J, Pomeranz BH, Corey PN. Incidence of adverse drug reactions in hospitalized patients—a meta-analysis of prospective studies. JAMA 1998;279(15):1200–5.

[45] Lee JW, Aminkeng F, Bhavsar AP, Shaw K, Carleton BC, Hayden MR, et al. The emerging era of pharmacogenomics: current successes, future potential, and challenges. Clin Genetics 2014;86(1):21–8.

[46] Carr DF, Alfirevic A, Pirmohamed M. Pharmacogenomics: current state-of-the-art. Genes 2014;5(2):430–43.

[47] Pirmohamed M. Personalized pharmacogenomics: predicting efficacy and adverse drug reactions. Ann Rev Genomics Hum Genet 2014;15:349–70.

[48] Urban TJ, Goldstein DB. Pharmacogenetics at 50: genomic personalization comes of age. Sci Transl Med 2014;6(220):1–9.

[49] Eloe-Fadrosh EA, Rasko DA. The human microbiome: from symbiosis to pathogenesis. Ann Rev Med 2013;64(64):145–63.

[50] Holmes E, Li JV, Marchesi JR, Nicholson JK. Gut microbiota composition and activity in relation to host metabolic phenotype and disease risk. Cell Metab 2012;16(5):559–64.

[51] Pflughoeft KJ, Versalovic J. Human microbiome in health and disease. Ann Rev Pathol Mech Dis 2012;7(7):99–122.

[52] Lindon JC, Nicholson JK, Holmes E. The handbook of metabonomics and metabolomics. Amsterdam; Oxford: Elsevier; 2007.

[53] Shah RR, Smith RL. Addressing phenoconversion: the Achilles' heel of personalized medicine. Br J Clin Pharmacol 2015;79(2):222–40.

[54] Joseph PG, Pare G, Ross S, Roberts R, Anand SS. Pharmacogenetics in cardiovascular disease: the challenge of moving from promise to realization concepts discussed at the Canadian Network and Centre for Trials Internationally Network conference (CANNeCTIN), June 2009. Clin Cardiol 2014;37(1):48–56.

[55] Maruthur NM, Gribble MO, Bennett WL, Bolen S, Wilson LM, Balakrishnan P, et al. The pharmacogenetics of type 2 diabetes: a systematic review. Diabetes Care 2014;37(3): 876–86.

[56] Perlis RH. Pharmacogenomic testing and personalized treatment of depression. Clin Chem 2014;60(1):53–9.

[57] Clayton T, Lindon J, Cloarec O, Antti H, Charuel C, Hanton G, et al. Pharmaco-metabonomic phenotyping and personalized drug treatment. Nature 2006;440(7087):1073–7.

[58] Fonville JM, Richards SE, Barton RH, Boulange CL, Ebbels TMD, Nicholson JK, et al. The evolution of partial least squares models and related chemometric approaches in metabonomics and metabolic phenotyping. J Chemom 2010;24(11-12):636–49.

[59] Clayton TA, Baker D, Lindon JC, Everett JR, Nicholson JK. Pharmacometabonomic identification of a significant host-microbiome metabolic interaction affecting human drug metabolism. Proc Natl Acad Sci USA 2009;106(34):14728–33.

[60] Coughtrie MWH. Sulfation through the looking glass: recent advances in sulfotransferase research for the curious. Pharmacogenomics J 2002;2(5):297–308.

[61] Schepers E, Meert N, Glorieux G, Goeman J, Van der Eycken J, Vanholder R. P-cresylsulphate, the main *in vivo* metabolite of p-cresol, activates leucocyte free radical production. Nephrol Dial Transplant 2007;22(2):592–6.

[62] de Loor H, Bammens B, Evenepoel P, De Preter V, Verbeke K. Gas chromatographic-mass spectrometric analysis for measurement of p-cresol and its conjugated metabolites in uremic and normal serum. Clin Chem 2005;51(8):1535–8.

[63] Phapale PB, Kim SD, Lee HW, Lim M, Kale DD, Kim YL, et al. An integrative approach for identifying a metabolic phenotype predictive of individualized pharmacokinetics of tacrolimus. Clin Pharmacol Ther 2010;87(4):426–36.

[64] Rahmioglu N, Le Gall G, Heaton J, Kay KL, Smith NW, Colquhoun IJ, et al. Prediction of variability in CYP3A4 induction using a combined H-1 NMR metabonomics and targeted UPLC–MS approach. J Proteome Res 2011;10(6):2807–16.

[65] Shin KH, Choi MH, Lim KS, Yu KS, Jang IJ, Cho JY. Evaluation of endogenous metabolic markers of hepatic CYP3A activity using metabolic profiling and midazolam clearance. Clin Pharmacol Ther 2013;94(5):601–9.

[66] Keun HC, Sidhu J, Pchejetski D, Lewis JS, Marconell H, Patterson M, et al. Serum molecular signatures of weight change during early breast cancer chemotherapy. Clin Cancer Res 2009;15(21):6716–23.

[67] Backshall A, Sharma R, Clarke SJ, Keun HC. Pharmacometabonomic profiling as a predictor of toxicity in patients with inoperable colorectal cancer treated with capecitabine. Clin Cancer Res 2011;17(9):3019–28.

[68] Winnike JH, Li Z, Wright FA, Macdonald JM, O'Connell TM, Watkins PB. Use of pharmacometabonomics for early prediction of acetaminophen-induced hepatotoxicity in humans. Clin Pharmacol Ther 2010;88(1):45–51.

[69] Kaddurah-Daouk R, Baillie RA, Zhu HJ, Zeng ZB, Wiest MM, Nguyen UT, et al. Lipidomic analysis of variation in response to simvastatin in the cholesterol and pharmacogenetics study. Metabolomics 2010;6(2):191–201.

[70] Kaddurah-Daouk R, Baillie RA, Zhu H, Zeng ZB, Wiest MM, Nguyen UT, et al. Enteric microbiome metabolites correlate with response to simvastatin treatment. PLoS One 2011;6(10): e25482.

[71] Trupp M, Zhu H, Wikoff WR, Baillie RA, Zeng ZB, Karp PD, et al. Metabolomics reveals amino acids contribute to variation in response to simvastatin treatment. PLoS One 2012;7(7):e38386.

[72] Ji Y, Hebbring S, Zhu H, Jenkins GD, Biernacka J, Snyder K, et al. Glycine and a glycine dehydrogenase (GLDC) SNP as citalopram/escitalopram response biomarkers in depression: pharmacometabolomics-informed pharmacogenomics. Clin Pharmacol Ther 2011;89(1):97–104.

[73] Abo R, Hebbring S, Ji Y, Zhu H, Zeng ZB, Batzler A, et al. Merging pharmacometabolomics with pharmacogenomics using '1000 Genomes' single-nucleotide polymorphism imputation: selective serotonin reuptake inhibitor response pharmacogenomics. Pharmacogenet Genomics 2012;22(4):247–53.

[74] Kaddurah-Daouk R, Boyle SH, Matson W, Sharma S, Matson S, Zhu H, et al. Pretreatment metabotype as a predictor of response to sertraline or placebo in depressed outpatients: a proof of concept. Transl Psychiatr 2011;1:1–7.

[75] Kaddurah-Daouk R, Bogdanov MB, Wikoff WR, Zhu H, Boyle SH, Churchill E, et al. Pharmacometabolomic mapping of early biochemical changes induced by sertraline and placebo. Transl Psychiatr 2013;3:e223.

[76] Zhu H, Bogdanov MB, Boyle SH, Matson W, Sharma S, Matson S, et al. Pharmacometabolomics of response to sertraline and to placebo in major depressive disorder—possible role for methoxyindole pathway. PLoS One 2013;8(7):e68283.

[77] Condray R, Dougherty GG, Keshavan MS, Reddy RD, Haas GL, Montrose DM, et al. 3-Hydroxykynurenine and clinical symptoms in first-episode neuroleptic-naive patients with schizophrenia. Int J Neuropsychopharmacol 2011;14(6):756–67.

[78] Lewis JP, Yerges-Armstrong LM, Ellero-Simatos S, Georgiades A, Kaddurah-Daouk R, Hankemeier T. Integration of pharmacometabolomic and pharmacogenomic approaches reveals novel insights into antiplatelet therapy. Clin Pharmacol Therap 2013;94(5):570–3.

[79] Ellero-Simatos S, Lewis JP, Georgiades A, Yerges-Armstrong LM, Beitelshees AL, Horenstein RB, et al. Pharmacometabolomics reveals that serotonin is implicated in aspirin response variability. CPT: Pharmacometrics Syst Pharmacol 2014;3 e125.

[80] Kapoor SR, Filer A, Fitzpatrick MA, Fisher BA, Taylor PC, Buckley CD, et al. Metabolic profiling predicts response to anti-tumor necrosis factor alpha therapy in patients with rheumatoid arthritis. Arthritis Rheum 2013;65(6):1448–56.

[81] Puskarich MA, Finkel MA, Karnovsky A, Jones AE, Trexel J, Harris BN, et al. Pharmacometabolomics of l-Carnitine treatment response phenotypes in patients with septic Shock. Ann Am Thorac Soc 2015;12(1):46–56.

[82] Ji Y, Schaid DJ, Desta Z, Kubo M, Batzler AJ, Snyder K, et al. Citalopram and escitalopram plasma drug and metabolite concentrations: genome-wide associations. Br J Clin Pharmacol 2014;78(2):373–83.

[83] Kema IP, de Vries EGE, Muskiet FAJ. Clinical chemistry of serotonin and metabolites. J Chromatograph B 2000;747(1-2):33–48.

[84] Yerges-Armstrong LM, Ellero-Simatos S, Georgiades A, Zhu H, Lewis JP, Horenstein RB, et al. Purine pathway implicated in mechanism of resistance to aspirin therapy: pharmacometabolomics-informed. Pharmacogenom Clin Pharmacol Ther 2013;94(4):525–32.

[85] Wikoff WR, Frye RF, Zhu H, Gong Y, Boyle S, Churchill E, et al. Pharmacometabolomics reveals racial differences in response to atenolol treatment. PLoS One 2013;8(3) e57639.

[86] Villasenor A, Ramamoorthy A, Silva dos Santos M, Lorenzo MP, Laje G, Zarate Jr. C, et al. A pilot study of plasma metabolomic patterns from patients treated with ketamine for bipolar depression: evidence for a response-related difference in mitochondrial networks. Br J Pharmacol 2014;171(8):2230–42.

[87] Patterson AD, Slanar O, Krausz KW, Li F, Hofer CC, Perlik F, et al. Human urinary metabolomic profile of PPARalpha induced fatty acid beta-oxidation. J Proteome Res 2009;8(9):4293–300.

[88] Everett JR, Loo RL, Pullen FS. Pharmacometabonomics and personalized medicine. Ann Clin Biochem 2013;50:523–45.

[89] Wang TJ, Larson MG, Vasan RS, Cheng S, Rhee EP, McCabe E, et al. Metabolite profiles and the risk of developing diabetes. Nat Med 2011;17(4) 448–U83.

[90] Wang-Sattler R, Yu Z, Herder C, Messias AC, Floegel A, He Y, et al. Novel biomarkers for pre-diabetes identified by metabolomics. Mol Syst Biol 2012:8.

[91] Shah SH, Sun J-L, Stevens RD, Bain JR, Muehlbauer MJ, Pieper KS, et al. Baseline metabolomic profiles predict cardiovascular events in patients at risk for coronary artery disease. Am Heart J 2012;163(5):844–850.

[92] Wang Z, Tang WHW, Buffa JA, Fu X, Britt EB, Koeth RA, et al. Prognostic value of choline and betaine depends on intestinal microbiota-generated metabolite trimethylamine-N-oxide. Eur Heart J 2014;35(14):904–10.

[93] Rizza S, Copetti M, Rossi C, Cianfarani MA, Zucchelli M, Luzi A, et al. Metabolomics signature improves the prediction of cardiovascular events in elderly subjects. Atherosclerosis 2014;232(2):260–4.

[94] Jove M, Mauri-Capdevila G, Suarez I, Cambray S, Sanahuja J, Quilez A, et al. Metabolomics predicts stroke recurrence after transient ischemic attack. Neurology 2015;84(1):36–45.

[95] Nicholson JK, Everett JR, Lindon JC. Longitudinal pharmacometabonomics for predicting patient responses to therapy: drug metabolism, toxicity and efficacy. Exp Opin Drug Metab Toxicol 2012;8(2):135–9.

[96] Suhre K, Wallaschofski H, Raffler J, Friedrich N, Haring R, Michael K, et al. A genome-wide association study of metabolic traits in human urine. Nat Genet 2011;43(6) 565–U97.

[97] Kettunen J, Tukiainen T, Sarin A-P, Ortega-Alonso A, Tikkanen E, Lyytikainen L-P, et al. Genome-wide association study identifies multiple loci influencing human serum metabolite levels. Nat Genet 2012;44(3) 269–U65.

[98] Shin S-Y, Fauman EB, Petersen A-K, Krumsiek J, Santos R, Huang J, et al. An atlas of genetic influences on human blood metabolites. Nat Genet 2014;46(6):543–50.

[99] Rueedi R, Ledda M, Nicholls AW, Salek RM, Marques-Vidal P, Morya E, et al. Genome-wide association study of metabolic traits reveals novel gene-metabolite-disease links. PLoS Genet 2014;10(2) e1004132.

Chapter 7

Population Screening for Biological and Environmental Properties of the Human Metabolic Phenotype: Implications for Personalized Medicine

Douglas I. Walker[1,2], Young-Mi Go[1], Ken Liu[1], Kurt D. Pennell[2] and Dean P. Jones[1]

[1]*Division of Pulmonary, Allergy and Critical Care Medicine, Emory University School of Medicine, Atlanta, GA, USA* [2]*Department of Civil and Environmental Engineering, Tufts University, Medford, MA, USA*

Chapter Outline

E. Holmes, J.K. Nicholson, A.W. Darzi & J.C. Lindon (Eds): Metabolic Phenotyping in Personalized and Public Healthcare. DOI: http://dx.doi.org/10.1016/B978-0-12-800344-2.00007-0

7.1 INTRODUCTION

Estimates from genome-wide association studies (GWASs) indicate that only 10–20% of diseases have a strong genetic component [1], with the remaining 80–90% having unknown etiology [2]. Genotyping enables the identification of single gene mutations that are highly penetrant [3]; however, disease etiology is often multifactorial and driven by a combination of genetic, environmental, and lifestyle factors [4]. A more complete understanding of how these factors contribute to disease susceptibility and progression is required for mitigating risk, developing effective treatment strategies and identifying at-risk populations. Currently, no unified method exists to characterize the sum involvement of lifestyle and environment in disease. Recent efforts in chemical profiling of biological specimens suggest that untargeted metabolomics provides a platform to develop and quantify biomarkers identifying external events impacting human health. Improvements in analytical structure and computational tools have enabled untargeted metabolic profiling to routinely detect 10,000–15,000 unique chemicals, which include endogenous biochemical molecules (eg, peptides, lipids, amino acids, nucleic acids, carbohydrates), dietary compounds (eg, vitamins, phytochemicals, food additives, minerals), microbiome co-metabolites and xenobiotics (eg, drugs, toxins, pollutants) [5]. Most importantly, metabolic profiling of biological samples can be completed in a cost-effective and moderately high-throughput manner, making population screening by metabolomics a reality for multiple applications, including Personalized Medicine, hazard identification, and advanced blood chemistry measurements. In order to obtain biologically relevant information from metabolic profiling studies, there is a need to understand how phenotypic and lifestyle traits contribute to expression of the metabolic phenotype at both individual and population levels.

7.2 DEFINING THE METABOLIC PHENOTYPE

The human metabolome is an expression of an individual's metabolic phenotype and includes all endogenous biological metabolites, the chemicals from human–environment interaction, and reactants arising from interaction of these compounds with enzymatic and bacterial processes occurring within multiple body components (Fig. 7.1). The true extent of the metabolic profile of a population is unknown, with current estimates suggesting greater than 10^6 metabolites in humans. Because of individual differences, chemical measurements in a population will vary in concentration, chemical properties, and diversity as a result of the unique chemical exposure experience and phenotype for each person. Thus, even with establishment of a sum human population reference metabolome, development of ways to use such a reference remains challenging.

Recent efforts attempting to curate a reference metabolome highlight the diversity present in the metabolic phenotype. The Human Serum Metabolome Database [6] was developed to provide information on endogenous small

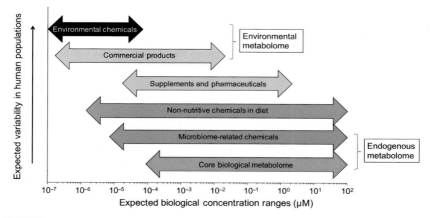

FIGURE 7.1 The human metabolome represents expression of an individual's metabolic pheno-type and is a combination of chemicals from core biological processes, the microbiome, dietary exposures, drugs, and exposure to xenobiotics from the environment and commercial products. Within a population, individual differences will result in chemical measurements that vary in con-centration and diversity because of unique chemical exposure experience and phenotypic character-istics for each person. Although a core metabolome consisting of the metabolites required for life will be detected universally in population screening, environmental chemicals, dietary compounds, and drugs in the metabolic phenotype are expected to vary greatly in concentration and presence. Furthermore, different methods of analysis are required to achieve chemical detection over a wide range of concentration magnitudes.

molecule substances and so far 4229 such metabolites have been identified in human serum samples. When predicted metabolites and dietary and environ-mental chemicals are included, a much greater number of metabolites were esti-mated. For example, the Human Metabolome Database (HMDB) [7] contains approximately 42,000 entries that are confirmed or expected in humans. The METLIN database [8] contains information on not only 240,000 chemicals, including metabolites, but also many synthetic chemicals not likely to be found in humans.

Untargeted metabolic profiling in human populations will also detect a large number of chemicals not readily identifiable because of the absence of matches in chemical databases and lack of reference spectra and analytical standards [9]. Environmental and dietary sources are suspected to be the main contributors to unknowns in the metabolome. When considering transformation products, intermediates, byproducts, and adducts, the number of possible chemical enti-ties is enormous. For example, a search of the PubMed Compound database over the molecular weight range of 50–2000 dalton returned 54×10^6 chemi-cals. Although many of these chemicals are synthetic and not relevant to human exposures, upward of 1 million chemicals are likely to contribute to the human metabolome. Plants are likely to contain 10^5–10^6 different small molecules.

Commercial products provide another source of chemical–human interaction contributing to the metabolic phenotype. Chemicals registered with the U.S. Environmental Protection Agency (EPA) include 10^5 unique species, with recent estimates from the Toxic Substance Control Act indicating that 70,000 are commonly used. To prioritize environmental chemicals for toxicity testing, a recent survey of 43,000 registered chemicals identified approximately 20,000 used directly in consumer products [10]. When considering the multiplicative effect of transformation and detoxification products of dietary- and environment-derived parent compounds, exogenous chemicals most likely exceed the endogenous contribution to the metabolic phenotype, further supporting the role of external factors in developing the human metabolic profile and importance in population-based studies.

7.3 THE HUMAN EXPOSOME

Establishing the role of environment in human health and disease will enable intervention with preventative measures and treatment therapies to reduce disease risk. Interaction of genetic and environment/lifestyle risk factors are suspected to be top contributors to the chronic disease burden; however, technologies available to measure genetic susceptibility far surpass technology for sequencing environmental exposures [4]. The need to include the environment in understanding human disease led Christopher Wild to introduce the concept of the exposome in 2005 [1], which he defined as "encompassing life-course environmental exposures (including lifestyle factors), from the prenatal period onwards." The exposome is envisioned as a complement to the genome, where a life course of exposure and interaction with the genome defines risk for disease development. Unlike the genome, however, exposures are transient and change on both short-term and long-term time scales, making quantitative assessment challenging. A more tangible definition of the exposome was proposed by Miller and Jones [11]: "The cumulative measure of environmental influences and associated biological responses throughout the lifespan, including exposures from the environment, diet, behavior, and endogenous processes." Exposures in this framework not only are limited to external chemical exposures but also include processes internal to the body (host factors) and wider socioeconomic influences [2,12]. Importantly, by redefining the exposome in terms of a cumulative measure of environmental influences and biological response, Miller and Jones acknowledged that the entirety of an individual's exposure does not need to be quantified, with exposure effect and maladaptations to external influences linking environment to health and disease. This concept emphasizes the need to identify biological markers such as plasma levels of environmental chemicals, transformation products, DNA adducts, epigenetic alterations, and protein measurements in exposome research that can be used as surrogate markers of environment, providing tangible measurements that can be used in disease prevention and management [11].

As discussed above, the metabolic phenotype represents an integrated chemical profile derived from environmental and endogenous processes; individual measurements provide a snapshot measure of an individual's exposome. Periodic chemical profiling of readily accessible biological fluids such as urine or blood has the potential to provide quantitative measurements of exogenous chemicals and biological response. In addition to chemical exposure, metabolic phenotyping also provides the ability to link external events such as stress and geographic location, to biochemical perturbations in the metabolic profile. Measurement of metabolic phenotypes in a population-wide manner will, therefore, enable a means of evaluating environment–human interaction on a wide scale, providing systematic information on exposures required for environment-wide association studies (EWASs) [13] to complement GWASs.

7.4 POPULATION SCREENING

To make large-scale population screening possible, cost-effective and high-throughput platforms are required to provide quantitative measurement of the metabolome. Because of the complexity of the human metabolome, no single platform will provide universal sequencing of all chemical species present in humans [4]; however, current instrumentation and workflows provide sufficient chemical coverage for population screening of metabolic phenotypes. Techniques using nuclear magnetic resonance (NMR) spectroscopy and mass spectrometry (MS) have been used for metabolic profiling of biological samples obtained from thousands of individuals to identify metabolic changes associated with aging, diet and the environment [14–17]. These approaches have been identified as key technologies in exposome research [12,18–20]. As a result of variations in instrument configuration, data processing approaches, and spectral stringency requirements, chemical coverage by these platforms varies widely; however, improved sensitivity obtained by MS-based metabolomics has led to its growing popularity. Increased use of MS for metabolomics has largely been driven by advances in data analytic tools [21–24], dedicated instrumentation [25], and high-resolution, high-speed MS [26–30], making possible routine measurement of 15,000–20,000 unique features in a 10-min analysis [20,28,31].

The main features of high-resolution metabolomics (HRM) are summarized in Fig. 7.2. The ultra-high resolution capabilities of Fourier transform instruments offers advantages in providing coverage of low-abundance environmental, dietary, microbiome, and drug-related metabolites. Comparison of MS platforms and additional analytical platforms are discussed elsewhere [32,33]. A rigorous analytical workflow using dedicated instruments is most advantageous. Triplicate analysis of each sample enhances reliability of detection and improves quantification by using a standard error of the mean instead of an individual value [28]. Simple extraction (addition of solvent with internal standards followed by removal of protein precipitate), instead of solid phase

---

High-resolution metabolomics for personalized medicine

Salient features of high-resolution metabolomics enabling large-scale population screening for personalized medicine

- **Replicate Injections:** By increasing technical replicates to three or more, the number of ions that can be reliably quantified is increased and effect sizes are reduced [28]. For example, an analyte present in only one sample out of 100 can be reliably measured if present in three replicates, even though it is absent from the preceding 99 samples.

- **Ultra high-resolution mass detection:** Mass spectrometers with sufficient mass resolution, sensitivity and dynamic range to detect high concentration endogenous metabolites and low abundance environmental chemicals in a single instrumental run [20].

- **Advanced data extraction:** Adaptive data extraction algorithms [22] provide detection of 10,000 to 20,000 unique chemical signals post-noise removal.

- **Throughput:** Increased mass spectrometer scan speed allow "fast" chromatography methods of 5 minutes or less. With dual column configurations, 24-hour, six days per week operation, capacity for 30,000 samples per year per instrument.

- **Cost:** Cost is primarily driven by instrument lease, maintenance and personnel. Lease/y is ~ $175,000; Personal cost/y ~$325,000; Supplies disposable and maintenance is ~$150,000. Including overhead of 50%, cost for 30,000 samples would approach ~$1 million.

FIGURE 7.2 High-resolution metabolomics for Personalized Medicine information box.

extraction with drying and reconstituting, decreases error due to variation in recovery efficiency of internal standards. Use of profile mode instead of centroid mode in MS preserves the ability to discriminate chemicals with very similar mass-to-charge ratio (m/z) and use of a data re-extraction routine with options to optimize parameter settings improve detection and quantitative accuracy [23,24].

Because of advances made in metabolomic platforms, it is now possible to implement metabolic phenotyping for screening of human populations. Using standardized sample preparation and instrument configurations, we believe that endogenous and xenobiotic chemical screening for 50,000–100,000 biological samples could be completed each year for approximately $5 million. However, for initial large-scale metabolic phenotyping, it is important to recognize a data-driven approach is required. The gross influence of environment and phenotypic character on the metabolome is poorly understood, and in many instances, evidence-based associations will have to be uncovered.

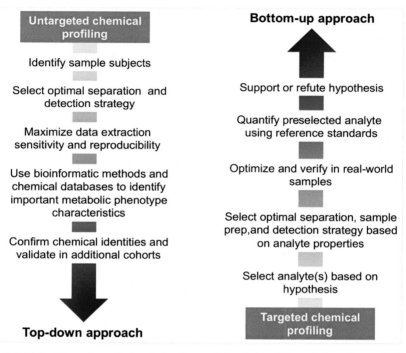

FIGURE 7.3   Comparison of untargeted and targeted chemical profiling techniques for population research. To derive the greatest benefit from population screening, untargeted chemical profiling is required. This is commonly referred to as a top-down analytical strategy and is a data-driven approach, where the goal is to maximize detection of chemical species for biomarker discovery and metabolome-wide association studies. Candidate metabolic biomarkers are then confirmed and validated within complementary populations. Targeted approaches utilize preselected analytical targets for measurement within a group of subjects and can be used to test hypotheses, estimate limited chemical exposures, and test for specific biomarkers of disease.

This will necessitate an initial exploratory phase to develop metabolic characteristics associated with an individual. Approaches in how to successfully apply a discovery approach for population screening can be learned from large-scale GWASs and genome-sequencing studies currently underway. For example, the International HapMap Project was designed to identify common genetic variants associated with disease and their response to drugs and environmental factors [34]. After completion of phases I and II, it became evident that, in many cases, the presence of common variants alone could not explain inheritable disease risk, and additional sequencing of low-frequency single nucleotide polymorphisms (SNPs) was required [35]. This highlights the need to apply metabolic phenotyping approaches in an agnostic manner for chemical analysis such as untargeted metabolomic profiling (Fig. 7.3). Untargeted profiling allows for maximum chemical coverage and reduces bias by not limiting analytical targets to known metabolites and pathways, and detected chemical signals are characterized after categorizing their importance. Although initial validation requires confirmation with co-elution and ion disassociation MS studies, development of cumulative detected metabolic feature databases and certified, well-characterized reference materials (ie, NIST SRM 1950 [36]) will provide direct identification and quantification based on instrument acquired information (mass, retention time, and intensity). This approach, termed "reference standardization," was recently used by Go et al. [20] to demonstrate an application in human populations.

An additional benefit of adapting untargeted metabolomic platforms to metabolic phenotyping is the ability to measure chemicals arising from environmental influences, providing quantitative information for exposome research. High-resolution, MS-based metabolomics platforms are now capable of detecting chemicals and their metabolites present at low nanomolar ranges from environmental, diet, and lifestyle exposures [20,37]. The ability to provide measures of a wide range of chemicals arising from external exposures in a systematic and quantitative manner greatly enhances exposure assessment, which traditionally relied on external monitoring, lifestyle factors, modeling, observational data, and targeted biomonitoring data. The simultaneous measurement of endogenous metabolites enables quantitative evaluation of the biological response through dose–response associations. A conceptual illustration of the exposure–bio-effect framework possible by metabolomics-based population screening is shown in Fig. 7.4. Although not directly establishing cause–effect relationships with distal health outcomes, it can provide information on exposure risk and whether it has occurred at a biologically relevant level.

Metabolic phenotyping using metabolomic platforms supports Personalized Medicine by providing comprehensive measure of tissue and/or biofluids chemistry for in-depth characterization of the metabolic phenotype. Translational applications include use as a diagnostic tool, identification of nutritional deficiencies, measurement of disease risk, tracking of therapeutic intervention efficacy, individualized exposure assessment, and metabolic health forecasting. Adaptation into a health care paradigm could be accomplished by utilizing easily accessible biofluids such as blood or urine. Incorporating a metabolomic panel, ie, "next generation" blood chemistry or urinalysis, into annual checkups will allow for periodic evaluation of endogenous, exogenous, and dietary health and would be complementary to additional omic measurements such as those obtained from the genome, epigenome, and proteome.

Before metabolic phenotyping can be applied in a Personalized Medicine context, a thorough understanding of how phenotypic characteristics, disease, and external factors influence the metabolome is required. Metabolome-wide association studies (MWASs) provide the framework for identifying specific metabolite associations with health endpoint or phenotypic traits [16,28,38]. In MWASs, large-scale metabolic phenotyping is related to a single or series of variables to identify metabolite changes specific to individuals expressing

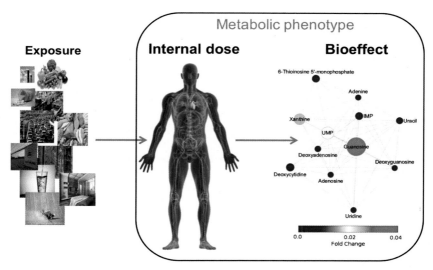

FIGURE 7.4 Conceptual framework for use of metabolic phenotyping in exposome research. Untargeted analytical techniques enable measurement of endogenous metabolism and plasma chemicals arising from environmental exposures, providing a means for chemical surveillance and bio-effect monitoring in general populations. The relationship between internal levels of exposure biomarkers and endogenous metabolism can then be related by testing for dose–response associations and changes consistent with early disease pathophysiology.

the variables of interest. MWASs can be employed for both discovery and diagnostic purposes and are analogous to GWASs, where genetic variants are tested as disease risk factors. Identification of metabolic phenotypes selected through MWASs provides insight into disease processes, dietary influences in the metabolome, exposure-related metabolic perturbations, and geospatial/cultural influences. Although these findings are initially correlative in nature, confirmation in validation cohorts and through elucidation of mechanistic processes responsible for the detected metabolic changes is expected to identify metabolic indicators of health.

Recent studies have applied metabolomic profiling tools to evaluate how the metabolic phenotype is influenced by numerous factors, including circadian rhythm, age, gender, disease, diet, climate, and chemical exposures. To incorporate metabolomics into a Personalized Medicine framework, there is a need to understand how variations in metabolic profiles arise and are related to health and lifestyle factors. In the following section, we review how these factors contribute to expression of the metabolic phenotype and their relevance to population screening and MWASs. Emphasis is placed on use of metabolic phenotyping for exposome research and evaluating chemical exposure bio-effects in human populations.

7.4.1 Diurnal Metabolic Phenotype

Metabolism has long been known to exhibit diurnal cycles [39,40]. Time of day, amount of sleep, timing of meals, and light–dark cycles are associated with metabolic fluctuations for control of energy expenditure and intake distribution during the waking period. Observational studies have indicated that diurnal changes influence xenobiotic metabolism [40], whereas loss of circadian control has been associated with onset of cancer and diabetes, suggesting diurnal changes are an important measurement of overall metabolic health [41]. Only a limited number of studies have applied metabolic phenotyping to evaluate diurnal changes in the metabolome and have primarily examined the effect of fasting status, amount of sleep, and the light–dark cycle in well-controlled environments. Metabolites exhibiting circadian oscillations in humans are largely associated with gene expression under circadian control, such as fatty acid and lipid metabolism, in addition to cholesterol and bile acid biosynthesis [42,43]. For example, in the study by Gooley and Chua [42], targeted profiling of 263 plasma lipids periodically collected from 20 healthy individuals was completed over a span of 28 h to assess interindividual circadian variations. Rather than restricting their study to population-averaged diurnal changes, the authors examined individual changes in lipid levels over the sampling time course to evaluate how individual fluctuations differ from the group. Results from Chua et al. [44] showed that lipids from all classes varied with the circadian pattern, and individual changes were greater than the group-rhythmic fluctuations; that is, the time of peak lipid levels varied greatly across the 20 individuals, even though the cohort was relatively homogeneous. Studies using metabolomic platforms aimed at metabolic intermediates have detected similar changes in lipid-related pathways [41,42] in addition to the amino acid glutamate, macronutrients, and hippurate [41,43]. Estimates from the previously discussed metabolomics studies indicated that up to 20% of metabolites could exhibit diurnal changes; however, the estimates were limited to characterized metabolites, and metabolic profiling was not completed in an untargeted manner.

Data on diurnal changes in the metabolome from studies using untargeted approaches are limited. In the study by Ang et al. [45], untargeted metabolomics using an MS-based platform was able to test for diurnal variation in eight healthy adults by using 1069 detected features. Measured metabolic phenotypes were compared at selected time points, identifying 203 chemical signals exhibiting a significant time-of-day variation. Since data were collected within an untargeted framework, the authors used several techniques to characterize the metabolites; however, only 34 of the 203 metabolites could be positively identified, representing just 17%. Similar to the previously discussed studies, the metabolites identified by Ang et al. [45] included fatty acids and lipids, in addition to tyrosine, cortisol, and methionine, and a large number of chemicals that remained unidentified when searched against publicly available databases.

Diurnal variation within the metabolic phenotype has the potential to influence metabolite expression and MWAS results. Furthermore, the pattern of change is strongly dependent on the individual and cannot necessarily be expected to change consistently across a population. Thus, diurnal rhythm in metabolic phenotyping is an important consideration that will need to be assessed in population screening. This will include identification of metabolic features, both known and unidentified, that exhibit large time of day variations. During validation of biomarkers identified from MWASs, time-dependent variability will also need to be evaluated to understand potential confounding factors when applied within a Personalized Medicine framework.

7.4.2 Sexual Dimorphisms in the Metabolic Phenotype

Sex is a strong determinant of the metabolic phenotype because of the fundamental biological differences between the males and females [46] on both genetic and metabolic levels [47]. Influences of the environment, disease, and treatment effectiveness exhibit gender stratification, requiring consideration of sexual dimorphism during population screening. Recently, a number of statistically well-powered studies have applied metabolomics to highlight sexual dimorphisms in the metabolic phenotype. For example, in the study by Dunn et al. [14], metabolic phenotyping of 1200 individuals from the United Kingdom was completed and evaluated for associations with gender, age, and health endpoints. Applying a two-way analysis of variance (ANOVA) using gender and age, the authors identified gender-dependent changes in both endogenous and exogenous metabolites. The differentially expressed metabolites included metabolic intermediates previously associated with sex, including 4-hydroxyphenyllactic acid, creatinine, citrate, urate, and tyrosine, in addition to novel associations such as an increased level of methionine sulfoxide in females, a marker of oxidative stress. Caffeine was also increased in females and highlights the role of lifestyle contributions to the metabolic phenotype, since increased caffeine levels are mostly likely caused by a combination of different intakes and gender-specific metabolism.

Metabolic phenotyping in complementary populations has shown similar sexual dimorphisms, including differences in central metabolic intermediates. In the study by Mittelstrass et al. [48], a comparison of 131 serum metabolites measured in a population of 1649 males and 1732 females identified 103 metabolites exhibiting statistically significant differences between the sexes. Importantly, sex-associated metabolites were present in all metabolic classes measured, including acylcarnitines, amino acids, phosphatidylcholines, lysophosphatidylcholines, sphingomyelins, and hexoses. The results from the study by Mittelstrass et al. [48] further supported the importance of sex stratification in metabolic phenotyping. Even though the study was limited to the common metabolic intermediates, broad differences were detected between males and females. Application of untargeted metabolomic approaches is expected to identify additional metabolites differentially expressed in males and females exhibiting sex-dependent expression. In the study by Krumsiek et al. [49], untargeted profiling of serum obtained from 903 female and 853 male participants detected 180 metabolic features exhibiting significant differences between the sexes. The majority of the metabolic associations were identified endogenous metabolites, including amino acids, vitamins/cofactors, carbohydrates, lipids, peptides, and nucleotides; however, 54 were unidentified metabolites, and 3 consisted of xenobiotics, which included 4-vinylphenol sulfate, piperine, and 2-hydroxyisobutyrate.

MWASs suggest that sex derived metabolic phenotype is not limited to differences in expected sex-dependent metabolites such as steroid hormones but also includes changes in core nutritive metabolism, chemicals from exogenous sources, and metabolic features with unknown function. Accounting for the differences in male and female metabolic phenotypes will be required for successful application of population screening in Personalized Medicine and exposome research. As was discussed by Krumsiek et al. [49], the identified metabolic pathways exhibiting sexual dimorphisms are related to sex-specific diseases such as coronary heart disease and gout. Thus, baseline differences due to sex are important to understanding initial metabolic changes leading to disease outcome for development of Personalized Medicine.

7.4.3 Age-Related Metabolic Phenotype

Over the lifespan of an individual, molecules, cells, and organs undergo a steady progression of damage occurring from the deterioration of homeostatic metabolic processes, resulting in loss of function, morbidity, and death [50]. Aging resulting from these changes is considered the most universal contributor to metabolic decline and related diseases. Current estimates suggest only 25% of lifespan is attributed to heredity [51], whereas lifestyle, environmental, and nutritional factors strongly influence the rate of aging. Estimating the contribution of age to the metabolic profile will inherently require deconvoluting the influence of diseases and the underlying processes contributing to disease onset. Confounding caused by the potential effect of age-related diseases and multimorbidity in populations representing young and old age groups is shown in Fig. 7.5. As an advanced age is reached, the likelihood of expressing clinical or preclinical manifestations of disease increases significantly [52].

In addition to looking at metabolic phenotype differences based on age grouping, a number of studies have used age-related traits, including telomere length, forced expiratory volume, forced expiratory vital capacity, hip bone mineral density, blood pressure, cholesterol, and dehydroepiandrosterone sulfate to identify metabolic associations with physiologic age [17,53]. For example, Menni et al. [17] measured 280 metabolites in 6605 individuals in a UK twin cohort study to identify metabolic correlations with age and clinical

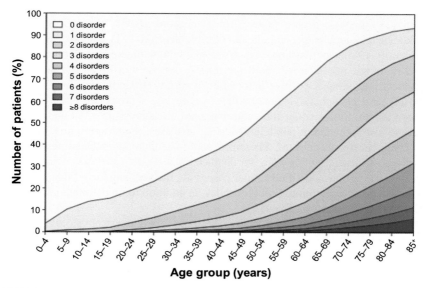

FIGURE 7.5 Prevalence of multimorbidities in different age groups for 1,751,841 patients. *Reproduced with permission from Elsevier [52].*

measurements known to be associated with aging. The authors identified 165 metabolites directly associated with age grouping when accounting for family relatedness, sex, and body mass index (BMI), suggesting that biochemical changes occurring during aging are expressed in the metabolic phenotype. Using the initially selected 165 metabolites, the authors were able to identify a subset of 22 measured chemicals that included endogenous and exogenous metabolites strongly correlating with age and with age-related clinical traits independently of age. The study by Menni et al. [17] largely consisted of females (93% of individuals), but other studies have identified similar metabolic shifts with aging in men. In Dunn et al. [14], two-way ANOVA of gender, age, and the gender–age interaction was completed by comparing the metabolic phenotype of persons <50 years old and >64 years old. Age-related changes were found to be both independent and dependent on gender identity, and similar to the results of Menni et al. [17] included changes in tryptophan metabolism, serine, citrate, and the food additive erythritol. Dunn et al. [14] also detected significant changes in the metabolites known to be associated with aging, further supporting the use of metabolic approaches to detect associations when applied to population screening. For example, cysteine levels were lower in the individuals >64 years old, which is consistent with evidence suggesting oxidative stress increases with age [54].

Untargeted metabolomics has also been used to discover metabolic phenotype associations with physiologic and chronologic age. In the study by

Zhao et al. [55], metabolic profiles, measured by untargeted, high-resolution MS from 423 otherwise healthy Native Americans were tested for association with leukocyte telomere length, an indicator of biological aging. The authors were able to identify 19 metabolites that were associated with telomere length independent of chronologic age. The metabolites primarily consisted of lipids and fatty acid products, including a positive association with anti-inflammatory fatty amides and glycerol phosphatidylethanolamines, suggesting a protective effect for telomere length. In this study, the authors limited the analysis to 1364 metabolites that are also included in publicly available databases, representing only 18% of the chemical signals detected. Therefore, additional unknown metabolites associated with telomere length were not characterized. Testing for associations with chronologic age has identified similar changes in metabolism, including fatty acid metabolites, phospholipids, and glycerides, in addition to unidentified chemicals providing no database matches [53].

In the previously described studies, both chronologic and physiologic aging were associated with changes in the metabolome. However, these changes are not caused by the natural aging process alone, as disease status, in addition to the cumulative effects of environment, diet, and lifestyle factors, also contribute to expression of the metabolic phenotype over a lifetime. Incorporating age-related changes in the metabolic phenotype through periodic measurement will be an important component when applying population screening to Personalized Medicine. Observing changes in the metabolome of an individual over time will support improved detection of disease risk and of poor health outcomes, providing increased opportunity for intervention and treatment.

7.4.4 Disease Metabolic Phenotype

The use of routine metabolic phenotyping in health care is expected to transform the understanding of disease, improve risk prediction, and increase efficacy of treatment. Therefore, population screening in healthy and diseased individuals has been one of the primary research focuses of metabolic phenotyping. For example, metabolic studies of neurodegenerative diseases [56,57], type II diabetes [58], cancer [59], human immunodeficiency virus (HIV) infection [26], and cardiovascular disease (CVD) [60] have identified a number of metabolic markers related to disease progression, risk, and the underlying derangement in metabolism characteristic of disease pathophysiology. The growing popularity of metabolomics in disease research and biomarker discovery can be demonstrated by searching PubMed for the terms "disease," "metabolomics," and "diagnostic," with the number of publications in which these terms were used increasing from 3 in 2002 to 549 in 2014 (Fig. 7.6). However, translating findings from metabolomics to routine clinical use has proven challenging. For successful application of metabolic phenotyping in Personalized Medicine, validation of disease biomarkers will have to be completed in a way that removes ambiguity in cause and effect, and this requires careful consideration of the

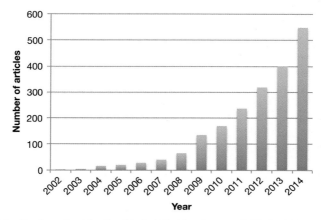

FIGURE 7.6   Number of articles listed in PubMed for the years 2002–2014 after searching for the terms "disease," "metabolomics," and "diagnostic." The growing popularity of metabolomics for disease research is evident in the growing number of publications each year over that time span, which increased from 3 in 2002 to 549 in 2014.

biomarker context with respect to a disease state and the underlying physiologic changes associated with it.

Metabolic markers have long been used to diagnose diseases and identify individuals requiring intervention. For example, elevated serum creatinine is used to evaluate and diagnose kidney disease, and hemoglobin A_{1C} is an important diagnostic tool for metabolic syndrome and diabetes. In the Personalized Medicine framework, metabolic phenotyping will supplement these traditional health indicators by enabling a more comprehensive measurement of blood chemistry. However, clinical diagnostic measurements often represent functional markers of disease pathophysiology—that is, changes in physiologic function occur as a result of the disease itself. With routine population screening, markers of metabolic changes that lead to eventual disease manifestation will be identified. By determining early biomarkers predictive of health outcome, lifestyle factors can be modified, progression can be more accurately monitored, and intervention can be started for prevention rather than treatment.

In the application of population screening for identification of disease risk, one must acknowledge the heterogeneity in disease processes, which are combinations of genetic pre-disposition, environmental factors, and lifestyle. Therefore, identification of predictive biomarkers will require an evidence-based approach, including large-scale population screening and animal models. The majority of studies completed to date are cross-sectional comparisons of healthy and diseased individuals, and only a limited number have applied metabolic phenotyping techniques to prospectively study metabolome associations with disease outcomes. Although cross-sectional studies provide insight into disease mechanisms and possible therapeutic targets, there are complications due to treatment effects, reverse causality, and an overall change in homeostasis as a result of altered metabolic functions. Prospective metabolic phenotyping studies overcome many of these limitations and improve the ability to identify biomarkers predicting disease risk and onset. For example, in the prospective metabolomics study of cardiovascular disease (CVD) by Wurtz et al. [60], NMR-based chemical profiling was completed in a discovery population of 7256 individuals (800 cardiovascular [CV] events during a 15-year follow-up) and two different validation cohorts of 2622 (573 CV events during a 20- to 30-year follow-up) and 3563 (368 CV events during a 11- to 13-year follow-up). Application of an MWAS framework to determine the metabolic associations with CV event identified phenylalanine (Phe) and three fatty acids, which acted as independent predictors of CVD onset 10–15 years later in both the discovery cohort and the two validation cohorts. Dietary sources of Phe association were explored by evaluating the correlation with plasma aspartame; however, a significant correlation was not present, which suggested increased expression results from an underlying biological mechanism related to CVD. Alternatives can be inferred from an MWAS of Phe performed in marmosets [28]. In this study, Phe was strongly correlated with other amino acids, dietary chemicals, and environmental chemicals containing an aromatic structure. Of interest, Phe was also strongly associated with lipids, especially in females. The results of the study by Wurtz et al. [60] highlighted the potential of population screening to identify novel disease biomarkers and use them in a predictive sense. The hazard ratio for future CVD was highest in the two youngest age groups (24–45 and 45–50 years), which are considered to be at lowest risk for CVD and have the most opportunity to decrease their risk factors prior to a CV event.

Population screening using accessible biological fluids such as blood or urine has the capability to provide a wealth of chemical information that can be used to evaluate an individual's health status. Unlike genetic testing for disease associations, chemical measurement with broad coverage enables detection of endogenous metabolites and xenobiotics such as dietary chemicals, chemicals from environmental exposure, and pharmaceuticals. Incorporating this chemical information into understanding how a metabolic phenotype transitions from healthy to diseased for screening and for identifying the prognosis will be a major development in Personalized Medicine. Application of such an approach represents a paradigm shift in focus from disease treatment toward risk identification and prevention.

7.4.5 Dietary and Nutritional Metabolic Phenotype

Metabolic profiling for nutritional and dietary metabolomics has received considerable attention recently [5,61,62]. The contributions of diet and nutrition to the metabolic phenotype are inherently complex because of external factors, including heterogeneity in intake, food preparation, diet composition, socioeconomic status, and availability, as well as host factors such as digestion, absorption, and clearance. Furthermore, specific nutrients from dietary sources will impact hundreds of highly regulated molecular systems, resulting in a series of diet-related changes to the metabolic phenotype. Therefore, diet is expected to contribute greatly to expression of the metabolic phenotype and heterogeneity when applying population screening techniques.

When considered as a source of chemical exposures, diet includes nutrients, non-nutritive chemicals, pesticides, preservatives, additives, and others. An average individual consumes approximately 40 of the required nutrients daily. Nutritive dietary organic chemicals can then be converted through intermediary metabolism to more than 1500 chemicals [5]; however, most chemicals in food are non-nutritive in nature. For example, biologists estimate that upward of 200,000 metabolites are present in the plant kingdom [63], which contribute significantly to the phytochemical body burden; 372 different polyphenols have been identified in different food types that are consumed regularly [64], and 26,619 chemical constituents have been curated from 907 different types of food [65]. Application of untargeted analytical techniques in food and agricultural sciences to characterize animal, plant, and derived products has further identified a large number of chemicals relevant to dietary exposures [66]. Combining information obtained from population screening with dietary exposures is expected to provide a more complete understanding of the role of non-nutritive dietary chemicals in health and enable unbiased nutritional assessment for an individual. Application of metabolic phenotyping within a Personalized Medicine framework will allow identification of individuals requiring dietary interventions, provide a more mechanistic understanding of biological response to foods (eg, food allergies), and improve food preparation and handling to avoid short-term and long-term health risks.

Current application of metabolic phenotyping to nutritional research has included studies to correlate specific dietary biomarkers with food intake, with the primary goal of characterizing diet and assessing nutritional status in a way that provides a less biased estimate than can be made from patient recall. Representative dietary biomarkers for different food classes are provided in Table 7.1. Although the results presented here are limited to previously published findings, it is expected that further characterization of the diet–metabolic phenotype interaction will identify additional biomarkers correlated with diet. For example, a number of studies have indicated that habitual diet (eg, high-fat, high-protein vs grain-based diets) influences the metabolic phenotype, resulting in specific metabolic chemicals capable of identifying primary food consumption. In the study by O'Sullivan et al. [67], profiling of urine was used to assess dietary patterns in 125 subjects. Following a 3-day dietary assessment and categorization based on established food groups, urinary profiles were compared. The results showed that the differences in the metabolic phenotype within different dietary groups were driven by food and nutrient intakes, including protein, sugar, starch, salt, magnesium, and alcohol. O'Sullivan et al. [67] used the metabolites differentiating the groups to identify healthy, unhealthy, and

TABLE 7.1 Representative Dietary Biomarkers for Identifying Food Consumption and Improving Food Recall in Population Studies from Scalbert, Brennan [37]

Food Category and Food	Metabolic Phenotype Biomarkers of Dietary Intake
Fruit	
Apple	Kaempferol, isorhamnetin, m-coumaric acid, phloretin
Orange	Caffeic acid, hesperetin, proline betaine
Grapefruit	Naringenin
Citrus	Ascorbic acid, β-cryptoxanthin, hesperetin, naringenin, proline betaine, vitamin A, zeaxanthin
Fruit (total)	4-O-Methylgallic acid, β-cryptoxanthin, carotenoids (mix), flavonoids (mix), gallic acid, hesperetin, isorhamnetin, kaempferol, lutein, lycopene, naringenin, phloretin, vitamin A, vitamin C, zeaxanthin
Vegetables	
Carrot	α-Carotene
Tomato	Carotenoids (mix), lycopene, lutein
Vegetables, leafy	Ascorbic acid, β-carotene, carotenoids (mix)
Vegetables, root	Ascorbic acid, α-carotene, β-carotene
Vegetables (total)	Ascorbic acid, α-carotene, β-carotene, β-cryptoxanthin, carotenoids (mix), enterolactone, lutein, lycopene
Fruit and vegetables (total)	α-Carotene, apigenin, ascorbic acid, β-carotene, β-cryptoxanthin, carotenoids (mix), eriodictyol, flavonoids (mix), hesperetin, hippuric acid, lutein, lycopene, naringenin, phloretin, phytoene, zeaxanthin
Cereal products	
Whole-grain rye	5-Heptadecylresorcinol, 5-pentacosylresorcinol, 5-tricosylresorcinol
Whole-grain wheat	5-Heneicosylresorcinol, 5-tricosylresorcinol, alkylresorcinols (mix)
Whole-grain cereals (total)	5-Heneicosylresorcinol, 3,5-dihydroxybenzoic acid, 3-(3,5-dihydroxyphenyl)-1-propanoic acid), 5-pentacosylresorcinol, 5-tricosylresorcinol, alkylresorcinols (mix)
Seeds	
Soy products	Daidzein, genistein, isoflavones (mix), O-desmethylangolensin
Meats	
Meat	1-Hydroxypyrene glucuronide, 1-methylhistidine
Meat, beef	Pentadecylic acid
Animal products (total)	1-Methylhistidine, 3-methylhistidine, margaric acid, pentadecylic acid, phytanic acid

(Continued)

TABLE 7.1 Continued

Food Category and Food	Metabolic Phenotype Biomarkers of Dietary Intake
Dairy products	
Milk, dairy products	Iodine, margaric acid, pentadecylic acid, phytanic acid
Fish	
Fatty	Docosahexaenoic acid (DHA), eicosapentaenoic acid (EPA), long-chain ω-3 polyunsaturated fatty acids (PUFAs), polychlorinated biphenyl (PCB) toxic equivalents, pentachlorodibenzofuran, PCB 126, PCB 153, ω-3 PUFAs
Lean	Long-chain ω-3 PUFAs
Beverage (non-alcoholic)	
Tea	4-O-Methylgallic acid, gallic acid, kaempferol
Coffee	Chlorogenic acid
Beverage (alcoholic)	
Wine	4-O-Methylgallic acid, caffeic acid, gallic acid, resveratrol metabolites
Alcoholic beverages (total)	5-Hydroxytryptophol/5-hydroxyindole-3-acetic acid, ethyl glucuronide

Used with permission from the American Society for Nutrition.

traditional diets, based on food contributing to total energy intake. The group consuming the unhealthiest diet, which included high intake of meat products, white bread, butter, and preserves (and corresponding lowest intake of fruit, vegetables, and whole-grain breads) had elevated trimethylamine-*N*-oxide (TMAO), which has been associated with increased risk of CVD [68]. The results from this study highlight the ability to assess diet-related contributions to the metabolic phenotype. Additional metabolomics studies have extended the relationship between diet and the metabolic phenotype to include dietary glycemic load [69], phytochemical dietary load [70], and Western versus Eastern diets [15].

Diet strongly contributes to the metabolic phenotype, health, and exposure burden. Therefore, the ability to provide a quantitative measure of dietary factors is an important component in Personalized Medicine and population screening. Although specific biomarkers of food intake have been identified [62], more general diet–metabolome interactions are still poorly understood. Application of untargeted metabolomics approaches in population screening is expected to provide additional insight into how dietary exposures are related to the metabolic phenotype. Understanding of diet–metabolome interactions will support identification of dietary risk factors in disease and allow interventions to reduce risk.

7.4.6 Geographic Metabolic Phenotype

Primary physical location will influence the metabolic phenotype by changing the composition of external factors contributing to the metabolome. For example, the type and the quantity of food intake are strongly influenced by culture and availability; climate, altitude, and length of light–dark cycles change total energy expenditure and metabolic rate; spatial–temporal variations in sources of environmental chemicals, chemical use, and availability of commercial products result in different exposures. Geographic variations among population groups linked to ethnicity and lifestyle are also closely associated with disease risk [15]. Therefore, understanding the role of geographic location in the development of metabolic phenotypes is an important consideration when applying population screening in support of Personalized Medicine.

Comparison of metabolic phenotypes from diverse locations has primarily identified diet-related chemicals contributing to the difference in populations. In the study by Holmes et al. [15], 24-h urinary metabolic profiles of 832, 1138, 496, and 2164 samples were obtained from individuals in China, Japan, the United Kingdom, and the United States, respectively. Metabolomic profiles obtained from the four populations identified well-defined Eastern and Western metabolic phenotypes, with the geographic metabolic differences more pronounced than those caused by gender. Regional differences were also identified, including the presence of varied metabolic patterns in Northern and Southern China. Metabolites discriminating between the different locations predominantly included metabolites of dietary origin and were correlated with energy intake, micronutrients, BMI, alcohol consumption, and dietary cholesterol. Similar results were obtained in the study of Walsh et al. [71], who evaluated the influence of multisite collection on the plasma and urinary metabolome of 219 adults from seven European countries. Regional differences were detected in diet-related metabolites, including TMAO, creatinine, *N*-methylnicotinate, and hippurate, and the authors attributed the differences to variations in intake of meat, fish, and polyphenol-rich food.

In addition to the direct effects of geography, as discussed above, environmental conditions such as temperature and altitude contribute to changes in metabolism in response to environmental stressors [72]. Increases in altitude and decreased atmospheric pressure have been associated with changes in oxygen-dependent metabolism and have been studied extensively for more than a century. More recent research has coupled metabolomic analyses of individuals in an induced high-altitude state to show changes of lactic and succinic acid with altered activity of hypoxia-inducible factor [73]. Environmental stresses and hot/cold response in extreme climates can exacerbate disease [74], alter the basal metabolic rate [75], and induce heat dissipation processes. Although these are important considerations, the associated metabolic changes have not been well characterized in human populations [72]. With recent advances in personal monitoring equipment, including digital sensors in smartphones and passive personal exposure monitors [76], it is now possible to track and collect a range of data about an individual's micro-environment and metabolic status. The combination of sensor data and metabolic phenotyping data will support improved identification of environmental stressors and health maintenance. Using currently available technology, heart rate, body temperature, cholesterol, glucose, blood pressure, and physical activity can now be routinely measured in real time. Combined with geographical information systems (GIS), weather and temperature information can provide daily, weekly, monthly, and yearly measurements of environmental factors available for integration with population screening data.

The varied nature of chemical exposures, spatial–temporal factors, disposition, changes in detoxification based on age, gender, weight, and health status result in a unique exposure history for each individual. Routine application of metabolic phenotyping in a population-wide framework will provide systematic information on exposures required for EWASs [13]. In the following sections, the use of metabolic phenotyping for assessing chemical exposures and contribution to disease burden is discussed. Strategies for incorporating chemical surveillance and bio-effect monitoring into a Personalized Medicine framework are also presented.

7.5   METABOLIC PHENOTYPING FOR ENVIRONMENTAL EXPOSURES

Contribution of low-dose, chronic environmental exposure to disease burden is poorly understood. To support detailed measurement of the environment for linking exposure events to health outcomes, analytical frameworks that provide quantitative assessment of chemical exposures in a high-throughput manner are required. Traditionally, the study of environmental risk factors in disease and health outcomes have relied on characteristics such as age, location, profession, and lifestyle risk factors determined through questionnaires to identify populations with a high probability of exposure. For example, residence near farmland, occupational use of solvents, and a self-reported smoking habit have been used to test associations between chemical exposure and health outcome [77]. Although these approaches enable identification of at-risk populations on a large scale, considerable uncertainty exists in estimating exposure, making it challenging to link disease outcomes with exposure in a dose-dependent manner. To overcome this limitation, targeted molecular epidemiology approaches that use molecular measurements of environment risk, including biomarkers of exposure, effect, and biological response, have been developed. Molecular measurement can be used to evaluate chemical internal dose, biologically effective dose, early biological response, and susceptibility, which can then be linked to disease or alternative biological endpoints. The use of biomonitoring, which estimates body burden of toxic chemicals through direct measurement in biological fluids, has especially contributed to understanding the potential association between exposure and diseases, such as Parkinson's disease (PD), Alzheimer's disease (AD), and type II diabetes [13,78,79]. However, biomonitoring is typically completed using targeted approaches, which are time consuming and often limited to specific classes of chemicals (eg, polychlorinated biphenyls [PCBs], brominated flame retardants), with targeted biomonitoring assays rarely exceeding 100 chemicals. Considering the large number of chemicals present in the environment, in diet, and in commercial products with potential for direct human exposure, it becomes apparent that the coverage by targeted biomonitoring alone is not sufficient for EWASs.

Metabolic phenotyping using high-throughput, untargeted metabolomics has been identified as one of the most promising analytical frameworks for providing detailed estimates of the exposome [4,11,18,19,25,28,37,80]. Because of the ability to detect chemicals from both endogenous and exogenous sources, a diverse series of exposure biomarkers, internal dose, and the biological response are obtained; each is an important functional input for exposome research. Although there are only limited examples of metabolic phenotyping for exposure surveillance and bio-effect monitoring, application in animal models and human populations provide insight into how it can be incorporated into population screening and Personalized Medicine. In the study by Park et al. [25], metabolic profiles obtained using untargeted metabolomics from seven mammalian species were compared to test whether humans would exhibit greater variation in plasma chemical content compared with animals housed in well-controlled research facilities. The resulting analysis identified 1485 metabolites present in all species, and an additional 2335 were variably detected in the different species. To differentiate between endogenous metabolites and environmental chemicals, plasma metabolomic profiles were evaluated for significance of within-species and between-species correlation to test for metabolite variations across the species. The resulting analysis identified two separate modules discriminating chemicals with small and large variations among the species, with the first module enriched in endogenous metabolites (644), and the second module consisted of environmental chemicals and endogenous metabolites (841) involved in xenobiotic metabolism. The results from Park et al. [25] supported the use of metabolic phenotyping for providing relevant biomarkers of exposure and response in population screening. Using an untargeted analysis, the authors were able to simultaneously measure endogenous metabolism and environmental chemicals in a high-throughput manner, which could then be used for biomonitoring and elucidation of biological response to support EWASs (Fig. 7.7). Similar results were obtained by Soltow et al. [37], Go et al. [81], and Go et al. [28], supporting the use of metabolic phenotyping for environmental chemical biomonitoring and exposome research.

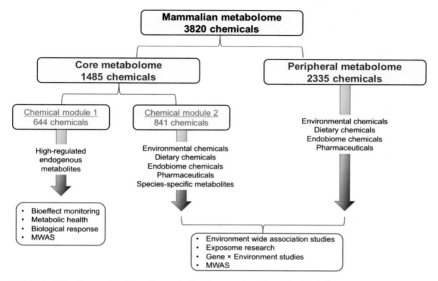

FIGURE 7.7 Categorization of metabolic phenotyping data obtained from seven mammalian species in Ref. [25] for bio-effect measurement and biomonitoring of environmental exposures. Comparison of metabolic profiles from the seven mammalian species identified a core metabolome consisting of chemicals common to the different species and a peripheral metabolome, which supported the core metabolome. Evaluation of the variations of metabolite levels and detection within the core metabolome identified two separate chemical modules. Chemical module 1 included endogenous metabolites having characteristics suitable for studies of biological responses to toxic exposure, evaluating metabolic health and metabolome-wide association studies. Module 2 included environmental chemicals, variable endogenous metabolites, dietary chemicals, metabolites derived from the microbiome, and pharmaceuticals. Metabolites in module 2 and the peripheral metabolome can be used to support environment-wide association studies and gene–environment studies in exposome research. *Used with permission from Elsevier.*

7.5.1 Population Screening for Hazard Identification

In addition to providing wider chemical coverage than available from targeted biomonitoring, population screening will assist in hazard identification for exposure to commercial chemicals. As discussed previously, current estimates suggest that upward of 100,000 chemicals are used in industry, commercial products, food preparation/storage, and pharmaceutical formulations. Although tremendous advances have been made in high-throughput toxicity screening (HTS) to assess chemical toxic effects [82], a large number in commercial use have not been thoroughly assessed for short-term and long-term toxicity to human populations. Furthermore, translating in vitro findings to quantitative risk assessment for human populations has proven challenging. For example, endpoints are typically assessed after exposure to a single chemical while not accounting for synergistic effects from mixtures; dose–response represents acute exposures and is not characteristic of the low-dose, chronic exposure typically observed in human populations; it is not possible to measure maladaptations following a history of exposure leading to mutagenesis and cancer; in utero toxicity resulting in developmental outcomes cannot be determined; and chemical availability, disposition, biotransformation, and contribution to health status cannot be evaluated in vitro. In many instances, animal model experiments are required to address the limitations of HTS and validate toxic mechanisms; however, it is not possible to complete these tests for all the chemicals (and mixtures) in current use. Metabolic phenotyping can assist in hazard identification by surveying populations for the occurrence of single chemicals and mixtures to which humans are routinely exposed. Prevalence can be used to prioritize for in-depth toxicity testing in model systems.

Although the use of untargeted metabolomic approaches will increase the ability to detect known and unknown chemical exposures, it must be acknowledged that complete coverage of all chemical exposures will not be possible. Many exogenous chemicals are rapidly transformed and excreted and can only be detected immediately following exposure. Metabolomic profiling sensitivity, methodology, sample processing, and biological fluid will influence the ability to detect chemical occurrence. Uncertainty will be present when providing absolute concentrations of chemicals detected by untargeted metabolomics, which is influenced by the availability of authentic reference standards, use of surrogate standards, and increasing analyte measurement number at the expense of sensitivity. However, the uncertainty in metabolic phenotyping is less than that present in current methods used to estimate exposure in the general population for chemicals without detailed measurement [83,84]. For example, in the study by Wambaugh et al. [84], chemicals with biomonitoring data available in the National Health and Nutrition Examination Survey were utilized to develop a model to predict exposure based on use, persistence, production, and physiochemical properties. Following parameter estimation using the training data set, exposure estimates for 7986 chemicals within the ACToR database was completed and then ranked on the basis of whether exposure occurs at concentrations consistent with the toxic response observed in HTS. However, confidence intervals for predicted concentrations were as high as four orders of magnitude, making accurate estimates in human population challenging.

To obtain adequate chemical exposure survey data in humans by population screening, measured absolute concentrations are required. Therefore, strategies enabling quantification of untargeted metabolic phenotyping for environmental chemicals will be needed to identify which exposure-related chemicals could pose a risk to health. One approach could include the use of metabolic phenotyping to identify chemicals of interest, upon which a targeted analytical technique [85] using authentic chemical standards could then be used to reanalyze all samples. However, targeted analytical techniques are labor intensive and time intensive, making it challenging and costly to analyze sample sets comprising 10^4–10^5 subjects. A second approach could include a series of analytical standards containing specified targets or use ^{13}C-labeled internal standards spiked into samples. While providing absolute concentrations, this method would result in loss of advantages gained using untargeted profiling, since chemical standards would have to be selected *a priori*. Recently, a quantification strategy relying on reference standardization was developed to provide absolute concentrations for a wide range of endogenous and exogenous chemicals detected by using untargeted metabolic phenotyping [20]. Using reference standardization, quantification after data acquisition is possible by referencing a pooled sample analyzed with each series of samples. The pooled reference, which is representative of the analytical samples, can be characterized and analytes quantified with traditional analytical chemistry techniques such as quantification by methods of addition [86] or standardization of a certified reference material such as NIST SRM 1950 [87]. Known concentrations within the reference standard can be used to determine a chemical response factor and calculate analytical sample concentrations based on single-point calibration. The benefit of this approach is that targeted quantification is only required in the reference sample; chemicals selected for quantification do not need to be selected *a priori*; and population-wide estimates of plasma chemical concentrations can be determined without having to reanalyze samples using a targeted approach.

By supporting quantitation of large numbers of chemicals detected in human samples, reference standardization can provide the systematic biological and environmental chemical measurements required for exposome research. For example, in the study completed by Go et al. [20], plasma from a healthy population of 153 individuals was analyzed by using untargeted metabolic profiling. Reference standardization was then applied to quantify a select number of metabolites and exogenous chemicals. Table 7.2 contains a representative list of chemicals derived from exogenous sources that were detected, confirmed, and quantified within the plasma samples. A thorough search of the literature indicated less than half had reported reference levels, with the remaining previously undetected in normal human populations. Similar results were obtained by Roca et al. [88] and Jamin et al. [89], and these results showed that untargeted chemical measurements were able to detect both common exposure biomarkers and environmental chemical metabolites previously uncharacterized in human populations. Thus, analytical platforms used in metabolic phenotyping now enable a more comprehensive measurement of internal chemical exposure in humans than was available from targeted biomonitoring techniques. The resulting survey of chemicals in human populations can be used to determine potential for exposure, identify chemicals to prioritize for toxicity screening, and test for links between environment, the genome, and health.

TABLE 7.2 Concentration of Selected Environmental Chemicals in a Cohort of 153 Healthy Individuals Detected and Quantified using High-resolution Metabolomics and Reference Standardization [20]

Metabolite	Source	Detected Median Concentration	Previously Reported Concentration Ranges
Caffeine (µM)	Dietary	22	26–129[a]
Chlorobenzoic acid (nM)	Insecticide	38	Not available
Chlorophenylacetic acid (nM)	Insecticide	28	Not available
Chlorsulfuron (nM)	Herbicide	0.6	Not available
Cotinine (nM)	Nicotine metabolite	3.9	6.7–13.5[a]
Dibutylphthalate (nM)	Plasticizer	21	15–989[b]
Dipropylphthalate (nM)	Plasticizer	57	Not available
Hippuric acid (µM)	Dietary	4.7	6–28[a]
Octylphenol (nM)	Commercial products	34	0.14–2.2[c,d]
Pirimicarb (nM)	Insecticide	0.8	Not available
Styrene (nM)	Industrial processes	5.3	0.4–5.3[a]
Tetraethylene glycol (nM)	Industrial processes	2.9	Not available
Triethylphosphate (nM)	Plasticizer/flame retardant	6.9	Not available
Triphenylphosphate (nM)	Plasticizer/flame retardant	22	Not available
Tris(2-chloropropyl) phosphate (nM)	Flame retardant	45	Not available
Xylylcarb (nM)	Insecticide	4.6	Not available

Used with permission from Oxford University Press.
[a]Ref. [7].
[b]TOXNET: US National Library of Medicine Toxicology Data Network. US National Institutes of Health. http://toxnet.nlm.nih.gov/ [accessed 01.10.15].
[c]Measured in urine, expressed as urinary nM/L.
[d]Ref. [113].

7.5.2 Metabolic Phenotyping in EWAS of Disease

Because of the ability to routinely measure a large number of environmental chemicals in the metabolic phenotype, EWAS can be completed with population screening data to identify whether chemical exposures are related to health outcome and disease. This can be accomplished through analytical strategies seeking to link environment to alternative endpoints (e.g., biological responses detected in metabolomics, complementary measured molecular markers) or identifying exposure-related associations with disease state. A number of strategies have been proposed for the use of metabolic phenotyping to identify environment–disease links [18,80,90,91]. For example, to avoid complication by factors related to reverse causality and identify exposures from the environment contributing to development of disease, a *meet-in-the-middle* (MITM) approach has been proposed [80,90,91]. With application of the MITM study design, the causal relationship between disease outcome and environment is assessed through a prospective search for intermediate biomarkers related to past exposure and associated with eventual disease development (Fig. 7.8). In human

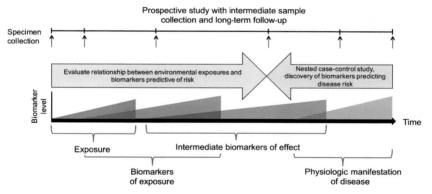

FIGURE 7.8 *Meet-in-the-middle* (MITM) approach to discover causal relationships between environmental exposures and disease outcome [80]. The MITM approach can be applied to prospective studies to link disease outcomes to intermediate biomarkers. In human populations, this is completed in three stages. The first requires epidemiologic evidence linking exposure to disease outcome. In a prospective study, this would require linking exposure and/or biomarkers of exposure to disease outcome through retrospective analysis of nested case–control studies. The second stage requires identifying associations between exposure and intermediate biomarkers. Intermediate biomarkers can be measured through metabolic phenotyping and metabolome-wide association studies (such as altered endogenous metabolite levels associated with exposure) or complementarily measured molecular markers (such as methylation levels or inflammatory response). The third stage requires identifying the relationship between intermediate biomarkers associated with environmental exposure and disease outcome. Within this framework, if there is an association between disease and exposure biomarkers in all three stages, a causal relationship between exposure and disease outcome is reinforced. Metabolic phenotyping applied to prospective studies supports the MITM principle by enhancing coverage of endogenous metabolites and xenobiotics, allowing more in-depth testing for exposure and disease associations.

populations, this is completed in three stages, which includes identification of (1) exposure and disease associations, (2) relationship between exposure biomarkers and intermediate biomarkers (such as alterations in the metabolic phenotype), and (3) relationship between intermediate biomarkers and disease outcome [80,91]. If associations between disease and exposure biomarkers are identified in all three stages, it reinforces the casual nature of the association. Measurement used in the MITM framework can also be determined through animal exposure and disease models, which assists in validating biological plausibility.

To provide a proof-of-concept implementation of the MITM approach, Chadeau-Hyam et al. [90] used metabolic phenotyping, dietary information and lifestyle information to test for environment-related biomarkers of colon and breast cancer outcomes in a nested case–control study design. Within the metabolic phenotype, the authors first tested for metabolic features distinguishing patients with breast cancer and those with colon cancer from healthy controls, followed by testing for dietary and lifestyle exposure associations. Although breast cancer exhibited no discernable differences between healthy and diseased individuals, colon cancer resulted in eight chemical regions associated with outcome. Comparison of chemical regions significant in colon cancer to those exhibiting associations with dietary exposure identified well-recognized risk factors (smoking) and protective factors (fiber intake), supporting the use of MITM for identifying putative markers linked to both exposure and disease outcomes. Metabolic phenotype markers determined through application of MITM and validated in model systems represent prime candidates for incorporation into population-based environmental epidemiologic studies and for use to identify disease risk factors in Personalized Medicine.

In addition to measurement for chemical surveillance and disease–environment biomarker identification, metabolic phenotyping can be applied to elucidate biological response to chemical exposure. This provides insight into toxicant targets, mechanisms, biomarkers of effect and chemical biological fate, including metabolism, distribution, and excretion. For example, in the study by Jeanneret et al. [92], metabolic profiling of an individual who received an extremely high dose of dioxin was completed to identify changes in endogenous metabolites correlated with exposure. The authors identified a series of steroid-related metabolites associated with high dioxin exposure, which were then tested as discriminatory metabolic phenotype biomarkers in 11 workers exposed to dioxin residues in a pesticide production plant and matched, healthy controls. A series of 24 metabolites distinguished exposed workers from controls, which indicated changes in the expression of cytochrome P450s, hepatotoxicity, bile acid biosynthesis, and oxidative stress; these results are consistent with the known aryl hydrocarbon receptor binding of dioxins. Importantly, the data showed that changes in the metabolome caused by a recent, acute dioxin exposure were still detected in the metabolic phenotype of individuals exposed 40 years earlier. Thus, in addition to biomonitoring, population screening by untargeted metabolomics provides the potential to identify biological markers indicating a history of exposure.

Metabolic phenotyping in tandem with traditional exposure assessment such as personal monitoring or targeted biomonitoring for exposure biomarkers can also be used to identify biological response to environmental stressors. For example, Wang et al. [93] utilized targeted analysis of urinary polycyclic aromatic hydrocarbons (PAH) metabolites to stratify individuals exposed to low and high levels of air pollution from industrial coking operations. The detected PAH biomarkers provided a means to evaluate dose-dependent changes in the urinary metabolome based on internal PAH exposure, which indicated association with muscular breakdown products, increased lipid peroxidation, and depletion of antioxidants. The presence of metabolites related to oxidative stress is consistent with PAH-related diseases, supporting the ability to detect early metabolic events in the exposure–disease continuum from chronic environmental exposures. Furthermore, by using individual biomarkers of exposure, the authors were able to identify biological associations changing in a dose-dependent manner, which avoids confounding caused by unknown/uncharacterized exposures and provides testable hypotheses for mechanistic study in experimental systems.

In addition to the two studies discussed above, metabolic profiling of human exposure to cadmium [94,95], pesticides [96], welding fumes [97], and arsenic [98] support the use of metabolic phenotyping for exposure-related biological changes, measurement of alternative toxicant endpoints, and insight into toxic mechanisms in human populations. Metabolic phenotyping can also be applied in an EWAS framework to systematically examine chemical associations with disease. For example, in an HRM study of PD progression [56], groups of subjects (80 from 146 patients with PD and 20 from controls) were matched by age, gender, smoking status, and pesticide exposure to identify metabolic phenotype related to PD and PD progression (slow vs rapid). Significant differences between the 80 patients and the 20 controls included matches to a polybrominated diphenyl ether (PBDE), tetrabromobisphenol A, octachlorostyrene, and pentachloroethane. The chemical feature corresponding to PBDE had a mean intensity 50% above controls, whereas the match to 2-amino-1,2-bis(p-chlorophenyl)ethanol, was more than 50% higher in individuals with rapid disease progression. Metabolic phenotyping of other aging-associated diseases have identified positive relationships between chemical exposures and disease. In a pilot study of neovascular age-related macular degeneration (NVAMD), an MWAS of 26 patients and 19 controls selected 94 metabolic features associated with NVAMD [99]. The identified features included a match to β-2,3,4,5,6-pentachlorocyclohexanol (β-PCCH), which is a hydroxylated metabolite of β-lindane originating from commercial grade insecticide formulations. To evaluate if the identity of β-PCCH was correct, the intensity was correlated against the measured metabolic phenotype to test for associations with additional exposure-related chemicals. Both the ^{37}Cl form of β-PCCH and an ambiguous match to 7-hydroxy-1,2,3,6,8-pentachlorodibenzofuran and 1,2,3,7,8-pentachlorod-ibenzodioxin were found to correlate with β-PCCH, suggesting a possible association for chemical exposures in NVAMD. Although the populations from these two studies are too small to draw firm conclusions and the identification of the chemicals associated with PD and NVAMD were not confirmed by comparison with authentic reference standards, the presence of disease–exposure associations provide possible targets for follow-up to determine if low-level environmental chemicals could be causally related to these diseases. Furthermore, these two studies indicate the potential of metabolic phenotyping for screening of exogenous chemicals in disease.

7.5.3 Chemical Exposure Measurement in Personalized Medicine

With the exception of occupational medicine, chronic environmental exposures are often not considered in the prevention and diagnosis of diseases. In many instances, chemical exposures represent a modifiable disease risk, and primary prevention can be achieved by reducing exposure risk in an individual. Before this can be applied in a Personalized Medicine context, a framework for providing exposure assessment is required, in addition to evidence supporting a causal relationship between exposure and health outcome. Metabolic phenotyping by untargeted metabolomic profiling meets the former requirement by providing a platform for universal chemical surveillance and bio-effect monitoring in human populations. Routine application is now possible at reasonable costs, making large-scale use of population screening for chemical surveillance feasible. This will enable a more thorough accounting of chemical exposures for an individual, but the relationship to diseases, particularly those with long latency periods, is currently unknown. This does not limit future use of metabolic phenotyping to predict disease risk, but it must be acknowledged that a discovery period will be required prior to deriving the full benefit of applying population screening techniques for chemical surveillance. Once the link between disease and environment is more fully elucidated, the advanced clinical and environmental measurements provided by untargeted metabolic profiling techniques, in addition to information provided by the genome [100,101], adductome [102], and proteome [103], will improve the ability of health care practitioners to include exposome-based measurements into predicting health outcome and designing preventive measures for chronic diseases.

7.6 CASE STUDY

By providing an untargeted measurement of the metabolic phenotype, population screening with high-throughput metabolomic platforms is expected to improve identification of environmental risk in disease, classification of nutritional status and provide early biomarkers of disease. Functional outputs from chemical profiling will have applications in public health, Personalized Medicine, and exposome research. In the following case study, metabolic phenotyping is applied to plasma samples obtained from 153 healthy individuals to provide proof-of-principle for high-resolution MS-based metabolomics in exposome research. First, the plasma chemical profiles were evaluated for the presence of a biomarker of exposure to the organochlorine insecticide dichlorodiphenyltrichloroethane (DDT). Following characterization of the distribution within this population of 153 individuals, an MWAS using a DDT exposure biomarker was evaluated for the presence of additional chemicals arising from environment exposure and associations with central metabolic processes. This case study is a proof-of-principle application of metabolic phenotyping for chemical biomonitoring and identifying biological response related to exposure, both important measurements in exposome research. Deployment of similar analytical frameworks in larger metabolic phenotyping studies is expected to provide insight into the distribution of chemical exposure biomarkers in the general population and identify if biologically relevant changes are occurring in response to exposure.

7.6.1   Introduction

Commonly, environmental epidemiology studies apply targeted analytical techniques to measure exposure biomarkers and molecular endpoints. The use of targeted measurements requires *a priori* knowledge of the biomarkers of interest, limiting the ability to detect possible confounding and unexpected chemical exposures. The use of untargeted metabolomic profiling techniques in population screening enables comprehensive measurement of a wide range of environmental, dietary, endogenous, and drug-derived chemicals and avoids limiting measurement to *a priori* selected analytes. With the data measured in metabolic phenotyping studies, it is then possible to perform an MWAS, which provides the framework for identifying specific metabolite associations with disease, health outcomes, and exposure. When applied to exposure biomarkers detected within the population being screened, MWAS can be used to test for dose-dependent associations within the metabolome, providing a means of describing toxicity, biological response, and metabolism *in vivo*. To illustrate the use of MWASs for characterizing possible co-exposures and exposure-related biological alterations, chemical associations with chlorophenylacetic acid (CPAA), a metabolite of DDT, was completed in a population of 153 healthy individuals. The results from this case study provide a proof-of-principle application of metabolic phenotyping for chemical biomonitoring and how population screening can be utilized to identify co-exposures and biological response to exposure.

DDT was originally used as an insecticide for mosquito control and agricultural purposes. Exposure to DDT has been linked to a wide range of diseases and health problems [79,104,105]. DDT has been classified as a persistent organic pollutant by the Stockholm Convention and remains widespread in the environment. Commonly, biomonitoring for DDT exposure is accomplished by measuring the most abundant isomer of DDT (*p,p'*-DDT) and its major metabolites *p,p'*-dichlorodiphenyldichloroethane (DDD) and *p,p'*-dichloro-2,2-bis(p-chlorophenyl) ethylene (DDE). Analysis is completed by in-depth sample cleanup/preparation with targeted quantification by gas chromatography MS [85,106]. Limited information is available on the more polar DDT residues such as 2,2-bis(chlorophenyl)acetic acid (DDA), 2,2-bis(chlorophenyl)-1-chloroethylene (DDMU), 2,2-bis(chlorophenyl) ethanol (DDOH), 2,2-bis(chlorophenyl)-1-chloroethane (DDMS), dichlorobenzophenone (DBP), and CPAA, which are still present in environmental media and represent possible chemical exposures [107].

7.6.2 Materials and Methods

7.6.2.1 Study Population

Metabolomic profiles were obtained from the study of Go et al. [20]. Briefly, a subset of plasma samples ($n = 153$) collected at Emory University was used for the metabolomics analysis. Subjects were healthy participants, between 30 and 90 years, who were studied to define a "normal" value or range of values in healthy people and to evaluate tests specifically designed to look for evidence of early multi-organ disease (ClinicalTrials.gov Identifier: NCT00336570). Both genders and individuals of different races and ethnicities were included. Confirmation and quantification of CPAA was completed using HRM and a reference standardization approach described in Go et al. [20].

7.6.2.2 High-Resolution Metabolomics

Human plasma samples were extracted and analyzed as described by Park et al. [25]. Briefly, extractions were performed with acetonitrile containing a mixture of internal standards and maintained in an autosampler at 4°C until injection. Liquid chromatography (LC) separation was achieved by using a C_{18} column (Targa, 2.1 × 10 cm; Higgins Analytical, Mountain View, CA) with a C_{18} guard column (Targa guard; Higgins Analytical) and an acetonitrile gradient, which was interfaced to a LTQ-Velos-Orbitrap mass spectrometer (Thermo Fisher, San Diego, CA) operated in positive electrospray ionization mode. Each sample was run in triplicate by using a 10 μL injection volume. Data were collected continuously over the 10-min elution time, and processed for peak extraction and quantification of ion intensities using xMSanalyzer [23] with apLCMS [108]. Tentative chemical identifications based on the KEGG (Kyoto Encyclopedia of Genes and Genomes) database [109] were completed by matching the accurate mass *m/z* feature to common adduct masses at a ±10 ppm mass error [defined as (detected m/z − theoretical m/z)/theoretical m/z × 10^6] threshold using the *feature.batch.annotation.KEGG()* function in xMSanalyzer.

7.6.2.3 Chlorophenylacetic Acid MWAS

Metabolic features associated with CPAA were selected using a Pearson correlation coefficient threshold of 0.3 and false discovery rate (FDR) threshold of

5% based on the method of Benjamini and Hochberg [110]. To evaluate network association of the significant features selected in the MWAS, a network correlation analysis was completed using the significant metabolic features and the raw metabolomics data. Multiple correlation thresholds and an FDR corrected *P*-value ≤0.05 were used to evaluate changes in the correlation structure. The correlation network was determined using the R package MetabNet [31], and metabolic pathway enrichment was assessed by *Mummichog* [111].

7.6.3 Results

7.6.3.1 Metabolomics Results

For 153 human plasma samples, 22,625 unique ions (based upon *m/z* features) were detected. Median relative standard deviation (RSD) for intensity among triplicate technical replicates was 9.1%, and total ion intensity for the entire sample set showed a RSD of 39.6%. Thus, the major variation in total signal intensity was caused by differences in the characteristics of individual samples resulting from total chemical content and not analytical variation. Data quality was also assessed by using the distribution of added, labeled internal standards. For $^{15}N^{13}C$-Met, the mean RSD was 6% and the median CV was 13.3%; for ^{15}N-Tyr, the mean CV was 8.9%, and the median RSD was 13.4%. Quantification of metabolic intermediates indicated expected ranges for amino acids and cofactors [20], suggesting that the population of individuals used for this analysis consisted of representative, healthy individuals.

7.6.3.2 Metabolic Association with CPAA

Annotation of the detected *m/z* features identified the M + H adduct (*m/z* = 171.0205, 1.17 ppm mass error) of CPAA at a retention time of 64 s. Confirmation by MS^2 matching to authentic reference standards indicated that this was a correct match. CPAA was detected in 146 of the samples, with intensity values varying over 50-fold (Fig. 7.9). Although no data existed in the literature giving representative ranges for blood levels of CPAA, the determined concentrations are similar in magnitude to blood concentration of DDE (~20 nM, 6.4 μg/L) measured in a representative US population [112]. Pearson correlation MWAS identified 71 *m/z* features that were significant with Pearson |r| ≥ 0.3 and FDR ≤ 5% (Fig. 7.10). The majority of the associations were positively correlated with CPAA, suggesting possible co-exposures, additional metabolites/ chemical products related to CPAA or increased enzymatic activity related to exposure.

  To identify the significant metabolites, we matched the *m/z* masses to the KEGG database by using common adducts for positive mode at a 10 ppm mass error threshold. Although these are only tentative matches and will need to be confirmed by comparison to authentic reference standards, the annotations are still useful for initial identification of metabolic associations with CPAA. Only 35 features matched metabolites present within the KEGG database.

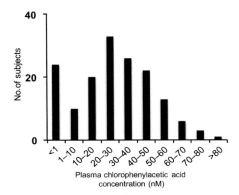

FIGURE 7.9 Distribution of plasma chlorophenylacetic acid levels measured in the 153 healthy individuals using high-resolution metabolomics and reference standardization [20]. *Used with permission from Oxford University Press.*

The remaining features had no probable matches, suggesting uncharacterized metabolic intermediates, unknown conjugates, or environmental chemicals not present in the KEGG database. After excluding nonclassified chemicals, the top three categories included pesticides (11%), lipids (9.3%), and carcinogens (9.3%). Phytochemical metabolites were also present, representing 8.6% of the matches. The identified metabolites were consistent with exposure to DDT and co-exposure to other pesticides (Table 7.3). Two metabolites of DDT were present within the significant features—chlorobenzoic acid, a bacterial degradation product of DDT, and DDMU, a degradation product of DDT. Only a limited number of the significant features were identified as endogenous metabolites and included metabolites involved in amino acid pathways, co-factor metabolism, lipids, and fatty acid intermediates. Metabolic intermediates in tyrosine metabolism were also present, including 5-O-(1-carboxyvinyl)-3-phoshoshikimate and fructose-1,6-bisphosphate. CPAA is converted to dihydroxyphenylacetate through an oxidoreductase enzyme, which is a metabolic intermediate in phenylalanine and tyrosine metabolism. Therefore, associations with tyrosine metabolism were expected, and the presence of metabolites related to tyrosine metabolism support the use of metabolic phenotyping to detect biologically relevant associations. Although it is not possible to determine if these metabolites are correlated as a result of co-exposures, co-transport on lipoproteins, or similar detoxification pathways, these results provide insight into the distribution of previously uncharacterized environmental chemicals in a representative healthy population.

7.6.3.3 Network Analysis

To identify metabolic pathway associations with the significant features, a correlation network analysis was completed using the 71 *m/z* features selected by

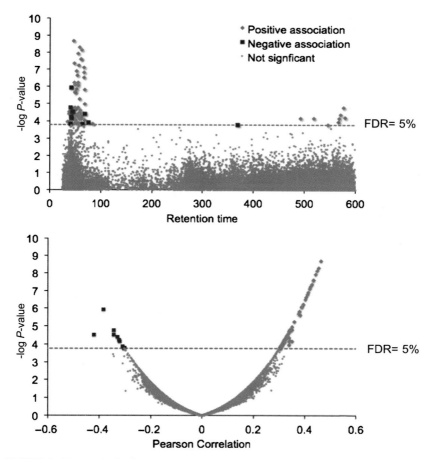

FIGURE 7.10   Metabolic feature association with plasma chlorophenylacetic acid (CPAA). Pairwise Pearson correlation analysis identified 71 metabolic features significantly associated with CPAA. Sorting by retention time indicated that these were primarily polar metabolites (A), since the use of C_{18} stationary phase with reverse phase chromatography would retain nonpolar metabolites such as lipids more strongly than polar chemicals. Correlations were predominantly positively associated with CPAA levels, suggesting possible co-exposures additional metabolites of dichlorodiphenyltrichloroethane, or increased enzymatic activity.

the MWAS and the raw metabolomics data. The resulting metabolic network maps are provided in Fig. 7.11 for correlation thresholds of $|r| \geq 0.7$ and 0.5 at FDR $\leq 5\%$. At the more stringent correlation threshold of 0.7, 154 features were found to be correlated with the CPAA-MWAS selected metabolic chemicals. The presence of metabolite clusters in Figure 7.11 suggests different patterns of metabolic associations and can be representative of chemicals from different sources, varied metabolic disposition, and dissimilar metabolic targets. After

TABLE 7.3 KEGG Database Matches Corresponding to Xenobiotic Chemicals for the Metabolic Features Associated with Plasma CPAA

Detected m/z	Metabolite Database Match	Chemical Formula	Source
150.9316	Dichloroacetate	$C_2H_2Cl_2O_2$	Halogenated organic acid
154.0043	Chloroallylaldehyde	C_3H_3ClO	Chloroalkane metabolism
157.0051	Chlorobenzoate	$C_7H_5ClO_2$	DDT metabolism
180.0437	Nitrosonaphthalene	$C_{10}H_7NO$	Combustion product
209.0271	Naphthalenesulfonic acid	$C_{10}H_8O_3S$	Combustion product
235.0915	Toxoflavine	$C_7H_7N_5O_2$	Bacterial toxin
258.0280	Chlorpropham	$C_{10}H_{12}ClNO_2$	Herbicide
284.0314	Proglinazine	$C_8H_{12}ClN_5O_2$	Herbicide
300.0228	Imidacloprid	$C_9H_{10}ClN_5O_2$	Insecticide
302.0181	Cloxyfonac	$C_9H_8ClO_4$	Growth regulator
303.0215	^{13}C Cloxyfonac	$^{13}C\ C_8H_8ClO_4$	Growth regulator
342.0134	Cyanofenphos	$C_{15}H_{14}NO_2PS$	Insecticide
343.0445	Pyraclofos	$C_{14}H_{18}ClN_2O_3PS$	Insecticide
344.0125	Ditalimfos	$C_{12}H_{14}NO_4PS$	Fungicide
346.0077	N-(6-Oxo-6H-dibenzo[b,d]pyran-3-yl)-2,2,2-trifluoroacetamide	$C_{15}H_8F_3NO_3$	Flavone
347.0114	^{13}C N-(6-Oxo-6H-dibenzo[b,d]pyran-3-yl)-2,2,2-trifluoroacetamide	$^{13}C\ C_{14}H_8F_3NO_3$	Flavone
348.0115	1-Chloro-2,2-bis(4'-chlorophenyl)ethane	$C_{14}H_{11}Cl_3$	DDT metabolism
365.9943	Trichloroethanolglucuronide	$C_8H_{11}Cl_3O_7$	Chlorinated solvent exposure

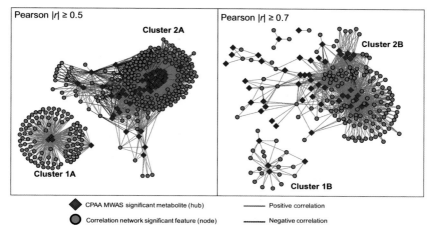

FIGURE 7.11 Correlation-based interaction network for 71 significant chlorophenylacetic acid (CPAA) metabolites. To test for metabolic pathway associations of metabolites identified in the CPAA metabolome-wide association study (MWAS), a correlation network was computed for different stringency thresholds (Pearson |r| ≥ 0.5; Pearson |r| ≥ 0.7) using the 71 metabolic features from the CPAA-MWAS as hubs and metabolites in the raw data as nodes. The formation of clusters within the interaction network can be used to define chemical modules (as discussed in Fig. 7.7) for evaluating biological response and the presence of xenobiotic chemicals. Network analysis identified two separate clusters present at both correlation thresholds. Cluster 1A and 1B primarily consisted of positive correlations, suggesting additional chemicals related to dichlorodiphenyl-trichloroethane (DDT) metabolism and arising from co-exposures. Cluster 2A and 2B contained both positive and negative associations, which is consistent with alterations in metabolism, and represents a possible biological response to DDT exposure. The interaction network obtained at Pearson |r| was used for pathway enrichment.

decreasing the correlation threshold to 0.5, the correlation structure of two clusters was maintained, further supporting disparate metabolic phenotype response to chemicals related to DDT exposure.

7.6.3.4 Metabolic Pathway Enrichment

To test for associations of the correlated *m/z* features with alterations in human metabolism, metabolic features significant at Pearson |r| ≥0.7 were tested for metabolic pathway enrichment. Pathways with greater than 4 metabolites detected in the network correlation analysis and *P*-value ≤0.05 were selected as significant and are shown in Fig. 7.12. The presence of glycolysis and gluconeogenesis pathways is consistent with the association of type II diabetes with DDT exposure, although it is not possible to determine if this is a reverse causality effect or an indicator of transition to clinical manifestation of disease state. In the long-term, exposure-associated alterations in pyrimidine and purine could also result in disruption of oxidative phosphorylation, in addition to diseases association with deficiencies in purine recycling, such as kidney failure or loss

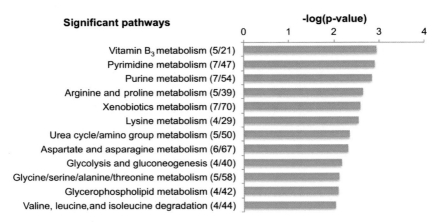

FIGURE 7.12 Pathway enrichment results for correlation interaction network with Pearson |r| $\geq$ 0.7. To identify alterations to metabolism associated with dichlorodiphenyltrichloroethane exposure, pathway enrichment was completed using the *Mummichog* software. Identified pathways significant at a permutation based *P*-value $\leq$0.05 and with more than four metabolites present in the correlation interaction network were related to chemical detoxification, essential and conditionally essential amino acids, co-factor metabolism, mitochondrial bioenergetics and nucleotide metabolism. These results support the use of metabolic phenotyping for identifying relevant environmental exposure biomarkers and testing for markers of biological response.

of immunity. The results from the metabolic pathway enrichment support the use of metabolic phenotyping to identify biologically relevant associations with chemical exposure, providing a platform for identification of alternative health endpoints and early biological indicators of exposure-related diseases.

7.6.4 Case Study Conclusions

Application of MWAS to the measurement of the metabolic phenotype using untargeted chemical measurement techniques demonstrated the capability of untargeted population screening for chemical surveillance and bio-effect monitoring. With a unified measurement method, a population of 153 subjects was screened for the presence of CPAA, an environmental chemical biomarker, and then tested for metabolic perturbations related to exposure. Annotation of the metabolites associated with CPAA identified additional chemical biomarkers and alterations in central metabolic pathways. Very little information is available on the biological consequences and extent of chronic, low-dose exposure to environmental chemicals. As is evident from this case study, population screening studies can be used to identify previously unknown chemical exposures and evaluate if the dose can be associated with alterations to metabolic pathways. This information can be used for exposome research, hazard identification, understanding of long-term health implications from chemical exposure, and understanding of the distribution of environmental chemicals in general populations.

7.7 CONCLUSION

Sequencing of the human genome has yet to identify a genetic link to many chronic diseases, and this has led to the expectation that a combination of genetic, environmental and lifestyle factors contribute to disease onset and progression. Unlike analytical platforms enabling population screening for genetic risk factors of disease, currently there is no platform permitting comprehensive measurement of environmental factors in humans. To adequately incorporate environment into a Personalized Medicine framework, there is a need to develop technologies that can provide quantitative measures of the influence of external factors on health. Untargeted measurement of the metabolic phenotype using high-throughput methods provides a platform for universal chemical surveillance and bio-effect monitoring, which vastly improves the chemical coverage available from targeted analytical techniques. Application of untargeted metabolic phenotyping in human populations enables a more comprehensive accounting of dietary, environmental, and pharmaceutical chemicals in biological fluids, in addition to providing measurement of endogenous metabolism. When used as a population screening tool for predictive health, metabolic phenotyping is expected to provide considerable insight into molecular markers of health and disease risk. Since expression of the metabolic phenotype includes endogenous metabolites, chemicals from human–environment interaction, and the reactant products arising from the interaction of these compounds with enzymatic and bacterial processes occurring within multiple body components, multiple factors influence the level, presence, and distribution of metabolites. Phenotypic traits, including circadian rhythm, gender, age, and disease status are associated with a number of changes to metabolite levels and will be important considerations when drawing conclusions for Personalized Medicine. External factors such as dietary patterns, external environment, geographic location, and chemical exposures are also strong contributors to the presence and distribution of metabolites. Providing an objective measurement of both phenotypic and environmental factors contributes to the usefulness of metabolic phenotyping in support of Personalized Medicine. The ability to provide sensitive measurements capable of detecting small changes in an individual's chemical profile will have important implications for disease prediction, progression monitoring, and preventive medicine. Therefore, the adoption and application of metabolic phenotyping in a population screening framework is expected to contribute to personalized health forecasting and provide valuable insight into the processes driving the development of diseases related to lifestyle and environmental factors.

ACKNOWLEDGMENTS

This work was supported by the Emory University HERCULES Exposome Research Center through NIEHS grant ES019776.

REFERENCES

[1] Wild CP. Complementing the genome with an "exposome": the outstanding challenge of environmental exposure measurement in molecular epidemiology. Cancer Epidemiol Biomarkers Prevention Publ Am Assoc Cancer Res Cosponsored by the Am Soc Preventive Oncol 2005;14(8):1847–50.

[2] Rappaport SM, Smith MT. Epidemiology. Environment and disease risks. Science 2010;330(6003):460–1.

[3] Vineis P, Schulte P, McMichael AJ. Misconceptions about the use of genetic tests in populations. Lancet 2001;357(9257):709–12.

[4] Rappaport SM, Barupal DK, Wishart D, Vineis P, Scalbert A. The blood exposome and its role in discovering causes of disease. Environ Health Perspect 2014;122(8):769–74.

[5] Jones DP, Park Y, Ziegler TR. Nutritional metabolomics: progress in addressing complexity in diet and health. Ann Rev Nutr 2012;32:183–202.

[6] Psychogios N, Hau DD, Peng J, Guo AC, Mandal R, Bouatra S, et al. The human serum metabolome. PLoS One 2011;6(2):e16957.

[7] Wishart DS, Jewison T, Guo AC, Wilson M, Knox C, Liu Y, et al. HMDB 3.0—the human metabolome database in 2013. Nucl Acids Res 2013;41(Database issue):D801–7.

[8] Smith CA, O'Maille G, Want EJ, Qin C, Trauger SA, Brandon TR, et al. Metlin: a metabolite mass spectral database. Ther Drug Monit 2005;27(6):747–51.

[9] da Silva RR, Dorrestein PC, Quinn RA. Illuminating the dark matter in metabolomics. Proc Natl Acad Sci USA 2015;112(41):12549–50.

[10] Dionisio KL, Frame AM, Goldsmith M-R, Wambaugh JF, Liddell A, Cathey T, et al. Exploring consumer exposure pathways and patterns of use for chemicals in the environment. Toxicol Rep 2015;2:228–37.

[11] Miller GW, Jones DP. The nature of nurture: refining the definition of the exposome. Toxicol Sci Off J Soc Toxicol 2014;137(1):1–2.

[12] Wild CP. The exposome: from concept to utility. Int J Epidemiol 2012;41(1):24–32.

[13] Patel CJ, Bhattacharya J, Butte AJ. An environment-wide association study (EWAS) on type 2 diabetes mellitus. PLoS One 2010;5(5):e10746.

[14] Dunn WB, Lin W, Broadhurst D, Begley P, Brown M, Zelena E, et al. Molecular phenotyping of a uk population: defining the human serum metabolome. Metabolomics 2015;11:9–26.

[15] Holmes E, Loo RL, Stamler J, Bictash M, Yap IK, Chan Q, et al. Human metabolic phenotype diversity and its association with diet and blood pressure. Nature 2008;453(7193):396–400.

[16] Holmes E, Wilson ID, Nicholson JK. Metabolic phenotyping in health and disease. Cell 2008;134(5):714–7.

[17] Menni C, Kastenmuller G, Petersen AK, Bell JT, Psatha M, Tsai PC, et al. Metabolomic markers reveal novel pathways of ageing and early development in human populations. Int J Epidemiol 2013;42(4):1111–9.

[18] Athersuch TJ, Keun HC. Metabolic profiling in human exposome studies. Mutagenesis 2015.

[19] Wild CP, Scalbert A, Herceg Z. Measuring the exposome: a powerful basis for evaluating environmental exposures and cancer risk. Environ Mol Mutagen 2013;54(7):480–99.

[20] Go YM, Walker DI, Liang Y, Uppal K, Soltow QA, Tran V, et al. Reference standardization for mass spectrometry and high-resolution metabolomics applications to exposome research. Toxicol Sci 2015;148(2):531–43.

[21] Smith CA, Want EJ, O'Maille G, Abagyan R, Siuzdak G. XCMS: processing mass spectrometry data for metabolite profiling using nonlinear peak alignment, matching, and identification. Anal Chem 2006;78(3):779–87.

[22] Yu T, Park Y, Li S, Jones DP. Hybrid feature detection and information accumulation using high-resolution LC–MS metabolomics data. J Proteome Res 2013;12(3):1419–27.

[23] Uppal K, Soltow QA, Strobel FH, Pittard WS, Gernert KM, Yu T, et al. Xmsanalyzer: automated pipeline for improved feature detection and downstream analysis of large-scale, nontargeted metabolomics data. BMC Bioinformat 2013;14:15.

[24] Libiseller G, Dvorzak M, Kleb U, Gander E, Eisenberg T, Madeo F, et al. Ipo: a tool for automated optimization of xcms parameters. BMC Bioinformat 2015;16:118.

[25] Park YH, Lee K, Soltow QA, Strobel FH, Brigham KL, Parker RE, et al. High-performance metabolic profiling of plasma from seven mammalian species for simultaneous environmental chemical surveillance and bioeffect monitoring. Toxicology 2012;295(1–3):47–55.

[26] Cribbs SK, Park Y, Guidot DM, Martin GS, Brown LA, Lennox J, et al. Metabolomics of bronchoalveolar lavage differentiate healthy HIV-1-infected subjects from controls. AIDS Res Hum Retrovirus 2014;30(6):579–85.

[27] Frediani JK, Jones DP, Tukvadze N, Uppal K, Sanikidze E, Kipiani M, et al. Plasma metabolomics in human pulmonary tuberculosis disease: a pilot study. PLoS One 2014;9(10):e108854.

[28] Go YM, Walker DI, Soltow QA, Uppal K, Wachtman LM, Strobel FH, et al. Metabolome-wide association study of phenylalanine in plasma of common marmosets. Amino Acids 2015;47(3):589–601.

[29] Kanu AB, Dwivedi P, Tam M, Matz L, Hill Jr. HH. Ion mobility-mass spectrometry. J Mass Spectrom JMS 2008;43(1):1–22.

[30] Makarov A, Denisov E, Kholomeev A, Balschun W, Lange O, Strupat K, et al. Performance evaluation of a hybrid linear ion trap/orbitrap mass spectrometer. Anal Chem 2006;78(7):2113–20.

[31] Uppal K, Soltow QA, Promislow DE, Wachtman LM, Quyyumi AA, Jones DP. Metabnet: an r package for metabolic association analysis of high-resolution metabolomics data. Front Bioeng Biotechnol 2015;3:87.

[32] Dieterle F, Riefke B, Schlotterbeck G, Ross A, Senn H, Amberg A. NMR and MS methods for metabonomics. Methods Mol Biol 2011;691:385–415.

[33] Scalbert A, Brennan L, Fiehn O, Hankemeier T, Kristal BS, van Ommen B, et al. Mass-spectrometry-based metabolomics: limitations and recommendations for future progress with particular focus on nutrition research. Metabolomics 2009;5(4):435–58.

[34] International HapMap C The International HapMap project. Nature 2003;426(6968):789–96.

[35] International HapMap C Altshuler DM, Gibbs RA, Peltonen L, Altshuler DM, Gibbs RA, et al. Integrating common and rare genetic variation in diverse human populations. Nature 2010;467(7311):52–8.

[36] Phinney KW, Ballihaut G, Bedner M, Benford BS, Camara JE, Christopher SJ, et al. Development of a standard reference material for metabolomics research. Anal Chem 2013;85(24):11732–8.

[37] Soltow Q, Strobel FH, Mansfield KG, Wachtman L, Park Y, Jones DP. High-performance metabolic profiling with dual chromatography-Fourier-transform mass spectrometry (DC-FTMS) for study of the exposome. Metabolomics 2013;9(1 suppl):S132–43.

[38] Nicholson JK, Holmes E, Elliott P. The metabolome-wide association study: a new look at human disease risk factors. J Proteome Res 2008;7(9):3637–8.

[39] Castro Cabezas M, Halkes CJ, Meijssen S, van Oostrom AJ, Erkelens DW. Diurnal triglyceride profiles: a novel approach to study triglyceride changes. Atherosclerosis 2001;155(1):219–28.

[40] Levi F, Schibler U. Circadian rhythms: mechanisms and therapeutic implications. Annu Rev Pharmacol Toxicol 2007;47:593–628.

[41] Dallmann R, Viola AU, Tarokh L, Cajochen C, Brown SA. The human circadian metabolome. Proc Natl Acad Sci USA 2012;109(7):2625–9.

[42] Gooley JJ, Chua EC. Diurnal regulation of lipid metabolism and applications of circadian lipidomics. J Genet Genomics 2014;41(5):231–50.

[43] Park Y, Kim SB, Wang B, Blanco RA, Le NA, Wu S, et al. Individual variation in macronutrient regulation measured by proton magnetic resonance spectroscopy of human plasma. Am J Physiol Regul Integr Comp Physiol 2009;297(1):R202–9.

[44] Chua EC, Shui G, Lee IT, Lau P, Tan LC, Yeo SC, et al. Extensive diversity in circadian regulation of plasma lipids and evidence for different circadian metabolic phenotypes in humans. Proc Natl Acad Sci USA 2013;110(35):14468–73.

[45] Ang JE, Revell V, Mann A, Mantele S, Otway DT, Johnston JD, et al. Identification of human plasma metabolites exhibiting time-of-day variation using an untargeted liquid chromatography–mass spectrometry metabolomic approach. Chronobiol Int 2012;29(7):868–81.

[46] Kim AM, Tingen CM, Woodruff TK. Sex bias in trials and treatment must end. Nature 2010;465(7299):688–9.

[47] Liu LY, Schaub MA, Sirota M, Butte AJ. Sex differences in disease risk from reported genome-wide association study findings. Hum Genet 2012;131(3):353–64.

[48] Mittelstrass K, Ried JS, Yu Z, Krumsiek J, Gieger C, Prehn C, et al. Discovery of sexual dimorphisms in metabolic and genetic biomarkers. PLoS Genet 2011;7(8):e1002215.

[49] Krumsiek J, Mittelstrass K, Do K, Stückler F, Ried J, Adamski J, et al. Gender-specific pathway differences in the human serum metabolome. Metabolomics 2015:1–19.

[50] Fontana L, Partridge L, Longo VD. Extending healthy life span—from yeast to humans. Science 2010;328(5976):321–6.

[51] Deelen J, Beekman M, Capri M, Franceschi C, Slagboom PE. Identifying the genomic determinants of aging and longevity in human population studies: progress and challenges. BioEssays: News Rev Mol Cell Dev Biol 2013;35(4):386–96.

[52] Barnett K, Mercer SW, Norbury M, Watt G, Wyke S, Guthrie B. Epidemiology of multimorbidity and implications for health care, research, and medical education: a cross-sectional study. Lancet 2012;380(9836):37–43.

[53] Jove M, Mate I, Naudi A, Mota-Martorell N, Portero-Otin M, De la Fuente M, et al. Human aging is a metabolome-related matter of gender. J Gerontol Ser A Biol Sci Med Sci 2015.

[54] Jones DP. Extracellular redox state: refining the definition of oxidative stress in aging. Rejuvenation Res 2006;9(2):169–81.

[55] Zhao J, Zhu Y, Uppal K, Tran VT, Yu T, Lin J, et al. Metabolic profiles of biological aging in American Indians: the strong heart family study. Aging 2014;6(3):176–86.

[56] Roede JR, Uppal K, Park Y, Lee K, Tran V, Walker D, et al. Serum metabolomics of slow vs. Rapid motor progression Parkinson's disease: a pilot study. PLoS One 2013;8(10):e77629.

[57] Trushina E, Mielke MM. Recent advances in the application of metabolomics to Alzheimer's disease. Biochim Biophys Acta 2014;1842(8):1232–9.

[58] Zhao J, Zhu Y, Hyun N, Zeng D, Uppal K, Tran VT, et al. Novel metabolic markers for the risk of diabetes development in american indians. Diabetes Care 2015;38(2):220–7.

[59] Kwon H, Oh S, Jin X, An YJ, Park S. Cancer metabolomics in basic science perspective. Arch Pharmacal Res 2015;38(3):372–80.

[60] Wurtz P, Havulinna AS, Soininen P, Tynkkynen T, Prieto-Merino D, Tillin T, et al. Metabolite profiling and cardiovascular event risk: a prospective study of 3 population-based cohorts. Circulation 2015;131(9):774–85.

[61] McNiven EM, German JB, Slupsky CM. Analytical metabolomics: nutritional opportunities for personalized health. J Nutr Biochem 2011;22(11):995–1002.

[62] Scalbert A, Brennan L, Manach C, Andres-Lacueva C, Dragsted LO, Draper J, et al. The food metabolome: a window over dietary exposure. Am J Clin Nutr 2014;99(6):1286–308.

[63] Goodacre R, Vaidyanathan S, Dunn WB, Harrigan GG, Kell DB. Metabolomics by numbers: acquiring and understanding global metabolite data. Trends Biotechnol 2004;22(5): 245–52.

[64] Neveu V, Perez-Jimenez J, Vos F, Crespy V, du Chaffaut L, Mennen L, et al. Phenol-explorer: an online comprehensive database on polyphenol contents in foods. Database J Biol Databases Curation 2010;2010:bap024.

[65] Foodb version 1.0 [Internet]; 2015 [cited October 19, 2015].

[66] Ibanez C, Simo C, Garcia-Canas V, Cifuentes A, Castro-Puyana M. Metabolomics, peptidomics and proteomics applications of capillary electrophoresis-mass spectrometry in foodomics: a review. Anal Chim Acta 2013;802:1–13.

[67] O'Sullivan A, Gibney MJ, Brennan L. Dietary intake patterns are reflected in metabolomic profiles: potential role in dietary assessment studies. Am J Clin Nutr 2011;93(2):314–21.

[68] Wang Z, Klipfell E, Bennett BJ, Koeth R, Levison BS, Dugar B, et al. Gut flora metabolism of phosphatidylcholine promotes cardiovascular disease. Nature 2011;472(7341):57–63.

[69] Barton S, Navarro SL, Buas MF, Schwarz Y, Gu H, Djukovic D, et al. Targeted plasma metabolome response to variations in dietary glycemic load in a randomized, controlled, crossover feeding trial in healthy adults. Food Funct 2015;6(9):2949–56.

[70] Walsh MC, Brennan L, Pujos-Guillot E, Sebedio JL, Scalbert A, Fagan A, et al. Influence of acute phytochemical intake on human urinary metabolomic profiles. Am J Clin Nutr 2007;86(6):1687–93.

[71] Walsh MC, McLoughlin GA, Roche HM, Ferguson JF, Drevon CA, Saris WH, et al. Impact of geographical region on urinary metabolomic and plasma fatty acid profiles in subjects with the metabolic syndrome across europe: the lipgene study. Br J Nutr 2014;111(3):424–31.

[72] O'Brien KA, Griffin JL, Murray AJ, Edwards LM. Mitochondrial responses to extreme environments: insights from metabolomics. Extreme Physiol Med 2015;4:7.

[73] Tissot van Patot MC, Serkova NJ, Haschke M, Kominsky DJ, Roach RC, Christians U, et al. Enhanced leukocyte hif-1alpha and hif-1 DNA binding in humans after rapid ascent to 4300 m. Free Radic Biol Med 2009;46(11):1551–7.

[74] McMichael AJ, Woodruff RE, Hales S. Climate change and human health: present and future risks. Lancet 2006;367(9513):859–69.

[75] Snodgrass JJ, Leonard WR, Tarskaia LA, Alekseev VP, Krivoshapkin VG. Basal metabolic rate in the yakut (sakha) of Siberia. Am J Hum Biol Off J Hum Biol Council 2005;17(2): 155–72.

[76] O'Connell SG, Kincl LD, Anderson KA. Silicone wristbands as personal passive samplers. Environ Sci Technol 2014;48(6):3327–35.

[77] Brown RC, Lockwood AH, Sonawane BR. Neurodegenerative diseases: an overview of environmental risk factors. Environ Health Perspect 2005;113(9):1250–6.

[78] Hatcher-Martin JM, Gearing M, Steenland K, Levey AI, Miller GW, Pennell KD. Association between polychlorinated biphenyls and Parkinson's disease neuropathology. Neurotoxicology 2012;33(5):1298–304.

[79] Richardson JR, Roy A, Shalat SL, von Stein RT, Hossain MM, Buckley B, et al. Elevated serum pesticide levels and risk for Alzheimer disease. JAMA Neurol 2014;71(3):284–90.

[80] Vineis P, van Veldhoven K, Chadeau-Hyam M, Athersuch TJ. Advancing the application of omics-based biomarkers in environmental epidemiology. Environ Mol Mutagen 2013;54(7):461–7.

[81] Go YM, Uppal K, Walker DI, Tran V, Dury L, Strobel FH, et al. Mitochondrial metabolomics using high-resolution fourier-transform mass spectrometry. Methods Mol Biol 2014;1198:43–73.

[82] Tice RR, Austin CP, Kavlock RJ, Bucher JR. Improving the human hazard characterization of chemicals: a tox21 update. Environ Health Perspect 2013;121(7):756–65.

[83] Wambaugh JF, Setzer RW, Reif DM, Gangwal S, Mitchell-Blackwood J, Arnot JA, et al. High-throughput models for exposure-based chemical prioritization in the expocast project. Environ Sci Technol 2013;47(15):8479–88.

[84] Wambaugh JF, Wang A, Dionisio KL, Frame A, Egeghy P, Judson R, et al. High throughput heuristics for prioritizing human exposure to environmental chemicals. Environ Sci Technol 2014;48(21):12760–7.

[85] Barr DB, Needham LL. Analytical methods for biological monitoring of exposure to pesticides: a review. J Chromatogr B 2002;778(1–2):5–29.

[86] Niessen WM, Manini P, Andreoli R. Matrix effects in quantitative pesticide analysis using liquid chromatography–mass spectrometry. Mass Spectrom Rev 2006;25(6):881–99.

[87] McGaw EA, Phinney KW, Lowenthal MS. Comparison of orthogonal liquid and gas chromatography–mass spectrometry platforms for the determination of amino acid concentrations in human plasma. J Chromatogr A 2010;1217(37):5822–31.

[88] Roca M, Leon N, Pastor A, Yusa V. Comprehensive analytical strategy for biomonitoring of pesticides in urine by liquid chromatography–orbitrap high resolution mass spectrometry. J Chromatogr A 2014;1374:66–76.

[89] Jamin EL, Bonvallot N, Tremblay-Franco M, Cravedi JP, Chevrier C, Cordier S, et al. Untargeted profiling of pesticide metabolites by LC–HRMS: an exposomics tool for human exposure evaluation. Anal Bioanal Chem 2014;406(4):1149–61.

[90] Chadeau-Hyam M, Athersuch TJ, Keun HC, De Iorio M, Ebbels TM, Jenab M, et al. Meeting-in-the-middle using metabolic profiling—a strategy for the identification of intermediate biomarkers in cohort studies. Biomarkers Biochem Indicators Exposure, Response Suscept Chem 2011;16(1):83–8.

[91] Vineis P, Perera F. Molecular epidemiology and biomarkers in etiologic cancer research: the new in light of the old. Cancer Epidemiol Biomarkers Prevention Publ Am Assoc Cancer Res Cosponsored Am Soc Preventive Oncol 2007;16(10):1954–65.

[92] Jeanneret F, Boccard J, Badoud F, Sorg O, Tonoli D, Pelclova D, et al. Human urinary biomarkers of dioxin exposure: analysis by metabolomics and biologically driven data dimensionality reduction. Toxicol Lett 2014;230(2):234–43.

[93] Wang Z, Zheng Y, Zhao B, Zhang Y, Liu Z, Xu J, et al. Human metabolic responses to chronic environmental polycyclic aromatic hydrocarbon exposure by a metabolomic approach. J Proteome Res 2015;14(6):2583–93.

[94] Ellis JK, Athersuch TJ, Thomas LD, Teichert F, Perez-Trujillo M, Svendsen C, et al. Metabolic profiling detects early effects of environmental and lifestyle exposure to cadmium in a human population. BMC Med 2012;10:61.

[95] Gao Y, Lu Y, Huang S, Gao L, Liang X, Wu Y, et al. Identifying early urinary metabolic changes with long-term environmental exposure to cadmium by mass-spectrometry-based metabolomics. Environ Sci Technol 2014;48(11):6409–18.

[96] Bonvallot N, Tremblay-Franco M, Chevrier C, Canlet C, Warembourg C, Cravedi JP, et al. Metabolomics tools for describing complex pesticide exposure in pregnant women in brittany (france). PLoS One 2013;8(5):e64433.

[97] Wei Y, Wang Z, Chang CY, Fan T, Su L, Chen F, et al. Global metabolomic profiling reveals an association of metal fume exposure and plasma unsaturated fatty acids. PLoS One 2013;8(10):e77413.

[98] Zhang J, Shen H, Xu W, Xia Y, Barr DB, Mu X, et al. Urinary metabolomics revealed arsenic internal dose-related metabolic alterations: a proof-of-concept study in a chinese male cohort. Environ Sci Technol 2014;48(20):12265–74.

[99] Osborn MP, Park Y, Parks MB, Burgess LG, Uppal K, Lee K, et al. Metabolome-wide association study of neovascular age-related macular degeneration. PLoS One 2013;8(8):e72737.

[100] Schwartz DA, Freedman JH, Linney EA. Environmental genomics: a key to understanding biology, pathophysiology and disease. Hum Mol Genetics 2004;2:R217–24.

[101] Sun YV. The influences of genetic and environmental factors on methylome-wide association studies for human diseases. Curr Genetic Med Rep 2014;2(4):261–70.

[102] Rappaport SM, Li H, Grigoryan H, Funk WE, Williams ER. Adductomics: characterizing exposures to reactive electrophiles. Toxicol Lett 2012;213(1):83–90.

[103] Go YM, Jones DP. Redox biology: interface of the exposome with the proteome, epigenome and genome. Redox Biol 2014;2:358–60.

[104] Cantor KP, Silberman W. Mortality among aerial pesticide applicators and flight instructors: follow-up from 1965–1988. Am J Ind Med 1999;36(2):239–47.

[105] Cohn BA, La Merrill M, Krigbaum NY, Yeh G, Park JS, Zimmermann L, et al. Ddt exposure in utero and breast cancer. J Clin Endocrinol Metabol 2015;100(8):2865–72.

[106] Sandau CD, Sjodin A, Davis MD, Barr JR, Maggio VL, Waterman AL, et al. Comprehensive solid-phase extraction method for persistent organic pollutants. Validation and application to the analysis of persistent chlorinated pesticides. Anal Chem 2003;75(1):71–7.

[107] Heberer T, Dünnbier U. Ddt metabolite bis(chlorophenyl)acetic acid: the neglected environmental contaminant. Environ Sci Technol 1999;33(14):2346–51.

[108] Yu T, Park Y, Johnson JM, Jones DP. APLCMS—adaptive processing of high-resolution LC/MS data. Bioinformatics 2009;25(15):1930–6.

[109] Kanehisa M, Goto S, Sato Y, Furumichi M, Tanabe M. Kegg for integration and interpretation of large-scale molecular data sets. Nucl Acids Res 2012;40(Database issue):D109–14.

[110] Benjamini Y, Hochberg Y. Controlling the false discovery rate—a practical and powerful approach to multiple testing. J Roy Stat Soc B Met 1995;57(1):289–300.

[111] Li S, Park Y, Duraisingham S, Strobel FH, Khan N, Soltow QA, et al. Predicting network activity from high throughput metabolomics. PLoS Comp Biol 2013;9(7):e1003123.

[112] CDC Fourth report on human exposure to environmental chemicals, updated tables, (February, 2015). Atlanta, GA: U.S. Department of Health and Human Services, Centers for Disease Control and Prevention; 2015.

[113] Qin Y, Chen M, Wu W, Xu B, Tang R, Chen X, et al. Interactions between urinary 4-tert-octylphenol levels and metabolism enzyme gene variants on idiopathic male infertility. PLoS One 2013;8(3).

Chapter 8

Handing on Health to the Next Generation: Early Life Exposures

Elaine Holmes[1], David MacIntyre[2], Neena Modi[3] and Julian R. Marchesi[2,4]

[1]*Department of Surgery and Cancer, Imperial College London, London, UK* [2]*Institute of Reproductive and Developmental Biology, Department of Surgery and Cancer, Imperial College London, The Hammersmith Hospital, London, UK* [3]*Department of Medicine, Imperial College London, London, UK* [4]*Centre for Digestive and Gut Health, Imperial College London, London, UK; School of Biosciences, Cardiff University, Cardiff, Wales, UK*

Chapter Outline

E. Holmes, J.K. Nicholson, A.W. Darzi & J.C. Lindon (Eds): Metabolic Phenotyping in Personalized and Public Healthcare. DOI: http://dx.doi.org/10.1016/B978-0-12-800344-2.00008-2
   213

8.1 WHY IT IS IMPORTANT TO HAND ON GOOD HEALTH TO THE NEXT GENERATION

The child is the father of the man.

William Wordsworth, 1802

The recognition of early life exposures as prime causal determinants of lifelong health and, indeed, the health of the next generation has tremendous potential to improve population well-being. The initial epidemiologic observations of David Barker and others showing a strong association between low birth weight and increased risk of premature death from cardiovascular disease [1,2] has stimulated an extraordinary outpouring of global research over three decades supporting this premise. The list of later health outcomes believed to have resulted from adverse early exposures now ranges from cardiovascular [3], autoimmune and obstructive pulmonary diseases [4], cancers [5], asthma [6], obesity [7,8], schizophrenia [9], and even kidney stones [10]. A growing body of evidence points to specific infant phenotypes, of which perhaps preterm birth and growth restriction are best recognized as predictors of compromised adult health [11]. The observation that women born preterm are substantially more likely to deliver preterm themselves [12] is but one piece of evidence supporting the view that adversity carries through into subsequent generations.

In parallel with these scientific insights, powerful postgenomic technologies provide tremendous opportunity to understand the molecular basis of early life exposures, including the biological substrates of deprivation and adversity, as causal factors leading to compromised adult health. Other technologies such as in vivo functional imaging, noninvasive monitoring, high-throughput analytical techniques employing tiny sample volumes, and the sciences of bioinformatics, epigenetics, and systems biology provide ability to include even the fetus and the newborn baby in biomedical research to a hitherto unprecedented extent.

This potential remains largely untapped, in part because social and political drivers are necessary in addition to scientific and technological advances. The UK Royal College of Paediatrics and Child Health has recently reported on research involving infants, children, and young people and made a series of recommendations in response to concerns that opportunities are not being seized to harness the potential to reduce the growing burden of noncommunicable, chronic diseases that have their origins in early development and lead to premature adult death [13]. The American Academy of Pediatrics detailed the role of what they term early life "toxic stress" on outcomes in adult life and highlighted the need for science-based strategies to build strong foundations for lifelong health [14].

A growing recognition by society of the moral and ethical imperative for closer integration between clinical practice and clinical research to bring about speedier reductions in care uncertainties [15,16] is helping to drive these changes, as is political acceptance of the need for evidence to inform policy [17,18]. However, the slow pace of change and the continuing under-representation of infants, children, and young people in medical research worldwide are salutary evidence of a compelling requirement for bold action to identify effective interventions, break down the divides between preclinical, clinical, and public health disciplines, and translate these into the delivery of strategic agendas and policy directives to improve maternal and child well-being.

8.2 THE INFLUENCE OF THE EARLY LIFE ENVIRONMENT

Environmental factors that operate in the in utero and neonatal periods of an individual's life cast a long shadow later into life. If we disregard the genetic determinants that we hand on to subsequent generations, we are still left with the capacity to shape the future health of an individual, even before he or she is born. Many actions we take during pregnancy, or in the first weeks after a baby is born, set lifelong patterns for metabolic behavior. Most mothers are anxious to provide a healthy start for their offspring, and for this reason, they often take care to adopt a particularly healthy diet and lifestyle when pregnant. There are obvious habits that, when carried out during pregnancy, will result in negative effects on the baby: for example, smoking is associated with stillbirth, prematurity, and low birth weight, and drinking excessive amounts of alcohol during pregnancy can lead to fetal alcohol syndrome, which results in learning disabilities and even physical deformities in some cases. Other medical advice during pregnancy includes limiting consumption of unpasteurized cheeses that may contain *Listeria* spp. and which can cause miscarriage or premature birth. These are obvious lifestyle modifications, yet beyond this lies another world of diet and lifestyle issues that can influence downstream health. We can make choices that directly impact on the metabolism of the baby, but it has also been shown that we can cause metabolic imprinting indirectly by influencing the gut microbiota. The gut microbiota or bacteria and other micro-organisms that live in the gastrointestinal (GI) tract help shape the developing immune system in an infant and form an inseparable partnership with us. Our microbiome contributes to gathering energy from the diet, maintaining a healthy immune system and fighting off pathogens, combating inflammation, and synthesizing vitamins. In fact, it is difficult to assess the influence of diet on human health without taking into consideration the interaction of diet with the microbiome. In the following sections, we explore some of the early life events that impact on human health, such as complications in pregnancy, diet and nutrition during pregnancy and infancy, mode of delivery, antibiotic administration, and infections (Fig. 8.1). We then consider the potential for metabolic phenotyping to identify early indices of downstream adverse health events against this background of the early life environment.

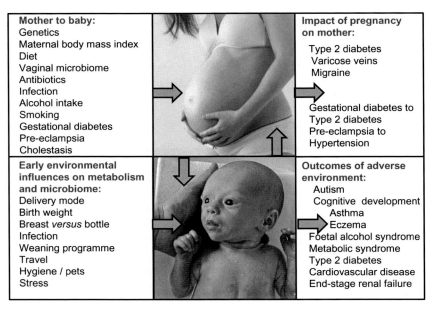

Mother to baby:	Impact of pregnancy on mother:
Genetics	
Maternal body mass index	Type 2 diabetes
Diet	Varicose veins
Vaginal microbiome	Migraine
Antibiotics	
Infection	
Alcohol intake	Gestational diabetes to
Smoking	Type 2 diabetes
Gestational diabetes	Pre-eclampsia to
Pre-eclampsia	Hypertension
Cholestasis	

Early environmental influences on metabolism and microbiome:	Outcomes of adverse environment:
	Autism
Delivery mode	Cognitive development
Birth weight	Asthma
Breast *versus* bottle	Eczema
Infection	Foetal alcohol syndrome
Weaning programme	Metabolic syndrome
Travel	Type 2 diabetes
Hygiene / pets	Cardiovascular disease
Stress	End-stage renal failure

FIGURE 8.1 Internal and external factors that affect the mother–infant interaction in pregnancy and early infancy and the downstream health consequences.

8.3 THE DEVELOPMENT OF METABOLISM IN NEONATES

Fetal metabolism relies on transplacental transfer of nutrients, predominantly glucose from the maternal circulation. Birth initiates a major shift in protein and amino acid metabolism in newborn infants in response to changes in thermogenesis, mobilization, and redox states of internal organs and tissues [19]. The continuous infusion of nutrients is interrupted at birth and replaced by feeding–fasting intervals with a transition to milk, which is high in fat and low in carbohydrate [20]. The infant adapts to this transition by hormone-mediated modifications of glucose and fatty acid metabolism. Fats become the main energy source of the neonate, providing up to 50% of the total calories in human milk. Fatty acids are required for brain development and are one of the main building blocks for cell membranes. Additionally, they solubilize and distribute hormones and vitamins in milk. Gluconeogenesis does not occur in the fetus but is evident soon after birth, with pyruvate making a significant contribution to glucose production in healthy full-term infants [21]. Brain development occurs at a rapid pace in infants, and whereas glucose is the sole energy source for the brain in adults, ketone bodies represent an important fuel source for neonates. Other key metabolic processes in the newborn infant involve remodeling of protein and changes in bile acid metabolism, with an increase in total bile acid concentrations over the first 4 weeks of life [22].

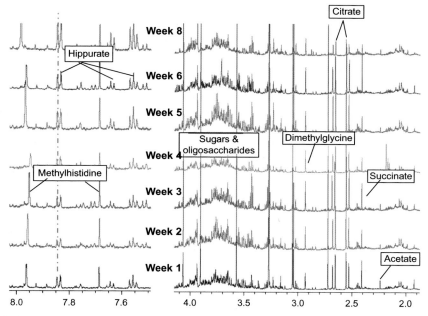

FIGURE 8.2   Changes in urinary composition over the first 3 weeks of life as detected by 600 MHz ¹H NMR spectroscopy. *Figure supplied by Francis Jackson.*

The developing metabolism of the neonate is reflected in the composition of body fluids, particularly urine. The kidneys mature over the first 1 to 2 years of life, particularly the renal tubules, and proteomic studies have shown that infant urine is enriched by more than 648 proteins, most of which are associated with cellular turnover and metabolism [23]. Preterm infants have been shown to have relatively elevated levels of insulin-like growth factor binding protein-1, -2, and -6; monocyte chemotactic protein-1; CD14; and sialic acid-binding immunoglobulin-like lectin-5 at birth compared with full-term infants, with these differences persisting for between 2 to 6 months after birth [24]. Similarly, metabonomic studies have shown substantial changes in urine content within the first few weeks of birth (Fig. 8.2), with the changes including a decrease in the urinary excretion of bile acids, N-methylnicotinamide, betaine, and taurine [25]. N-methylnicotinamide is derived from tryptophan, which is present in higher concentrations in the fetal umbilical vein and artery than in maternal blood. Taurine serves many metabolic functions and is required for neuronal development as well as acting as a renal osmolyte. Total body content of taurine is known to increase over the gestational period. Moreover, taurine metabolism has also been shown to differ in germ-free rodents compared with conventional rodents, thereby suggesting a role for gut bacteria in taurine metabolism. Differences between preterm and full-term infants may also assist in understanding the sequelae of metabolic changes in urinary composition. Studies based on nuclear magnetic resonance (NMR) spectroscopy of urine by Foxall et al. as early as 1995 showed increased renal osmolytes (trimethylamine-*N*-oxide and myo-inositol) in urine from preterm babies [26]. Gu et al. showed a transition in composition from newborn to 1 year old, consisting of a decrease in urinary excretion of betaine, succinate, citrate, and glycine, coupled with increased urinary excretion of creatine and creatinine [27]. Increased creatine and creatinine excretion over the first year of life mainly reflects the increasing muscle mass of the infant, whereas age-related changes in tricarboxylic acid cycle (TCA) intermediates (citrate and succinate) reflect a change in energy balance [28] and have been previously observed in germ-free rodents upon colonization. In contrast to the study by Guneral, other studies have reported increased urinary excretion of TCA intermediates as infants get older, together with higher concentrations of gluconeogenic amino acids [29]. Differences in the excretion of amino acid metabolism and the urea cycle in infants born preterm or those with intrauterine growth restriction (IUGR) in comparison with infants born at full term and normal weight have been reported [30], underscoring the relationship between urine composition and the development of the infant. Other research has shown differences in 1-carbon and choline metabolism between IUGR, or small-for-gestational-age babies, and normal-weight infants [31]. Thus, development of metabolism reflected in urinary composition occurs in utero and ex utero and continues to develop throughout infancy.

8.4 THE MARRIAGE OF THE MICROBIOME AND HUMAN METABOLISM

Colonization of the mammalian gut with its microbial ecosystem is an evolution-driven process to allow for a lifelong symbiotic relationship with its host. Co-evolution of host and microbiome has influenced both host and microbiome functionality such that metabolic complementarity exists within the microbiota and the intestinal bacteria significantly extend host metabolic capacity by providing critical biosynthetic pathways for the host. Other bacteria that colonize the infant gut stimulate the intestinal environment to produce nutrients for themselves. For example, Gordon et al. have shown that *Bacteroides thetaiotaomicron*, which colonizes the small intestine during the weaning period, promotes production of fucosylated glycans by the epithelial cells for its own nutritional requirements [32]. The process of colonization of the gut is very dynamic and random in the first 2 years [33], and thus, many bacterially encoded functions do not become established in this period. Although this randomness makes it difficult to determine what functions are being established and which ones are important, it may make it easy to influence the host's biology. A recent study in newborn infants showed that administration of probiotics in the first 12 months of life was able to reduce the incidence of eczema later in life [34]. This report shows that although the type of bacteria colonizing the gut in early life may seem random, the composition and succession of the bacteria that colonize is, in fact, important and has long-term consequences for the host. Furthermore, colonization of the gut was traditionally thought to begin after delivery, at which time babies receive their first exposure to the microbiota through the maternal birth canal or feces, or in the case of cesarian section (C-section), via skin contact and from the immediate environment [35]. However, mounting evidence is beginning to suggest that the environment in utero is not sterile, as originally thought, but harbors bacteria from the mother's GI tract [36]. Bacteria such as *Enterococcus* spp. and *Lactobacillus* spp. from the urogenital or GI tract have been isolated from the umbilical cord, meconium, fetal membranes, and even amniotic fluid. Since the actual numbers of these bacteria have been low in comparison with their presence on mucosal surfaces in the GI tract, it has been suggested that they serve to prime the immune system rather than colonize the neonate. The diabetes status of the mother has also been shown to influence the composition of the meconium. In mothers with different forms of diabetes mellitus, there was a strong link between the types of bacterial phyla in the meconium and the diabetes status of the mother. The children of mothers with diabetes show enrichment of the *Bacteroidetes* (phyla) and *Parabacteroides* (genus) [37].

Although direct contact between microbes and the growing embryo is limited during pregnancy, body weight and weight gain during pregnancy have been associated with low levels of bifidobacteria and *Bacteroides* spp. in the offspring but high numbers of staphylococci and members of the *Enterobacteriaceae*, with *Escherichia coli* showing a positive association with weight gain during pregnancy [38].

Accumulating epidemiologic evidence indicates that alterations of the intrauterine environment during fetal development or in the perinatal period can result in increased risk of susceptibility to a wide range of metabolic, neurodevelopment, and psychiatric diseases later in life in addition to altering the odds ratio of developing metabolic syndrome and related conditions later in life. Thus, the mode of delivery is known to shape the early infant gut microbial landscape, and the delivery method also transfers a metabolic signature to the umbilical cord tissue. A gas chromatography–mass spectrometry (GC-MS) method was used to characterize umbilical cord tissue from infants delivered naturally via spontaneous or induced labor, and from infants delivered by C-section. The stress associated with labor was particularly reflected by the concentrations of sugars such as fructose, mannose, and glucose [39]. Although birth by C-section has been associated with reduced fetal stress, epidemiologic studies lean toward negative associations of C-section with long-term health outcomes for cardiovascular disease, asthma, and metabolic syndrome [40,41]. Whereas the microbiome of babies delivered vaginally is dominated by *Lactobacillus*, *Prevotella*, and *Atopobium* species, babies delivered by C-section have a microbiome that more closely resembles that of the maternal skin community, with staphylococci being a dominating member [42]. In a cross-sectional study of over 2000 children assessed at age 8, the authors attempted to establish if there was an association between C-section and asthma and/or atopy [43]. Findings showed that the adjusted odds ratio for diagnosis of asthma was 1.41 in the C-section group; for atopy, it was 1.67, where there was a family history of these conditions. In countries such as Cyprus, where the rate of C-sections is as high as 50%, the rise in asthma has also mirrored this trend. As babies born via C-section are not exposed to the mothers' vaginal microbiome, it has been hypothesized that the neonatal cytokine and Th1/Th2 helper cell balance are different from those found in vaginally delivered babies. However, the literature has conflicting reports, with almost as many studies finding no association between C-section delivery and asthma or atopy.

A significant proportion of urine composition is shaped by gut bacteria. These are acquired at birth, and the intestinal tract is dominated first by facultative and aerotolerant bacteria such as staphylococci, streptococci, and species from *Enterobacteriaceae* in the first few days of life, succeeded by increasingly stringent anaerobic species such as bifidobacteria and lactobacilli [44]. Metabolic phenotyping studies have been used to characterize changes in urine composition over time.

Typically, around 10–30% of the urinary metabolites we detect in urine are synthesized, with at least some input from gut microbiota. Commonly excreted metabolites that require microbial modulation of either dietary components or human metabolites include hippurate, phenylacetylglutamine, 4-cresyl sulfate, hydroxyphenylpropionic acid, and various indoles.

Metabolic profiling has been used to characterize the phenotype of germ-free rodents exposed to the laboratory environment, and it has been found that it takes approximately 3 weeks for the urine composition to normalize. For example, the urine NMR spectra of germ-free rodents have very few signals in the region of 6.0–9.0 ppm, indicating that there are not many aromatic molecules present. Within the first 48 h, the animals begin to excrete phenylacetylglycine (equivalent to the phenylacetylglutamine in humans) followed by 4- and 3-hydroxy-phenylpropionate, and finally hippurate at around 19–21 days. Hippurate, the glycine conjugate of benzoic acid (a microbial product of plant-based phenols), remains the most prominent gut microbe–host co-metabolite in urine [45]. Urinary concentrations of trimethylamine-N-oxide, a bacterial degradation product of choline metabolism, have also been found to increase in tandem with hippurate levels, whereas the relative proportions of TCA cycle intermediates (succinate, citrate, 2-oxoglutarate) decreased. Similar findings have been reported in the urinary profiles of rodents following exposure to the antibiotics vancomycin [46], gentamicin [47], enrofloxacin [48], and penicillin [49,50], inducing depletion of the indoles, phenyl compounds, and bile acids components, with composition returning to baseline levels between 2 and 3 weeks after dosing. Many of the metabolites we find in urine and feces are influenced by the enzymatic activity of gut bacteria. For example, immediately after dosing with enrofloxacin, there was evidence of a decrease in alanine, isoleucine, leucine, valine, and threonine in feces, reflecting loss of microbial proteases, an ability to catabolize proteins, and a decrease in short-chain fatty acids (SCFAs) consistent with a decrease in lactate-utilizing bacteria [48]. One of the most striking changes in the urine composition is a decrease in trimethylamine-*N*-oxide, indicative of loss of microbial choline catabolism. Taurine excretion also decreases following antibiotic administration, consistent with loss of microbial cysteine dioxygenases, whereas creatine and creatinine increase in relative proportions, since they undergo less microbial degradation. Interestingly, the recolonization process following antibiotic exposure has been shown to be sensitive to the microenvironment with cage effects observed in terms of urinary composition after antibiotic exposure [51]; this would imply that initial colonization of infants may be a rather more plastic process than originally thought, since the local environment is known to shape microbial colonization. The consequence of harboring gut bacteria and the exact composition of that microbial community extend well beyond its influence on the urinary profile and beyond the local environment of the intestine. Organs such as the liver, heart, and kidneys in germ-free animals are associated with modified metabolic profiles, including increased tauroconjugation of bile acids [52], and the brains of germ-free rats contain relatively higher concentrations of choline, taurine, and *N*-acetyl aspartate compared with those of conventional animals (Jonathan Swann, personal communication). These findings all underscore the close relationship between the human host and the resident microbial communities in developing and controlling chemical communication across and between multiple tissues and organs.

8.5 CONDITIONS IN PREGNANCY AFFECTING OFFSPRING

8.5.1 The Vaginal Microbiome in Pregnancy

To understand the potential implications of the vaginal microbiome in pregnancy outcomes, it is useful to consider its role in reproductive health maintenance in the nonpregnant state, where most research has been focused. The presence of *Lactobacillus* species in the vagina are widely considered to be associated with a "healthy" state, wherein they inhibit the growth of pathogenic bacteria through the synthesis of lactic acid and hydrogen peroxide (H_2O_2) and the release of antibacterial bacteriocins [53]. Seminal work by Ravel et al. revealed that in nonpregnant women of reproductive age, the vaginal microbiome structure can be classified into five community state types (CSTs): Some women may have their microbiome dominated by *Lactobacillus crispatus* (CST I), *L. gasseri* (CST II), or *L. iners* (CST III); and some women may have depleted *Lactobacillus* species and increased numbers of strict anaerobic bacteria (CST IV) or *L. jensenii* (CST V) [54]. The type IV community state was also shown to be more likely to occur in black women, suggesting that ethnic differences may exist. This is particularly pertinent when considering that preterm birth rates are higher in the black population. Surprisingly, some vaginal bacterial CSTs are dynamic and can rapidly change from one type to another, whereas other CSTs exhibit relative stability [55]. For example, type I communities enriched for *L. crispatus* rarely change to another CST, whereas CST III (*L. iners*)–dominant communities more often transition to type II or IV communities. Menstruation and menopause have both been shown to have a profound impact on community composition, suggesting a functional role for hormonal signaling in the determination of the vaginal microbial structure.

Throughout pregnancy, the uterus is protected from ascending infection by the cervical mucus plug and a relatively benign vaginal microbiome that is dominated by *Lactobacillus* species. Similar to the nonpregnant state, culture-dependent methods have shown that specific *Lactobacillus* species are major determinants of the stability of the vaginal microbiota throughout pregnancy. Whereas *L. crispatus* appears to promote stability of the normal vaginal microbiota, *L. gasseri* and *L. iners* predispose to the occurrence of an abnormal microbiome [56]. Recent culture-independent 16S rRNA gene sequence–based studies in both cross-sectional [57] and longitudinal [58] pregnant patient populations concur with these findings, showing that the vaginal microbiome in pregnancy is reduced in both overall diversity and richness and is typically associated with more stable CSTs. In contrast, the postpartum period is associated with rapid loss of *Lactobacillus* spp. and an increase in species diversity [59].

Species of *Lactobacillus* secrete lactate, acetate, and other end products of anaerobic metabolism, together with species-specific metabolites and bacteriocins, which collectively influence growth of other organisms. The vaginal microbiota and their secretion products interact with and evoke a local tissue immune response. This complex biochemical interplay between the microbiota and the host is reflected in the metabolic milieu of the cervicovaginal fluid. NMR metabolic profiling of nonpregnant vaginal fluid samples has identified unique metabolic end products associated with different vaginal bacterial communities [55]. Lactic acid concentrations were shown to be higher in community states dominated by lactobacilli species, whereas acetate levels were comparatively increased only in *Streptococcus*-dominated communities. Both acetate and succinate concentrations were increased in communities enriched for *Prevotella, Atopobium*, and other anaerobic bacteria. Such changes in biochemical endpoints of fermentation represent a direct measurement of how shifts in the overall microbiome can perturb the local tissue response and alter the chemical composition of the vagina. Work by Witkin et al. has shown that lactic acid enhances the release of interleukin-1β (IL-1β) and IL-8 from vaginal epithelial cells, suggesting a synergistic relationship between the host and the microorganism, which may be dependent on both intrinsic (genetic) and extrinsic (environmental) factors [60]. These authors also demonstrated that elevated levels of D-lactic acid and an increased ratio of D- to L-lactic acid promote the expression of vaginal extracellular matrix metalloproteinase inducer, a key enzyme involved in the activation of matrix metalloproteinase-8, which may be involved in cervical ripening and fetal membrane remodeling [61].

8.5.2 The Gut Microbiome in Pregnancy

Koren et al. recently undertook an examination of the gut microbiota throughout pregnancy and revealed a remarkable shift in the gut microbiota from the first trimester to the third trimester, which corresponds with increased adiposity gain and insulin resistance observed during gestation [62]. Although advancing pregnancy is associated with reduction in microbial richness, there is a marked increase in the diversity of the gut microbiome among pregnant women in the third trimester, which involves an overall increase in the proportion of *Proteobacteria* and *Actinobacteria*. Stool collected during the third trimester showed greater levels of energy loss and also contained the highest concentrations of proinflammatory cytokines. Fecal transplants of the third trimester microbiota to germ-free mice induced greater adiposity and insulin insensitivity compared with first trimester microbe transplants. These findings provide striking evidence that host–microbe interactions not only modulate key energy transitions during pregnancy but also play an important role in modulating the host inflammatory response mechanism. This role would have major implications in various aspects of pregnancy where inflammatory mediation is a key causal regulator (eg, onset of preterm and full-term parturition). Zhang et al. proposed that dysbiosis during pregnancy could relate to adverse outcomes such as maternal–fetal immune rejection, cardiovascular maladaptation, and metabolic syndrome [63]. In line with this hypothesis, there are indications that probiotics confer some measure of protection against pre-eclampsia, gestational diabetes, allergic conditions, and maternal and infant weight gain [64]. The maternal microbiota is a factor that contributes to shaping the gut microbial ecosystem in the infant, and although the maternal gut microbiome increases in *Proteobacteria* and *Actinobacteria* during pregnancy, overweight pregnant mothers have been shown to carry increased numbers of staphylococci and *Enterobacteriaceae* compared with normal-weight pregnant women [38]. The dynamic remodeling of the gut microbiome during pregnancy results in changes in the levels of SCFAs, the levels of which correlate with maternal weight gain, maternal serum leptin levels, and infant weight and length [65]. Several studies have evaluated the changes in metabolic phenotype as pregnancy progresses, and in keeping with earlier discussions, several of these changes during pregnancy relate to altered gut microbial function and to lipid metabolism [66]. Over the course of the pregnancy, metabolites such as trimethylamine (bacterial degradation of choline and carnitine), acetate, and formate, all potentially have been associated with fetal growth rate [67], and taurine and histidine have been associated with fetal adiposity [68]. Taurine has been reported to be a candidate biomarker for early-onset pre-eclampsia that can be detected toward the end of the first trimester [69].

8.5.3 Premature Birth

Preterm birth (delivery before 37 weeks of gestation) represents the single largest cause of infant death and ill health across both developed and developing nations. It is also associated with adverse health outcomes in adulthood, with around 50% of surviving infants left with significant motor and sensory deficits, learning disabilities, and respiratory disorders [70]. Preterm birth occurs in approximately 10% of pregnancies, and its incidence is rising globally. Placental dysfunction and poor vascularization of the placenta is associated with preterm birth and neurologic morbidity. The causes of preterm birth are multifactorial, but infection is associated with at least 40% of cases, particularly those classified as extreme preterm labor (PTL) (<28 weeks), which carry the worst neonatal outcomes [71]. Infection and consequent inflammation are, so far, the only pathologic conditions for which a firm causal link with PTL has been established.

Infection of the choriodecidual space and the amniotic fluid is the most widely studied etiological factor underlying PTL, with vaginal colonization by micro-organisms associated with increased risk, especially in those women who deliver at <30 weeks' gestation [72]. In the second trimester, intra-amniotic infection can be detected in around half of all women with asymptomatic cervical dilatation, and it is proposed that cervical shortening may allow micro-organisms from the vagina to gain access to the chorioamniotic membranes [73]. Once present, bacteria or their products (eg, endotoxins) are recognized by the innate immune response system of the maternal host via toll-like receptors (TLRs) expressed in gestational tissues. TLR binding activates proinflammatory transcription factors such as nuclear factor kappa B (NFκB) and activator protein 1 (AP-1), initiating a local immune response that modulates immune cell trafficking and leukocyte infiltration [74]. This activation leads to the production of cytokines and matrix metalloproteases, which break down the collagen component of the cervix and fetal membranes and induce the production of prostaglandins that stimulate the myometrial contractions driving PTL [75].

As previously discussed, a *Lactobacillus* species–dominated vaginal microbiome is often considered representative of the healthy pregnant state. In contrast, dysbiosis of the vaginal microbiome is linked with an increased risk of preterm birth [76]. Bacterial vaginosis (BV) is characterized by reduced numbers of *Lactobacillus* species, a higher pH, and an increased abundance of potential pathogenic bacteria, including *Gardnerella vaginalis*, group B streptococci, *E. coli*, *Mycoplasma hominis*, *Peptostreptococcus* species, and *Bacteroides* spp. There exists a greater than two-fold increased risk of preterm birth in women with BV during pregnancy [77]. Moreover, the incidence of BV during pregnancy is associated with increased neonatal morbidity, including respiratory distress at delivery, requirement of intermittent positive pressure ventilation, and admission to the neonatal intensive care unit [78]. There also exists a correlation between the presence of *Ureaplasma* species in the vaginal microbiome, preterm birth, and poor neonatal outcomes [79]. Infection of the chorioamnion with *U. urealyticum* is associated with low birth weight, and respiratory tract colonization in preterm infants via vertical transmission is highly correlated with the development of chronic lung disease [80]. Similarly, vertical transmission of *M. hominis* and glucose nonfermentative gram-negative rods are independent risk factors for preterm births at <33 weeks' gestation [81].

8.5.4   Preterm Prelabor Rupture of the Membranes

The underlying etiology of preterm birth is undoubtedly multifactorial, yet around 30–40% of all cases are preceded by preterm prelabor rupture of the fetal membranes (PPROM) [82]. The composition of the vaginal microbiome is therefore likely to be a critical determinant in the cause of PPROM, the risk of subsequent ascending infection, and the ultimate outcome of the neonate. The first exposure of the neonate to the microbiota typically occurs during passage through the vagina, and thus the composition of these bacteria is likely an important factor for subsequent neonatal colonization and immune development. Consistent with this, the vaginal microbiota is associated with the development of neonatal sepsis following PPROM [83] and the colonization of the respiratory tract of premature infants with chronic lung disease [80].

Although a clear link has been established between bacterial infection and preterm birth, results of antibiotic treatment as an attempt to prevent PTL have been disappointing [84]. The large multicentre randomized control trials ORACLE and ORACLE II focused on the use of antibiotics in PPROM in over 4000 women and spontaneous PTL with intact membranes in over 6000 women. These trials identified that the use of erythromycin (a broad-spectrum bacteriostatic macrolide antibiotic) in singleton pregnancies with PPROM offered some neonatal benefit in the prevention of short-term respiratory function, chronic lung disease, and major neonatal cerebral abnormality, but no neonatal benefits with antibiotic treatment were detected in women with intact membranes [85]. As a result of these trials, erythromycin has been adopted as the treatment of choice for PPROM in many obstetric units in the United Kingdom. However, concerns have been raised over the widespread use of broad-spectrum antibiotics for all patients with PPROM [86]. Destruction of commensal bacteria crucial for the early colonization of the neonate and the potential to create resistant bacterial strains represent possible unintended side effects of this approach and may do more harm than good in the long term. We have started to address this question by examining the vaginal microbiome at 28 weeks' gestation in women who subsequently delivered at term ($n = 25$) and women who presented at 28 weeks with PPROM ($n = 15$). As seen in Fig 8.3A, those women who delivered at term (shaded red) had vaginal microbiomes that were overwhelmingly *Lactobacillus* spp. dominated (>95% of total species).

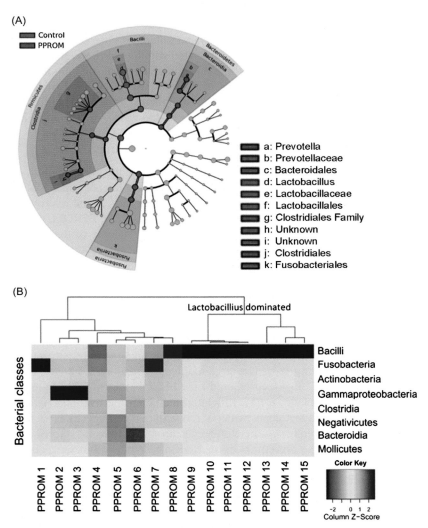

FIGURE 8.3 Comparison of the vaginal microbiome at 28 weeks' pregnancy in women who go on to deliver at term or present with PPROM, using 454 FLX pyrosequencing of V1-V3 regions of the 16S rRNA gene. (A) Cladogram describing taxa differences between the vaginal microbiome at 28 weeks of pregnancy in women who went on to deliver at term ($n = 25$, shaded red) and women who presented at 28 weeks with PPROM ($n = 15$, shaded green). Different abundance values of discriminatory taxons (according to LEfSe analysis) are listed where the color denotes the class with higher median for both the small circles and the shading. (B) Hierarchical clustering analysis of 16S rRNA gene deep sequencing class data of the vaginal microbiome–enabled stratification of patients with PPROM into *Lactobacillus*-dominant (PPROM 9-15) and *Lactobacillus*-depleted heterogeneous groups (PPROM 1-8).

Those women who presented with PPROM could be stratified into two groups: (1) low diversity, *Lactobacillus* spp. dominated (consistent with women who delivered at term); and (2) high diversity, *Lactobacillus* depleted, with high numbers of *Fusobacterium* spp., *Bacteroidia* and/or *Clostridia*. This separation implies different underlying mechanisms of PPROM in these women and thus indicates the need for differing treatment strategies. The ability to stratify patients with PPROM on the basis of their microbiome may offer a targeted treatment approach whereby only those patients with vaginal dysbiosis would be selected for antibiotic treatment. This could help prevent unnecessary antibiotic treatment and potentially protect the "healthy" existent microbiome of other patients. A limitation of next-generation sequencing analyses of the vaginal microbiome is that the time from sampling to data acquisition and interpretation is in the order of days to weeks. This is, therefore, not a practical approach for informing the clinical decision-making process. To address this, we have tested the hypothesis that the complex biochemical interplay between the microbiota and the host is reflected in the urinary metabolic profiles of urine, which itself can be viewed as the time-averaged representation of the metabolic changes of the individual and the direct interaction between metabolic activities of microbes and the host [87]. Thus, the analysis of urinary metabolites may provide a rapid and effective method of monitoring vaginal microbiota–host interaction, identifying women at risk of PPROM, and facilitating the stratification of patients groups who may require or respond appropriately to a given treatment regime. Our pilot study has shown that metabolic profiling of urine samples can readily assist the identification of PPROM patients with either a mixed or lactobacilli-dominated microbiome (Fig. 8.3B). By integrating urinary NMR data with the vaginal microbiome data, we also showed that it is possible to identify specific urinary metabolites with concentrations that correlate with the presence of specific abnormal bacterial classes (Fig. 8.4). For example, there exists a correlation between urinary concentrations of glucose and para-cresol with high abundance of clostridia in the vaginal microbiome. It is unclear if high glucose concentrations in urine facilitate the colonization and growth of vaginal clostridia. Alternatively, the presence of clostridia in the vaginal microbiota may be associated with their colonization of the gut microbiome in these patients. Clostridia are one of the few classes of anaerobic bacteria that possess an ability to produce p-cresol, a phenolic compound produced by the degradation of tyrosine via para-hydroxyphenylacetate (p-HPA) [88]. A number of studies have shown p-cresol is bacteriostatic and inhibits the growth of other bacteria and thus the production of p-cresol provides the bacterium with a competitive advantage over other microbiota. Further work is required to investigate the causal link between changes in the microbiome and metabolites in the urinary metabolome; however, metabolic profiling of urine may provide a novel approach for the rapid assessment of the vaginal microbiome and its impact on the maternal host system response.

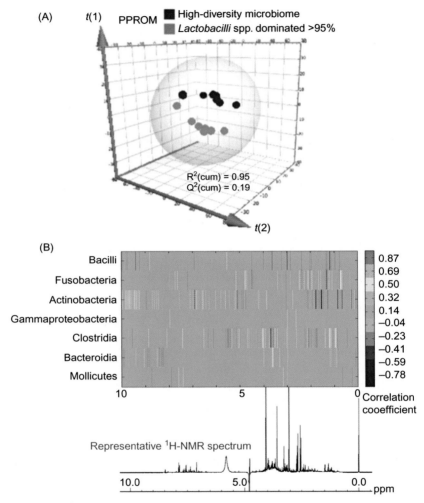

FIGURE 8.4   The urinary metabolome reflects system response changes that are associated with the vaginal microbiome. (A) PLS-DA of ¹H-NMR–derived urine metabolite profiles showing separation of profiles of patients with PPROM based on their vaginal microbiome. (B) Correlation coefficients between ¹H-NMR urine spectra and bacterial classes present in the vaginal microbiome. Specific peaks are increased (yellow-red) or decreased (blue) in the presence of specific vaginal bacterial classes.

8.5.5   Asymptomatic Bacteriuria

The bacterial environment in the urinary tract undergoes dramatic changes during pregnancy and asymptomatic bacteriuria, defined as a culture of 10^5 organisms/mL urine in the absence of symptoms, has been associated with

PTL, IUGR, PPROM, and pre-eclampsia [89]. Urinary metabolite profiles have been shown to identify and differentiate urinary tract infections [90] based on changes in the excretion patterns of nonspecific markers of bacterial metabolism such as acetate and lactate but also metabolites, including trimethylamine, scyllo-inositol, and 4-aminohippurate. Diagnosis of urinary tract infections (UTIs) generally takes 24–28 h, and antibiotics are often administered prior to obtaining the outcome. Burrillo et al. have recently devised an adaptation to the Gram staining method used to identify bacteria, by coupling this with a matrix-assisted laser desorption ionization–time of flight mass (MALDI-TOF) spectrometric method, which is able to detect the presence and type of bacteria present in urine [91]. In some communities, the presence of asymptomatic bacteriuria in pregnant women has been found to be as high as 18.8%, with staphylococci and *E. coli* representing the most common isolates found [92]. Thus, it is recommended that screening be conducted in all pregnant women. New metabolic profiling techniques based on rapid evaporative ionization mass spectrometry (REIMS) have been shown to provide a culture independent tool for the identification of clinically relevant bacteria [93] and may provide a more rapid and cheaper solution to population screening programs for asymptomatic bacteriuria (see Chapter 4 for more details).

8.5.6   Intrahepatic Cholestasis

Intrahepatic cholestasis of pregnancy (ICP) affects 7 in every 1000 pregnant women in the United Kingdom and is a metabolic disorder that typically develops in the third trimester of pregnancy [94]. The prevalence is population dependent, and a higher incidence is found in South Asians and South Americans (up to 15%) [94]. The condition is associated with maternal liver dysfunction and high blood bile acid levels causing symptoms of itch, premature labor, and sometimes stillbirth. Obese women are more prone to the condition, and both affected mothers and children are at increased risk of longer-term metabolic diseases such as diabetes and fatty liver. Although the causes of ICP are not well understood, genetics (eg, mutations in phospholipid transporters) and hormone levels, particularly estrogens, have been associated with development of the condition, and oral contraceptive use is also linked to its occurrence. Normal pregnancy is hypercholanemic [95,96], but for some women, the demands of pregnancy initiate a stronger response during this "metabolic switch" in late second and third trimesters, and physiologic homeostasis is disturbed, which results in metabolic diseases such as gestational diabetes and cholestasis. ICP is independently associated with further impairments of glucose homeostasis, gestational diabetes, and hypercholesterolemia [97]. Although ICP is not generally associated with immediate serious health consequences in the mothers, the likelihood of developing cirrhosis, pancreatitis, or gallstones later on in life is increased. ICP is metabolically characterized by alterations in bile acid metabolism, and a marked increase in serum cholic acid conjugated with taurine and glycine is known to occur [98]. Ursodeoxycholic acid is the most effective treatment for ICP [93]. Bile acids undergo enterohepatic recirculation and are modified by gut bacteria. Some research has suggested that gut bacteria may play a role in the onset of ICP, proposing inflammation-mediated leaky gut, with absorption of bacterial endotoxin as a mechanism [99]. The association between lipid and sterol metabolism and gut microbial metabolism has been well studied; for example, germ-free mice have significantly reduced transcriptional activity and reduced fatty acid metabolic processes in the gut [100]. It has also been shown that germ-free mice have 42% less body fat than conventional mice [101]. ICP is also associated with functional variants in the FXR bile acid sensor [102]. However, as yet, the role of the gut microbiome in bile acid metabolism in pregnancy has not been extensively studied. Ley et al. transplanted fecal samples from women in their first and third trimesters to germ-free mice and showed that mice implanted with third trimester feces gained became more obese and demonstrated reduced insulin sensitivity compared with those implanted with first trimester fecal samples [62]. There is clear scope for application of microbiomics and metabolic phenotyping to study the metabolic consequences and mechanisms of ICP.

8.5.7 Maternal Obesity

Many of the immune and metabolic changes seen during pregnancy reflect metabolic syndrome, which is a group of obesity-related metabolic abnormalities that can manifest as hyperlipidemia, insulin resistance, and increased adiposity. Adverse fetal and maternal outcomes have been associated with overweight and obesity in pregnancy, including increased risks for preterm delivery, preeclampsia, fetal death, and low Apgar scores [103]. In clinically obese mothers, the relative risk of giving birth to babies with low Apgar scores rises to 31% higher than in normal weight mothers [104]. In a recent US-based study on 2155 pregnant women, 29% of the women were clinically obese [105]. The infants of these women were significantly more likely to require ventilation on delivery. Epigenetic studies have shown that hypermethylation occurs at several DNA locations. This effect was stronger than any association found with paternal obesity and thus an in utero mediation of the effect was proposed [106]. Obesity during pregnancy has also been shown to increase long-term cardiovascular disease risk in the mothers [107]. Conversely, pregnancies occurring after bariatric surgery for weight loss were found to be associated with lower risk of gestational diabetes and lower incidence of large-for-gestational-age infants, although there was a positive correlation with small-for-gestational-age delivery and a trend toward a higher rate of neonatal deaths [108]. In addition to adverse perinatal outcomes, maternal obesity in pregnancy is also linked to increased risk of childhood, adolescent, and adult obesity, as well as metabolic syndrome in the offspring [109]. Although various studies report on different downstream health impacts of being born to obese mothers, there is widespread consensus that the in utero environment is a dominant determinant of long-term health, particularly with respect to cardiovascular disease [110]. Identifying metabolic pathways that are susceptible to fetal programming of disadvantageous body composition would open up new avenues for therapeutic intervention with potential for a massive gain in socioeconomic benefit. Inflammation during the pregnancy has been implicated as a driver for in utero programming of metabolic disease [111].

During a normal pregnancy, the immune status changes to adapt to the fetus. Initially, inflammation is triggered on implantation of the blastocyst, and this gives way to immunosuppression in order to protect against rejection of the fetus. In the later stages of pregnancy, the body again enters into an inflammatory state. Obesity in pregnancy is associated with a higher incidence of infection, and it has been found that obese pregnant women have impaired immune cell regulation and cytokine production compared with non-obese pregnant women, particularly with regard to IL-6 and tumor necrosis factor alpha (TNF-α) [7]. One proposal is that leptin, which is higher in obese individuals, induces the production of inflammatory cytokines such as IL-12, IL-10, and TNF-α [112]. In normal pregnancy the placenta can sense and adapt to the maternal inflammatory environment, and maternal inflammation does not always translate to fetal inflammation [113]. However, maternal obesity enhances the inflammatory response of the placenta. Adipose tissue in obese pregnant individuals tends toward inflammation, with an increase in CD68+ and CD14+ differentiated macrophages, which can affect the neurologic and hematopoietic systems [7]. Kozyrskyj et al. identified that high concentrations of *Lactobacillus* spp. and low concentrations of *Bacteroides* spp. colonizing the infant gut within the first 3 months in the postpartum period predicted the risk for infant and child overweight and proposed that the overweight pregnant mother may vertically transfer a specific microbial composition to the infant [114]. They also showed that boys are more susceptible to transfer of the obesity trait, which is consistent with the literature on later-life development of metabolic syndrome and other chronic conditions. SCFAs are produced by the gut microbiota and act as an energy source for colonic cells. Acetate, in particular, has been directly associated with maternal weight gain and inversely associated with maternal leptin levels and newborn body weight, underscoring the influence of the gut microbiome on pregnancy outcomes [65]. Much of the epidemiologic and clinical studies in humans are backed by animal studies. Diet-induced obesity in a range of animal species, including rodents and sheep, have shown that high-fat diets are associated with hyperglycemic and hyperinsulinemic fetuses and with reduced placental efficiency [115]. Placental fatty acid transport proteins, PPARγ, and toll-like receptors (TLR2, TLR-4), all of which are associated with inflammation were found in a sheep model of maternal obesity.

Thus, obesity in pregnancy may "prime" the basal level of inflammation such that a relatively benign inflammatory stimulus during pregnancy may result in major activation of inflammation, leading to various poor outcomes for mother and infant, and this may be partially modulated by the gut microbiome.

8.5.8 Gestational Diabetes

Gestational diabetes mellitus (GDM) is a complication of pregnancy, which affects approximately 4–7% of all pregnancies in Europe but can be as high as 18% in some European populations [116,117], and in some communities, for example, in northern India, the prevalence can be as high as 41.9% [118]. The propensity for developing GDM is higher with increasing body mass index (BMI) and maternal age and is associated with higher risk of infants developing macrosomia, higher incidence of birth defects and IUGR, higher likelihood of premature births, and greater percentage of intrauterine fetal deaths [119,120]. Overweight women are at a higher risk of developing GDM among other complications. In normal pregnancies, insulin resistance increases toward the end of the second trimester to levels comparable with type 2 diabetes (T2D), but most women compensate for this by higher insulin secretion from β-cells. Hormones released by the fetal placental unit and the accumulating maternal fat contribute to the increased insulin resistance. Development of GDM predisposes women to develop T2D, with up to 70% estimated to develop T2D within 15 years [117]. Increased pregnancy levels of inflammation are thought to play a role in the pathology of T2D. Although the inflammatory mechanisms associated with GDM are not fully understood, the NF-κB and I kappa B kinase (IKKβ) and c-Jun NH2-terminal kinase (JNK) pathways are thought to be involved in the development of insulin resistance [121].

Several studies using metabonomic technology to characterize GDM have been conducted, mostly with the aim of identifying earlier or more sensitive biomarkers of the condition. Huynh et al. provided a summary of metabolic phenotyping studies of GDM conducted in serum, urine, and amniotic fluid [122]. Serum characteristics that are associated with GDM include increased cholesterol, fatty acids, triglycerides, valine, pyruvate, lactate, and glucose with decreased concentrations of betaine; proline; trimethylamine-N-oxide; creatine; myristic, palmitic, linoleic, and arachidonic acids; and glutamate [122,123]. GDM-associated changes in the urine profile include decreased excretion of trimethylamine-N-oxide, 3-hydroxyisovalerate, 2-hydroxyisobutyrate, N-methylnicotinamide, and N-methyl-2-pyridone-5-carboxamide, with increased excretion of hippurate and 4-hydroxyphenylacetic acid. With regard to amniotic fluid, fewer studies have been conducted, and the changes attributed to GDM are higher concentrations of glucose and lower concentrations of amino acids, creatine, and glycerophosphocholine. In a multiplatform study combining NMR spectroscopy, LC-MS, and capillary electrophoresis mass spectrometry (CE-MS), the most striking changes in plasma composition was found to be in the depletion of lysophosphoglycerides in GDM. Tauro-conjugated bile acids were also present in lower levels in patients with GDM than in individuals going through a healthy pregnancy [124]. In urine, decreased excretion of carnitine was associated with GDM, whereas the excretion of other amino acids was found to be increased. Modulation of carnitines is a characteristic of fatty liver and several other hepatic pathologies.

The majority of the studies, to date, have been conducted on small groups of individuals, with almost a complete lack of validation cohorts, as many of the studies have not been carried out in fasted conditions. Therefore, although interesting associations between health during pregnancy and metabolism of both mother and infant have been flagged, a more systematic approach is required to translate the current observations into a tool that would have clinical relevance. Nevertheless, the technology is a promising approach to defining a characteristic panel of biomarkers that can be evaluated early in the pregnancy journey.

Since the microbiome is implicated in inflammation driven by the gut microbiota in pregnancy, several studies have proposed the use of probiotics to ameliorate this effect [125]. However, an analysis of several reported probiotic trials concluded that just one study had shown a reduction in the rate of GDM when probiotics were administered early in pregnancy [126]. They also reported that probiotic intake did not have an impact on the risk of miscarriage or neonatal death. Thus, more trials and validation of the effects of probiotic administration during pregnancy are warranted. One group has shown that the bacterial composition of the meconium in infants born to mothers with GDM is enriched by the same bacterial taxa as the maternal fecal microbiome, thereby giving rise to a potential channel for vertical transfer of susceptibility to GDM [37]. As mentioned previously, several urinary metabolites such as trimethylamine-*N*-oxide, hippurate, and 4-hydroxyphenylacetic reflect gut microbial activity and underscore the potential role of the gut microbiome in GDM pathology. A less tangible, but nevertheless intriguing, potential connection with the gut microbiome is that GDM has been associated with increased risk of autism after adjustment for BMI and other confounders [127]. There is, in fact, some evidence to support the suggestion that children with autism have a dysregulated microbiome, with specific changes in the clostridial species, leading to altered urinary excretion or cresols and other phenolic compounds [128].

Differences in the composition of colostrum and milk between healthy mothers and those with GDM point to a more persistent metabolic signature of GDM. A panel of 27 proteins has been shown to be predictive of GDM, suggesting that GDM has implications in modulating proteins involved in nutrition and immunity [129]. Further evidence of postpartum glucose dysregulation is the presence of increased glycosylation of lactoferrin in milk, which may influence the innate protective function of milk [130]. In a study that explored the metabolism of infants born to mothers with GDM, lower serum levels of glucose were found, along with relatively higher concentrations of lipoproteins, alanine, valine, and lysine, but no overt differences in clinical outcomes were detected in the infants [131]. However, early differences in metabolism may be predictive of disrupted metabolism in later life, and follow-up studies are required to investigate this hypothesis.

8.5.9 Pre-eclampsia

Pre-eclampsia is defined by high blood pressure during pregnancy and is accompanied by increased excretion of protein in urine. The condition usually comes to light in the second half of the pregnancy and is detrimental to both mother and fetus. Like GDM and IUGR, it is associated with placental dysfunction, wherein, typically, there is poor placental infusion and a tendency toward hypertension. Women who manifest pre-eclampsia double their risk of future heart disease and stroke [132]. Approximately 5% of pregnant women in the United Kingdom will develop pre-eclampsia, and the condition manifests in as many as 10% of all pregnancies worldwide [133].

In terms of the metabolic characteristics of pre-eclampsia, some of the profile overlaps with those of GDM, which is perhaps not surprising because they share a subset of clinical features and outcomes. Similarities include elevated serum levels of linoleic and other fatty acids and bile acids and increased ketoacids in comparison with healthy pregnancy serum [134]. Elevation of eicosanoids and other inflammatory markers is a key feature of the pre-eclampsia serum signature, and taurine is also found in higher concentrations [134]. Taurine has many biological functions and is known to be involved in neuronal processes, to act as a cellular osmolyte, and to be involved in cell membrane stabilization. More importantly, it is believed that taurine plays a role in placental trophoblast development and may be involved in remodeling of the spiral arteries early in pregnancy [69]. Diminished trophoblast invasion is consistent with reduced taurine in placental tissue, and it is known that taurine administration to spontaneously hypertensive rats elicits a reduction in blood pressure consistent with the antihypertensive effects of taurine [135].

A decrease in 5-hydroxytryptophan and adipic acid have been reported to differentiate between pre-eclampsia and healthy pregnancy samples in metabolic profiling studies carried out on blood samples by using LC-MS. One MS study reported a differential plasma profile based on four metabolites (alanine, hydroxyhexanoylcarnitine, phenylalanine, and glutamate) for women who went on to develop pre-eclampsia; the samples were obtained in the first trimester of pregnancy, indicating that metabolic dysregulation may precede the development of hypertension [136]. In a similar study using NMR spectroscopy as the analytical vehicle, citrate, glycerol, hydroxyisovalerate, and methionine were found to differ in the first trimester (11–13 weeks' gestation) in the plasma samples of women who later developed pre-eclampsia [137]. The same group showed that there was also a metabolic distinction between women who developed pre-eclampsia early and late in pregnancy, with glycerol and methylhistidine (both lower in late pre-eclampsia) carrying the most weight in the differential profile [138]. Models of reduced oxygen tension in placental villous explants have shown differences in the regulation of osmolytes such as threitol and 2-deoxyribose and in fatty acids such as hexadecanoic acid and in prostaglandin metabolism [139,140].

From several of the metabolic studies, pieces of evidence suggesting gut microbial involvement in the expression of pre-eclampsia are accumulating. For example, Austdal et al. found lower concentrations of 4-cresylsulfate, phenylacetylglutamine, and hippurate and higher concentrations of dimethylamine and trimethylamine-*N*-oxide in the urine of women with pre-eclampsia [141]. Formate and acetate excretion was also inversely correlated with pre-eclampsia, whereas branched chain amino acids and urea were found in higher concentrations in urine collected from the pre-eclampsia group. Markers of lipotoxicity have been associated with pre-eclampsia [142].

More direct evidence of the bacterial origins of pre-eclampsia is the fact that some of the placentas delivered from women with pre-eclampsia have been shown to harbor bacteria, whereas no bacteria have been found in control placental tissue [143]. In one such study, bacteria were identified in the placental tissue of 12% of women with pre-eclampsia, and the organisms isolated included *Listeria, Salmonella, E. coli* (associated with GI infection); *Klebsiella pneumonia, Anoxybacillus* (respiratory tract infections), *Variovorax, Prevotella, Porphyromonas,* and *Dialister* (periodontitis) [143]. In line with this, periodontitis is known to be associated with pre-eclampsia [144] and merits further investigation to determine whether this association could be an etiologic factor in the development of pre-eclampsia.

8.6   POSTNATAL FACTORS THAT INFLUENCE DOWNSTREAM HEALTH

8.6.1   Interaction Between the Infant Microbiota and Mother's Milk

The infant's ability to feed, parental beliefs, socioeconomic status, and environment all influence the way an infant is fed at birth and throughout infancy. For reasons of infant or maternal health as well as choice, artificial milk substitutes have been given to babies for centuries. The World Health Organization (WHO) recommends that infants be exclusively breastfed for the first 6 months of their lives [145]. This recommendation is mainly based on the fact that maternal milk confers immunity on the infant, and evidence has shown that breastfed infants have a significantly reduced risk of GI infections [145].

For the first few months of life, milk provides the entire nutrition for the infant and comprises macronutrients, including protein, lactose, total fat, saturated fatty acids, mono-unsaturated fatty acids, polyunsaturated fatty acids, linoleic acid (18:2 n-6), alpha-linolenic acid (18:3 n-3), arachidonic acid (20:4 n-6), docosahexaenoic acid (22:6 n-3), trans fatty acids; vitamins (retinol/carotene, D, E, C, thiamin, riboflavin, niacin, B_6, B_{12}, folate, pantothenate, and biotin); and minerals (calcium, phosphorus, magnesium, sodium, potassium, chloride, copper, iron, zinc, manganese, selenium, and iodine) [146]. In addition to these components, there are many other chemicals present in milk, and these change over time in response to the developing nutritional needs of the infant as well as the mother's diet.

Oligosaccharides, termed *human milk oligosaccharides* (HMO), are present at between 5–15 g/L of human milk. They comprise hundreds of types of molecules, composed of combinations of five monosaccharides (for a more complete review, see Ref. [147], Fig. 8.5). The types of HMOs can be host specific [148], and since they act as nutrient sources for the microbes and are, in fact, the baby's first prebiotics [147], different babies will be exposed to different HMOs. Moreover, it has been hypothesized that they also play roles beyond generating or promoting growth of members of the genus *Bifidobacterium*, generally associated with beneficial outcomes. The extra functions they have been proposed to provide to the baby include acting as antiadhesive antimicrobials and preventing pathogenic microbes from colonizing the gut [149]; modulators of intestinal epithelial cell

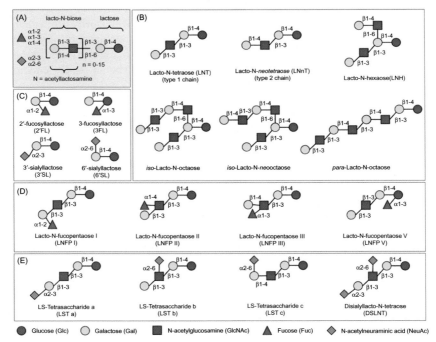

FIGURE 8.5 HMO blueprint and selected HMO structures. (A) HMOs follow a basic structural blueprint. (Monosaccharide key is shown at the bottom of the figure.) (B) Lactose can be fucosylated or sialylated in different linkages to generate trisaccharides. (C) Lactose can be elongated by addition of either lacto-N-biose (type I) or N-acetyllactosamine (type II) disaccharides. Addition of disaccharides to each other in the β1-3 linkage leads to linear-chain elongation (para-HMO); a β1-6 linkage between two disaccharides introduces chain branching (iso-HMO). (D) Elongated type I or II chains can be fucosylated in different linkages to form a variety of structural isomers, some of which have Le blood group specificity. (E) The elongated chains can also be sialylated in different linkages to form structural isomers. Disialylated lacto-N-tetraose (bottom right) prevents NEC in neonatal rats.

responses [150]; immune modulators [151]; protection against NEC [152]; and nutrients for brain development [153]. The composition of human milk has been characterized by several analytical platforms, including NMR spectroscopy, GC-MS, LC-MS, and CE-MS. The metabolic composition is most dynamic in the first 3 weeks, and a clear metabolic trajectory can be seen irrespective of the technology used. The compositional differences between milk samples taken at various time after birth can be visualized used multivariate modeling as shown in Fig. 8.6. As with most applications, the merits of each platform are different, and a combination of spectroscopic platforms offers the most comprehensive characterization of the biochemical components present in milk. There are questions as to whether it is effective to supplement breast milk with prebiotics to correct for any aberrant microbiome development associated with, for example, preterm birth.

Artificial milk undergoes constant improvements, with many manufacturers adding compounds known to have beneficial effects. For example, short-chain

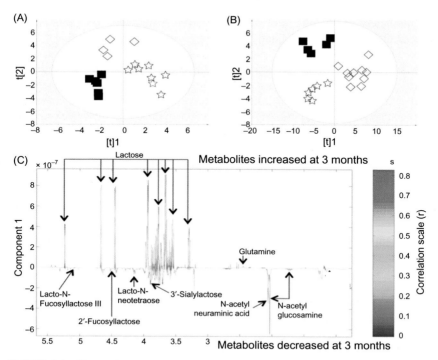

FIGURE 8.6   Changes in maternal milk composition in the period after delivery, as detected in the PLS-DA scores plots modeled on (A) GC-MS data, and (B) CE-MS data. White diamonds represent samples taken during days 3–5 post-partum, black squares show samples taken days 6–10 post-partum and white stars represent samples taken over a period beyond 10 days post-partum. The differential composition between milk obtained 7 days and 3 months after delivery is shown in panel (C) based on a ^{1}H NMR spectroscopic PLS-DA model, showing decreasing oligosaccharide concentration over time and increasing lactose concentration. *Figure supplied by Nicholas Andreas.*

galacto-oligosaccharides (GOSs) and long-chain fructo-oligosaccharides (FOSs) are added to the formula of most manufacturers' products, and others have incorporated long-chain polyunsaturated fatty acids such as docosahexaenoic acid and arachidonic acid, in response to research that has shown improved intelligence scores in infants receiving increased intakes of these compounds [154]. However, despite decades of research into improving the infant formula, the vast complexity of human breast milk cannot be captured by a standardized formula. The composition of milk constantly changes according to the needs of the infant and is dependent on time since last feed, length of lactation, age, and other factors, such as gender [155].

8.6.2 Weaning Food

The scope of metabonomic and microbiomic applications in nutrition is a vast field and is explored only briefly here. The tripartite association among metabolism, the diet, and the gut microbiota is key to the early origins of many diseases, and at present the exact relationship, and host–microbe signaling is incompletely understood at best. Various nutrimetabonomic (the study of the interaction between the diet and metabolism using metabolic profiling) studies have shown specific metabolic effects of specific dietary components, and a few studies have addressed the systemic metabolic consequences of complete diets such as the Mediterranean diet [156]. More recently, both nutrigenomic and nutrimetabonomic studies have explored the possibility of implementing personalized nutrition approaches, and research applications have been summarized in several seminal texts [157–159]. There are not many nutritional studies in children reported so far and even fewer in infants, but early dietary patterns are thought to be key to setting up long-term dietary habits and also metabolic imprinting. Some studies involving early nutrition point to some foods and nutrients being detrimental to metabolism or intestinal health. For example, iron fortification in Kenyan infants to reduce anemia was shown to cause intestinal inflammation and increase translocation and colonization of harmful bacteria in the gut [160]. Of the metabolic profiling studies conducted in children to date, several point toward the production of gut microbe–host co-metabolites as being a differentiating axis between specific diets. For example, in a study conducted in 8-year-old children given either milk or meat-enriched protein sources, the milk diet was found to lead to higher excretion of hippurate [161]. Similarly, high-calcium diets can influence the mucosal-associated gut bacteria in weaned pigs, and high-cereal diets have been shown to upregulate TLR-2 expression [162].

Several studies investigating weaning diets have shown the importance of the first encounter with diet. The pig is often used as a model for human nutrition studies because of the similarity in the GI systems. In one study investigating the effect of weaning diet on the metabolic profile, the metabolic phenotype of the pigs reflected whether the piglets were weaned onto a soy-based or an egg-based diet even after the piglets were all switched to a fishmeal-based diet for 4 weeks [163]. Thus, the metabolism of the pigs after acclimatizing to an adult diet held the metabolic memory of the first diet the pigs were exposed to weeks after they were changed to a standardized diet. Whether the piglets were weaned onto soy or egg was reflected in the biochemical composition of both urine and plasma, and the metabolites that differed between the two groups included metabolites such as enriched indoleacetylglycine, hippurate, and hydroxyphenylacetylglycine in the soy-based diet, which are all synthesized by the dual action of the mammal and its microbiota. Conversely, tricarboxylic acid cycle intermediates, particularly citrate, were enriched in the egg-based diet. Thus, it would appear that the initial diet influences the bacteria that colonize the gut and that, once established, these bacteria persist and are to some extent oblivious to subsequent changes in diet.

As mentioned, it would appear that metabolic syndrome and diabetes have their origins in early life, and perhaps even in utero. Thus, understanding the relationship between early life environment and later development of diabetes is paramount to combating these conditions, and there is an imperative to study the gene–environment interactions that contribute to their onset. The first signs of β-cell autoimmunity can occur in the first year of life. One of the critical questions is to ascertain whether the early postpartum period and infancy offer an opportunity for metabolic rescue of a poor in utero environment. Some research suggests that dietary supplementation with prebiotics or probiotics (eg, *Lactobacillus jensenii*) may help maintain intestinal homeostasis and promote growth and health [164]. The notion of functional foods ties is rather better with Eastern philosophies such as Traditional Chinese Medicine, which does not make a distinction between foods and medicine but instead views the patient and their environment as interconnected and attempts to treat conditions at a systemic level.

Probiotic interventions have been suggested for managing preterm birth, in pediatric and neonatal critical care, and as a protective measure against downstream effects where there has been incidences of GDM, obesity, or preeclampsia during pregnancy [165,166]. Given the complexity of the interactions of microbes with human metabolism, it is unlikely that a "one size fits all" approach to nutritional intervention will work, and we face the almost overwhelming task of characterizing the effects of diet on host–microbiome interactions. However, as we move forward into the twenty-first century with longitudinal cohorts and advanced technology, we are perfectly placed to embark on this high-energy, high-reward journey.

8.6.3 Infant Obesity

Infant and childhood obesity correlate with chronic conditions such as metabolic syndrome and cardiovascular disease, which manifest later in life. Childhood obesity is growing at an alarming rate and currently affects around 43 million preschool children [167]. There is some proof that the gut microbiota contribute to the early obesogenic environment, and some research has shown that specific metabolites such as plasma lysophosphocholine (14:0) correlate with rapid growth in infancy and with childhood obesity [168]. In a study in 4- to 5-year-olds, *Enterobacteriaceae* was present in higher proportions in the fecal samples of overweight children [169]. Koleva et al. presented a summary of studies looking at bacterial predictors in infants who later showed weight gain and/or obesity and cited colonization of *Bacteroidetes* in the first few months of life as an important determinant of later weight gain [167], with fewer fecal bifidobacteria and a greater proportion of *Staphylococcus* spp. at 6 months of age correlating with overweight in children at 7 years.

A particularly controversial topic, covered in more detail in later sections of this text, is the potential association between antibiotic exposure in infants and late weight gain. Although there are contradictory findings regarding the link between antibiotic exposure and weight gain, some studies find a clear and direct association. In one particular study, this association between antibiotics given in the first year of life and excess weight and central adiposity in preadolescence was only found to be significant in boys [170], which is consistent with sex-dependent differences in the development of metabolic syndrome in adults born preterm [171].

8.6.4 Malnutrition

Although overnutrition is a key driver of pathology in the developed world, undernutrition remains one of the biggest childhood killers in the world, accounting for approximately 45% of all deaths in children under 5 years [172]. Undernutrition can manifest in various forms, including acute forms: kwashiorkor (wasting with edema) and marasmus (wasting with loss of fat and muscle), and chronic conditions such as stunting. Just as transplantation of an obese microbiome into germ-free rodents can promote weight gain, transplantation of the microbiome from infants with kwashiorkor has been shown to induce weight loss in germ-free rodent models [173]. Metabolic dysregulation accompanying the transplantation of a kwashiorkor microbiome included changes in urinary excretion of methylamines (product of bacterial degradation of choline), amino acids, allantoin creatinine and creatine (muscle metabolism), and sebacic acid can be sensitive diagnostic factors [174]. Both undernutrition and overnutrition of the fetus have been shown to increase the risk of chronic pathologies such as metabolic syndrome, cardiovascular disease, and dyslipidemia in later life [175]. Children born to famine-exposed mothers have been shown to be at increased risk of diabetes mellitus and impaired fasting glucose in adulthood, and as with several obesity-related conditions, the effects were greater in men than in women [176]. Other studies have shown that the risk for developing T2D is up to 125% higher in individuals who have had early-life exposure to famine compared with those without famine exposure [177]. Both maternal undernutrition and overnutrition may regulate the expression of genes involved in lipid and carbohydrate metabolism, with further effects on metabolism being induced by epigenetic factors and by modulation of the microbiome, which is passed from mother to infant [178].

Studies in animal models have been used to advance the understanding of the mechanistic consequences of malnutrition. Pathways involving energy metabolism (tricarboxylic acid cycle intermediates), carbohydrate metabolism, protein metabolism (amino acids), purine metabolism, and muscle turnover (creatine) were all shown to be perturbed in rats in response to protein-energy malnutrition [179]. In a separate famine model involving malnourishment of pregnant female rats resulting in fetal retardation of neurodevelopment, a characteristic metabolic profile was seen in the amniotic fluid, involving the release of increased concentrations of glycine, inositol, putrescine, and rubidium, whereas the concentrations of methionine, dopa, tryptophan, zinc, cobalt, and selenium decreased [180].

Several texts have demonstrated the effect of malnutrition on the microbiome and allude to the potential of manipulating the microbiome to achieve or support recovery from the malnourished state. For example, several opinion papers have described the opportunities offered by probiotics to combat malnourishment, enteric infections, and exposures to toxins, which populations in developing countries are often exposed to [181,182]. Malnutrition has been shown to induce dysbiosis in animal models, and tryptophan restriction, through lack of dietary protein, has been shown to result in vitamin B_3 deficiency, leading to a reduction in the production of ileal epithelial antimicrobial peptides and thus to diarrhea [172]. Kane et al. summarized several microbial studies conducted in malnourished infants and children. Their observations included higher amounts of *Proteobacteria*, specifically *Campylobacter* spp., and in general a lower diversity of bacterial species in malnourished children. Higher diversity is associated with improved resistance to pathogens. Conversely, *Bifidobacterium longum* was enriched in children who were not malnourished [172].

A body of research focused on the use of functional foods to correct malnutrition has come up with mixed results. Although the microbiome and metabolism respond to high-protein diets, there is an indication that the changes are temporary and quickly revert once the diet is withdrawn or that the microbiome is only partially ameliorated by nutritional intervention [173,183].

8.6.5   Infection and Diarrhea

Diarrhea is a condition involving the passage of loose, watery stools more frequently than normal and is associated with various medical conditions, the most common being GI infection. This condition can quickly lead to dehydration, and chronic diarrhea is associated with malnutrition in the developing world. Inadequate sanitization, unclean drinking water, and cramped living conditions promote the spread of enteric infections, which reduce the barrier function of the intestine. Multiple episodes of diarrhea in children in their first 2 years of life has been associated, on average, with a 8 cm reduction in height and a 10-point lower IQ level [184]. In later life, stunted growth has been shown to predispose to obesity and its co-morbidities.

In one MS-based study in malnourished Ugandan children, aged 6 months to 5 years, high plasma levels of nonesterified fatty acids, ketones, and even-chain acylcarnitines were found, together with high levels of ghrelin, growth hormone, cortisol, IL-6, peptide YY (PYY), and glucagon-like peptide 1. Conversely, plasma albumin, amino acids, C3 carnitine, insulin, adiponectin, and leptin were low. Administration of a high-protein diet was found to reverse these metabolic differences between malnourished and healthy children [185].

In a study comparing children under 5 years of age with enteric diarrhea in two locations in Colombia, although age, location, and gender were all found to affect the fecal microbiome, *Bifidobacterium* and *Lactobacillus* species were found to be inversely correlated with incidence of diarrhea leading to the proposal to introduce probiotics or a diet rich in bifidogenic components as a prophylactic measure [186]. In a separate study of almost 1000 children with moderate to severe diarrhea across Africa and Southeast Asia, *Escherichia/Shigella*, *Granulicatella* species, and *Streptococcus mitis/pneumoniae* groups were positively associated with diarrhea [187].

A study of acute diarrhea in dogs combined metabonomics and microbiomics and found that acute diarrhea was characterized by a lower diversity of gut bacteria with overrepresentation of *Clostridium* spp. and underrepresentation of *Bacteroidetes* and *Faecalibacterium*, corresponding to lower fecal concentrations of propionate [188]. Serum levels of indoles and kynurenic acid were also decreased in the acute diarrhea group. Thus, again, the importance of early acquisition and continued maintenance of a "healthy" microbiome is highlighted.

In Section 8.6.1, we discuss the fact that human milk contains bioactive factors such as oligosaccharides, secretory immunoglobulins, and lactoferrin, which have no direct nutritional benefit but contribute to survival and healthy development by either promoting healthy bacteria or protecting against pathogens. Breastfeeding has been shown to be inversely correlated with the number of episodes of diarrhea [189]. Clinical studies have shown that probiotics have some efficacy in the treatment of infants with necrotizing enterocolitis and for treating allergic disorders in children [190], but studies to date are limited and based on inadequate sample sizes. Fecal transplantation of the gut microbiota has been used with some success in several diarrhea-associated conditions, the most well known of these being persistent *Clostridium difficile* infections in adults [191]. However, this procedure has also been used for treating neonatal infections such as severe pseudomembranous enteritis. In one case study, a 13-month-old infant with persistent diarrhea and malnutrition, who was treated unsuccessfully with several courses of antibiotics, underwent fecal microbial transplantation, after which the diarrhea resolved [192].

8.6.6   Antibiotics, Cognition, and the Gut–Brain Axis

Fleming is widely acknowledged as the father of modern antibiotics after his discovery of penicillin. However, the growing tsunami of antibiotic resistance, associated with antibiotic overuse, threatens to move medicine back to the nineteenth century if we cannot find ways to overcome antibiotic resistance. Antibiotics are designed to destroy pathogens and can indiscriminately wipe out bacteria, even when the agents are targeted toward certain bacterial classes. This arbitrary targeting means that good, as well as bad, bacteria are destroyed and the intestinal environment is disturbed. Antibiotic resistance is growing and can be passed from mother to infant. However, there is also evidence to support the development of antibiotic resistance in neonates, independent of maternal influence. For example, a study in healthy twin infants with the aim of exploring the fecal antibiotic resistome of healthy amoxicillin-exposed and antibiotic-naive twins and their mothers followed up the subjects for 12 months after delivery. A diverse fecal resistome developed in the infants by 2 months of age and was distinct from the maternal resistome [193]. A proportion of the infants studied developed resistance to clinically important broad-spectrum β-lactam antibiotics, including piperacillin-tazobactam, aztreonam, and cefepime, even when their mother did not display resistance to these antibiotics.

As discussed earlier in this chapter, around 10–30% of the metabolites visible in a ^{1}H NMR spectrum of urine involve a microbial contribution to their synthesis. Administration of antibiotics to animals and humans results in a dramatic change in the spectral profile (Fig. 8.7), with disappearance of hippurate, phenylacetylglutamine (humans) or glycine (rodents), 4-cresyl glucuronide, and sulfate, and indoleacetyl glycine, as described in Section 8.4. Once administration of antibiotics has ceased, it takes approximately 3 weeks for the metabolic profile to return to normal, around the same time as it takes for a germ-free animal to develop a profile that approximates a conventional animal. Stool composition alters metabolically and recovers within a similar time frame, and the metagenomic profile or the resident bacteria in the gut follow a similar temporal trajectory. Apart from the obvious drawbacks of acquired resistance, the overuse of antibiotics has been associated with obesity and the potential to cause depression, inflammatory diseases, and even neurobehavioral conditions such as autism. Generally, in adults, the microbiota returns to equilibrium in a matter of weeks after a dose of antibiotics. However, for the first 12–24 months of life, the infant microbiome is much more plastic, and stochastic events can lead to differential colonization of the intestine after antibiotic administration. For example, there is a correlation between childhood inflammatory bowel disease and the number of courses of antibiotics taken, particularly when children under the age of 12 months have received more than two courses of antibiotics that target anaerobic bacteria [194,195]. There are growing concerns that maternal use of antibiotics during pregnancy can also affect the offspring, and in a New York cohort of 436 mother–child dyads followed

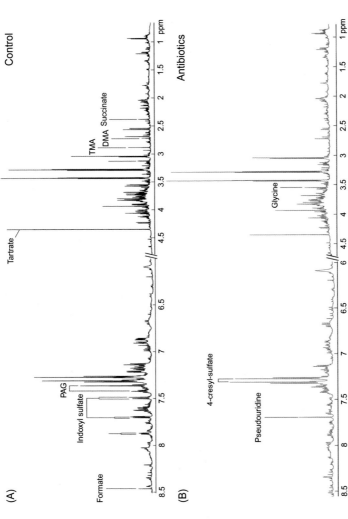

FIGURE 8.7 The effects of antibiotic treatment on the urine composition of a mouse: (A) Before treatment. (B) 48 h after treatment, showing decreased urinary excretion of hippurate, dimethylamine, trimethylamine, succinate, formate, and indoxyl sulfate as visualized using 600 MHz ^{1}H NMR spectroscopy following antibiotic administration. *Figure supplied by Dr. Jonathan Swann.*

up until 7 years after birth, children exposed in utero to antibiotics in the second and third trimesters of pregnancy were 84% more likely to be obese, with similar statistics for percentages body fat and waist circumference percentages [196]. An association between childhood obesity and birth by C-section was also noted in this cohort, underscoring the early life influence of the microbiome. In a separate study conducted in Taiwan, a positive relationship between acetaminophen and/or antibiotic exposure during the first year of life and the subsequent development of atopic dermatitis, asthma, and allergic rhinitis was observed [197]. Antibiotics, predominantly from *Streptomyces* spp., are widely used in agriculture to promote growth in animals bred for meat. The literature over the last decade is peppered with examples of the modulated associations between the microbiota and either weight gain or weight loss. However, the mechanism for microbial modulation of weight, as well as its role in obesity, remains elusive. Since there is no consensus on the microbes that trigger obesity despite numerous attempts at replication of studies such as those described by the Gordon laboratory [32,100], one must assume that the relationship between host and microbiome in influencing adiposity is a complex and a multifactorial one. Modulation of the microbiota by antibiotics is known to modulate bile acid synthesis and glucose metabolism in mice. In humans, oral vancomycin reduced fecal microbial diversity and decreased the *Firmicute/Proteobacteria* ratio [198]. As with the rodent models, the antibiotic decreased insulin sensitivity and bile acid dehydroxylation. Amoxycillin, on the other hand, did not affect any of these parameters. With regard to the connection between antibiotics and obesity, some researchers have even suggested that chronic exposure to low levels of antibiotics present in water bodies, coming from changing farming practices, could be responsible for the epidemic rise in obesity in the United States [199]. Increasing levels of antibiotics in water bodies have been proposed to contribute to the growing phenomenon of antibiotic resistance. Antibiotic resistance genes (ARGs) are continuously discharged into water bodies via human and animal waste. Freshwater reservoirs provide potential vectors for ARGs may create opportunities for gene transfer to human and animal pathogens via the food chain or drinking water. Although the levels of quantifiable ARGs were generally low, sulfonamide genes were detected in all sampled lakes in one study carried out in Switzerland [200]. Similar findings have been reported for *Acinetobacter* spp., where it has been concluded that although there is a low prevalence of acquired antibiotic resistance in water, there is potential to acquire and disseminate resistance via drinking water [201].

Against this background of the metabolic consequences of antibiotic use, one question of global interest is whether exposure to antibiotics in infancy predisposes individuals to obesity or metabolic syndrome in later life. In a large secondary analysis of 74,946 children aged 5–8 years, it was established that antibiotic exposure in the first 12 months of life was associated with increased BMI in boys [202]. In another study of over 11,000 children in the United Kingdom, it was shown that antibiotic administration prior to 6 months of age was consistently associated with increased BMI from 10–38 months [203]. Thus, it has been suggested that modulation of the infant microbiome by antibiotics can program the adult body composition, but the evidence is not clear. For example, in one study, amoxicillin was administered to rats prior to weaning, and although the authors observed a transient change in the microbiome and an increase in food consumption, these effects did not persist for long after cessation of antibiotic treatment [204].

In a study of 397 pregnant women, use of antibiotics other than penicillins was associated with lower birth weight. In this study, it was shown that there was an association between antibiotic intake and several differentially methylated regions of genes, of which one was correlated with birth weight, suggesting that antibiotics can influence methylation on imprinted genes [205].

In addition to the association between antibiotic exposure and body weight, antibiotic exposure in utero and early life has been linked to neurodevelopment and cognition. Autism spectrum disorder (ASD), first described by Kanner in 1943, represents a group of developmental conditions characterized by impairment in verbal and nonverbal communication and social interactions. It is also associated with repetitive behavior and inability to adopt imaginative play [206]. Autism is part of the pervasive developmental disorders family and typically manifests during the first 3 years of life, being present from birth or manifesting later after apparently normal neurobehavioral development (regressive autism). The etiology of ASD is not known, but there is a general consensus that both genetics and environmental factors contribute to development of ASD. Hypotheses regarding the etiology of ASD include exposure to chemicals (eg, pesticides) in utero, oxidative stress, vaccination or its thiomersal carrier, genetic causes (particular association with the X chromosome), advanced maternal age, planned C-section, low Apgar scores, hyperbilirubinemia, and birth defects [207,208]. Inflammation has also been implicated as a trigger for autism, potentially mediated by a defective placenta, an immature blood–brain barrier, the immune response of the mother to infection while pregnant, a premature birth, encephalitis in the child after birth, or a toxic environment either in utero or in the first few months of life [209]. Theories regarding the measles, mumps, and rubella vaccine have been largely discredited, but for every suggested cause, there is typically a range of contradictory evidence. It is therefore evident that ASD represents a complex set of interrelated conditions and is likely to originate from a highly multifactorial etiology. This complexity is further supported by the vast array of co-morbidities that typically accompany autism, in particular GI dysfunction. Understanding the role of the gut microbiota in autism would undoubtedly offer new therapeutic avenues, regardless of whether the intestinal dysfunction is causal or merely coincidental.

Given that postnatal microbial colonization of the GI tract overlaps with a critical period of brain development and maturation, it is feasible that changes in the composition of the microbiota during infancy may potentially affect consequences of brain function later in life. In this context, it is interesting to note

(1) onset of autism often follows treatment with antibiotics; (2) many children with late-onset autism (18–24 months of age) have a history of extensive antibiotic use; and (3) certain antibiotics such as vancomycin have been reported to decrease symptoms of autism, with a relapse following cessation of treatment. In addition, a 10-fold increase in certain bacteria in stool samples from children with autism compared with those from healthy controls has been observed. There have been allegations that the onset of autism, or ASD, could be associated with antibiotic use and a dysfunctional microbiome in infancy, particularly since studies have linked GI symptoms with regressive autism. The whole field of autism has been tainted with poorly designed studies, wild theories, and confounding factors such as strong dietary preferences, dietary restrictions, and administration of numerous supplements. Several studies have identified higher incidences of ear infections (and subsequent courses of antibiotics) and fevers in children with ASD [210]. There is a correlation between the introduction of amoxicillin in the 1980s and the increase in the number of children diagnosed with ASD, and this correlation cannot be accounted for by improved diagnosis of ASD. Conversely, other studies have provided evidence that antibiotic therapy can, in fact, improve behavior in children with autism. In a single-blind placebo trial of D-cycloserine in 10 children, there was a significant reduction in withdrawal and an improvement in social interaction [211].

The link between the microbiota and ASD is further strengthened by the fact that GI dysfunction is reported in a high percentage of children diagnosed with autism, although symptoms vary from diarrhea to constipation. Increased intestinal permeability has been associated with individuals with ASD (36.7%) in comparison with healthy controls (4.8%), but a casein-free diet was found to correct this [212]. Interestingly, family members of children with autism also showed increased intestinal permeability.

Several studies have characterized the microbiome of cohorts of children with autism and found higher levels of clostridia species [213,214], in particular *C. histolyticum*. Clostridia are known to produce toxins and often flourish in a perturbed gut ecosystem. Metabolic profiling studies have also generated indirect evidence of a shift in clostridial species. 4-Cresol is a gut microbial metabolite that can be synthesized by several bacteria, including *C. difficile* and *C. scatalogens*. Urinary excretion of 4-cresyl sulfate (the main phase II conjugate produced in the liver) is increased in children with autism and, to a lesser extent, in their siblings [127]. Other studies have identified increased amounts of urinary 3-(3-hydroxyphenyl)-3-hydroxypropionic acid in children with ASD, another clostridial metabolite, which is thought to be a metabolite of the tyrosine analog 3-hydroxyphenylalanine [215]. In addition to an increased presence of clostridia, a shift at the phylum level toward an increased *Bacteroidetes/Firmicute* ratio has been reported in the feces of children with ASD, along with high levels of *Desulfovibrio* spp. [214]. In autistic children with GI symptoms, bacteria of the *Sutterella* genus, predominantly *Sutterella wadsworthensis* and *S. stercoricanis* were prevalent in the epithelial mucosa [216].

Unfortunately, although there are tantalizing connections between ASD and a dysfunctional microbiome, studies to date suffer from a low sample size and poor comparison between children with autism who do or do not have GI dysfunction, and no causal link can be drawn. One possibility is that before the blood–brain barrier fully closes, infants are more susceptible to toxic metabolites produced by an aberrant microbiome, but several researchers argue against this hypothesis [217].

The metabolic phenotype of children with autism has been found to differ from that of nonaffected children with bacterial products of choline degradation such as methylamine and dimethylamine, hippurate, phenylacetylglutamine, reinforcing the possible involvement of the microbiota in autism. In contrast, Lis et al. reported lower urinary levels of hippurate and 4-hydroxyhippurate in children with autism compared with age-matched controls, as measured by ion exchange chromatography [218]. Increased excretion of nicotinic acid metabolites (N-methyl-2-pyridone-5-carboxamide, N-methyl nicotinic acid, and N-methyl nicotinamide) has also been associated with autism. Nonaffected siblings of children with ASD are often used as the control group for comparison, since their environment is better matched to that of the cases. However, several studies have indicated that the metabolism of siblings is also different from nonrelated age matched controls and, as mentioned, family members of individuals with ASD have been shown to have an increased prevalence of leaky gut.

Another suspect in the etiology of ASD is acetaminophen (paracetamol). The drug Calpol is commonly administered for fever to infants and children and can activate the endocannabinoid system, which is known to be capable of modulating brain function during development [219]. The primary route of acetaminophen excretion in children is sulfation. Sulfur deficiency is another common feature of ASD, and low levels of reduced plasma glutathione and sulfur-containing amino acids is consistent with this observation [220]. Since the primary route of acetaminophen excretion in children is sulfation, it may serve to further deplete the already low sulfur pools in the body, which are challenged by high levels of cresol production by the clostridia-rich microbiota. There is some evidence that dietary supplementation with glutathione may improve the levels of trans-sulfuration metabolites to some extent [221].

There is a strong unmet need for prospective studies that profile at-risk infants, for example, those born to families with one or more siblings with ASD, who have been shown to be more at risk for developing ASD, particularly in the case of boys [222] or preterm infants [223]. The metabolic phenotypes of affected children with ASD are distinct from nonaffected individuals, but the children in most studies tend to fall into the age range of 3–10 years. What is required is an early marker for autism risk. It has been proposed that the ability to mitigate autism or its symptoms is directly correlated with the age when behavioral therapy is given [223]. Since the microbiome is passed from mother to baby, there is also a possibility that the risk for autism could be predicted during pregnancy.

8.7 PROPOSED STRATEGY FOR A METABONOMIC FRAMEWORK FOR MONITORING MATERNAL AND INFANT HEALTH

In most developed nations, the structure for antenatal care typically begins with a booking visit and an ultrasonography scan toward the end of the first trimester, with subsequent visits spaced approximately 4 weeks apart until week 28 when the frequency of appointments typically have a fortnightly pattern [121]. This frequency is largely to ensure detection of pregnancy complications such as gestational diabetes, cholestasis, pre-eclampsia, or IUGR, which tend to manifest toward the end of the second or the third trimester. On the basis of medical history and clinical parameters, women are stratified into high risk or low risk, but the tools for stratification are relatively imprecise and ineffective. There is a clear unmet medical need for improved risk stratification. During pregnancy, most women tend to be heavily invested in the health of their baby and, as a consequence, their own health. In general, the pregnancy journey offers unparalleled opportunities for detailed sampling of biometrics and biofluids, enabling detailed profiling of the maternal journey leading up to birth of the infant. This sampling provides not only an immediate opportunity to relate pregnancy events to neonatal outcomes such as prematurity and acute bacterial infections but also a resource for collecting metabolic data that may be predictive of later life events, for example, in investigating the proposed association between low birth weight and metabolic syndrome. Moreover, as discussed in previous sections, maternal health in pregnancy is closely related to the development of clinical outcomes in later life, with gestational diabetes increasing the risk of developing T2D and pre-eclampsia being a risk factor for developing hypertension. Therefore, maternal biofluid samples may be predictive of downstream disease risk, and more research is warranted to pursue this hypothesis. We propose that an integrated mapping of the urine, fecal, and plasma metabolite profiles and the fecal microbiome during the pregnancy journey will ultimately enable cost-effective monitoring of individual pregnancy journeys and provide rich biobanks for progressing our understanding of the role of the microbiome in human development and health.

8.7.1 Challenges of Metabolic Profiling of Infant Biosamples

Obtaining appropriate urine, serum, and fecal samples for metabolic profiling from infants can present a significant analytical challenge. Samples volumes are, by nature, smaller than those obtained from adults, and parental consent for sampling can be difficult to obtain for both healthy and preterm infants. Although fecal samples are relatively easy to obtain noninvasively directly from the diaper, the length of time a sample has been exposed to room temperature and contamination from urine and potentially fecal matter are difficult to ascertain. The cleanest way to catch urine samples is to secure a plastic sampling bag to the infant by using an adhesive. Often, parents find this procedure unacceptable because the bag can irritate the baby's skin. Therefore, the most common way to obtain a sample of infant urine is to either place a cotton wool swab in the diaper and extract the urine by squeezing the cotton wool or by squeezing the diaper directly. These methods do not, however, prevent contamination of the sample by fecal material, and additionally chemicals from the diaper and the cotton wool, such as propylene glycol, acrylate, and tertbutanol, have been shown to leach into the sample (Fig. 8.8). To minimize the amount of blood taken from the infant, blood collection is often achieved by collecting a single drop of blood

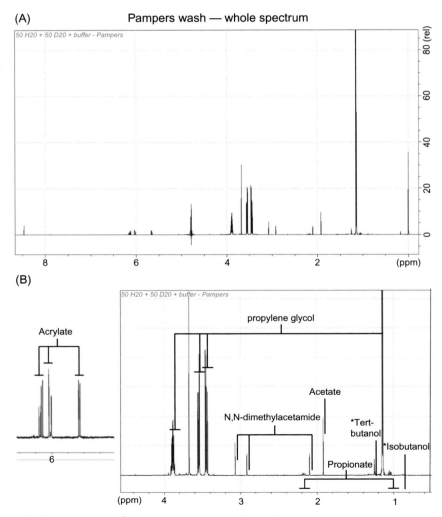

FIGURE 8.8   600 MHz ^{1}H NMR spectrum of water flush of a diaper showing chemical contaminants that leach into urine samples. *Figure supplied by Dr. Michael Kyriakides.*

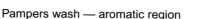

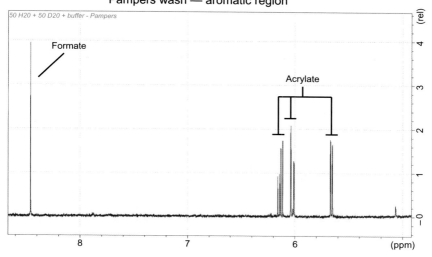

FIGURE 8.8   (Continued)

from a heel prick directly onto filter paper. Although chemical information on numerous compounds found in blood can be obtained, many of the biochemical components, including volatiles and eicosanoids, are lost. Therefore, sample handling and analytical spectroscopy protocols have been optimized to accommodate low sample volumes (Elaine Holmes, to be published).

Compositional analysis of infant biofluids, particularly serum, and also urine has been somewhat limited, and most metabolic profiling studies have focused on adults and children rather than neonates and infants for a variety of practical and ethical reasons. Likewise, only a few metabolic profiling studies that comprehensively describe the dynamic changes in biofluids throughout the course of pregnancy have been published. Thus, one clear research need is to provide detailed metabolic and microbial trajectories for healthy pregnancies and neonates, which can be used as a database or baseline on which systemic differences related to pathologic perturbations can be mapped and understood.

8.8   SUMMARY

The legacy of health begins before birth, with multiple factors shaping the combinatorial outcome of metabolic and microbial development in the infant. These factors are layered across the genetic, epigenetic, metagenomic, and environmental exposures of the parents. Understanding the influence and interactions of these factors can help us shape a better future with regard to the health of the next generation. For example, there is potential to modify the vaginal microbiome associated with adverse pregnancy outcomes by antibiotic or prebiotic/

probiotic intervention; better nutritional strategies can help improve the development of a healthy metabolism and the friendly gut microbiome; and the long-term effects of antibiotic administration in preterm infants can potentially be ameliorated by manipulation of the microbiome. Metabolic profiling has a strong role in discovering predictive biomarkers of downstream health and disease, not only in its own right as a method for elucidating chemical patterns associated with specific conditions but also in its application as a functional genomic tool for assessing the metabolic activity of the developing gut microbiome. In general, humans, particularly mothers, are more strongly invested in the health of themselves and their offspring than at any other point in life, and coupled with a well-structured antenatal program, this offers unparalleled research opportunities for understanding the origins of modern diseases.

REFERENCES

[1] Barker DJ, Osmond C, Golding J, Kuh D, Wadsworth ME. Growth in utero, blood pressure in childhood and adult life, and mortality from cardiovascular disease. Br Med J 1989;298(6673):564–7.

[2] Barker DJ. Fetal origins of coronary heart disease. Br Med J 1995;311(6998):171–4.

[3] Alexander BT, Dasinger JH, Intapad S. Fetal programming and cardiovascular pathology. Compr Physiol 2015;5(2):997–1025.

[4] Krauss-Etschmann S, Meyer KF, Dehmel S, Hylkema MN. Inter- and transgenerational epigenetic inheritance: evidence in asthma and COPD? Clin Epigenetics 2015;7(1):53.

[5] de Oliveira Andrade F, Fontelles CC, Rosim MP, de Oliveira TF, de Melo Loureiro AP, et al. Exposure to lard-based high-fat diet during fetal and lactation periods modifies breast cancer susceptibility in adulthood in rats. J Nutr Biochem 2014;25(6):613–22.

[6] Hartwig IR, Sly PD, Schmidt LA, van Lieshout RJ, Bienenstock J, Holt PG, et al. Prenatal adverse life events increase the risk for atopic diseases in children, which is enhanced in the absence of a maternal atopic predisposition. J Allergy Clin Immunol 2014;134(1):160–9.

[7] Rizzo GS, Sen S. Maternal obesity and immune dysregulation in mother and infant: a review of the evidence. Paediatr Respir Rev. 2014;16:251–7.

[8] Correia-Branco A, Keating E, Martel F. Maternal undernutrition and fetal developmental programming of obesity: the glucocorticoid connection. Reprod Sci 2015;22(2):138–45.

[9] Debnath M, Venkatasubramanian G, Berk M. Fetal programming of schizophrenia: select mechanisms. Neurosci Biobehav Rev. 2015;49:90–104.

[10] Howles SA, Edwards MH, Cooper C, Thakker RV. Kidney stones: a fetal origins hypothesis. J Bone Miner Res. 2013;28(12):2535–9.

[11] Wu G, Imhoff-Kunsch B, Girard AW. Biological mechanisms for nutritional regulation of maternal health and fetal development. Paediatr Perinat Epidemiol 2012;26(Suppl. 1):4–26.

[12] Bhattacharya S, Raja EA, Mirazo ER, Campbell DM, Lee AJ, Norman JE, et al. Inherited predisposition to spontaneous preterm delivery. Obstet Gynecol. 2010;115(6):1125–33.

[13] Modi N, Clark H, Wolfe I, Costello A, Budge H, writing group of the Royal College of Paediatrics and Child Health Commission on Child Health Research Goodier R, et al. A healthy nation: strengthening child health research in the UK. Lancet 2013;381(9860):73–87.

[14] Garner AS, Shonkoff JP, Committee on Psychosocial Aspects of Child and Family Health; Committee on Early Childhood, Adoption, and Dependent Care; Section on Developmental

and Behavioral Pediatrics. Early childhood adversity, toxic stress, and the role of the pediatrician: translating developmental science into lifelong health. Pediatrics 2012;129(1):e224–31.

[15] Modi N, Clark H, Wolfe I, Budge H, writing group of the Royal College of Paediatrics and Child Health Commission on Child Health Research. A healthy nation: strengthening child health research in the UK. Lancet 2013;381(9860):73–87.

[16] Denne SC, Hay Jr. WW. Advocacy for research that benefits children: an obligation of pediatricians and pediatric investigators. JAMA Pediatr 2013;167(9):792–4.

[17] World Health Organization EVIPNet (Evidence Informed Policy Network); this was initiated by WHO and the Ministries of Health in 25 countries in 2005; Vision: a world in whichpolicy-makers use the best scientific evidence, contextualized to the reality of their nation, to inform policy-making and policy implementation). <http://www.who.int/evidence/about/evipnet/en/>

[18] Oxman AD, Lavis JN, Lewin S, Fretheim A. SUPPORT tools for evidence-informed health policymaking (STP) 1: what is evidence-informed policymaking? Health Res Policy Syst 2009;7(Suppl. 1):S1.

[19] Ward Platt M, Deshpande S. Metabolic adaptation at birth. Semin Fetal Neonatal Med 2005;10(4):341–50.

[20] Pégorier JP, Girard J. Liver metabolism in the fetus and neonate: in principles of perinatal—neonatal metabolism. New York: Springer-Verlag; 1998. p. 601–26.

[21] Kalhan SC, Parimi P, Van Beek R, Gilfillan C, Saker F, Gruca L, et al. Sauer estimation of gluconeogenesis in newborn infants. Am J Physiol Endocrinol Metab 2001;281(5):E991–7.

[22] Sasaki H. Development of bile acid metabolism in neonates during perinatal period part 1: bile acid levels in sera of mothers, fetuses and neonates. Pediatr Int 1984;26(2):150–60.

[23] Wang Z, Li M. Evolution of the urinary proteome during human renal development and maturation. Adv Exp Med Biol 2015;845:95–101.

[24] Charlton JR, Norwood VF, Kiley SC, Gurka MJ, Chevalier RL. Evolution of the urinary proteome during human renal development and maturation: variations with gestational and postnatal age. Pediatr Res 2012;72(2):179–85.

[25] Trump S, Laudi S, Unruh N, Goelz R, Leibfritz D. ^{1}H-NMR metabolic profiling of human neonatal urine. MAGMA 2006;19(6):305–12.

[26] Foxall PJ, Bewley S, Neild GH, Rodeck CH, Nicholson JK. Analysis of fetal and neonatal urine using proton nuclear magnetic resonance spectroscopy. Arch Dis Child Fetal Neonatal Ed 1995;73(3):F153–7.

[27] Gu H, Pan Z, Xi B, Hainline BE, Shanaiah N, Asiago V, et al. ^{1}H NMR metabolomics study of age profiling in children. NMR Biomed 2009;22(8):826–33.

[28] Guneral F, Bachmann C. Age-related reference values for urinary organic acids in a healthy Turkish pediatric population. Clin Chem 1994;40(6):862–8.

[29] Moltu SJ, Sachse D, Blakstad EW, Strømmen K, Nakstad B, Almaas AN, et al. Urinary metabolite profiles in premature infants show early postnatal metabolic adaptation and maturation. Nutrients 2014;6(5):1913–30.

[30] Dessì A, Atzori L, Noto A, Visser GH, Gazzolo D, Zanardo V, et al. Metabolomics in newborns with intrauterine growth retardation (IUGR): urine reveals markers of metabolic syndrome. J Matern Fetal Neonatal Med 2011;24(Suppl. 2):35–9.

[31] Fanos V, Van den Anker J, Noto A, Mussap M, Atzori L. Metabolomics in neonatology: fact or fiction? Semin Fetal Neonatal Med 2013;18(1):3–12.

[32] Hooper LV, Xu J, Falk PG, Midtvedt T, Gordon JI. Proc Natl Acad Sci USA 1999;96:9833.

[33] Palmer C, Bik EM, DiGiulio DB, Relman DA, Brown PO. Development of the human infant intestinal microbiota. PLoS Biol 2007;5(7):e177.

[34] Niers L, Martín R, Rijkers G, Sengers F, Timmerman H, van Uden N, et al. The effects of selected probiotic strains on the development of eczema (the PandA study). Allergy 2009;64(9):1349–58.

[35] Mackie RI, Sghir A, Gaskins HR. Developmental microbial ecology of the neonatal gastrointestinal tract. Am. J. Clin. Nutr. 1999;69:S1035–45.

[36] Thum C, Cookson AL, Otter DE, McNabb WC, Hodgkinson AJ, Dyer J, et al. Can nutritional modulation of maternal intestinal microbiota influence the development of the infant gastrointestinal tract? J Nutr. 2012;142(11):1921–8.

[37] Hu J, Nomura Y, Bashir A, Fernandez-Hernandez H, Itzkowitz S, Pei Z, et al. Diversified microbiota of meconium is affected by maternal diabetes status. PLoS One 2013;8(11): e78257.

[38] Santacruz A, Collado MC, García-Valdés L, Segura MT, Martín-Lagos JA, Anjos T, et al. Gut microbiota composition is associated with body weight, weight gain and biochemical parameters in pregnant women. Br J Nutr. 2010;104(1):83–92.

[39] Hashimoto F, Nishiumi S, Miyake O, Takeichi H, Chitose M, Ohtsubo H, et al. Metabolomics analysis of umbilical cord blood clarifies changes in saccharides associated with delivery method. Early Hum Dev. 2013;89(5):315–20.

[40] Huang L, Chen Q, Zhao Y, Wang W, Fang F, Bao Y. Is elective cesarean section associated with a higher risk of asthma? A meta-analysis. J Asthma 2015;52(1):16–25.

[41] Hyde MJ, Modi N. The long-term effects of birth by caesarean section: the case for a randomised controlled trial. Early Hum Dev 2012;88(12):943–9.

[42] Dominguez-Bello MG, Costello EK, Contreras M, Magris M, Hidalgo G, Fierer N, et al. Delivery mode shapes the acquisition and structure of the initial microbiota across multiple body habitats in newborns. Proc Natl Acad Sci USA. 2010;107(26):11971–5.

[43] Kolokotroni O, Middleton N, Gavatha M, Lamnisos D, Priftis KN, Yiallouros PK. Asthma and atopy in children born by caesarean section: effect modification by family history of allergies – a population based cross-sectional study. BMC Pediatr 2012;12:179.

[44] Adlerberth I, Wold AE. Establishment of the gut microbiota in Western infants. Acta Paediatr 2009;98(2):229–38.

[45] Nicholls AW, Mortishire-Smith RJ, Nicholson JK. NMR spectroscopic-based metabonomic studies of urinary metabolite variation in acclimatizing germ-free rats. Chem Res Toxicol 2003;16(11):1395–404.

[46] Yap IK, Li JV, Saric J, Martin FP, Davies H, Wang Y, et al. Metabonomic and microbiological analysis of the dynamic effect of vancomycin-induced gut microbiota modification in the mouse. J Proteome Res 2008;7(9):3718–28.

[47] Lenz EM, Bright J, Knight R, Westwood FR, Davies D, Major H, Wilson ID. Metabonomics with ^{1}H-NMR spectroscopy and liquid chromatography-mass spectrometry applied to the investigation of metabolic changes caused by gentamicin-induced nephrotoxicity in the rat. Biomarkers. 2005;10(2–3):173–87.

[48] Romick-Rosendale LE, Goodpaster AM, Hanwright PJ, Patel NB, Wheeler ET, Chona DL, et al. NMR-based metabonomics analysis of mouse urine and fecal extracts following oral treatment with the broad-spectrum antibiotic enrofloxacin (Baytril). Magn Reson Chem 2009;47(Suppl. 1):S36–46.

[49] Sun J, Schnackenberg LK, Khare S, Yang X, Greenhaw J, Salminen W, et al. Evaluating effects of penicillin treatment on the metabolome of rats. J Chromatogr B Analyt Technol Biomed Life Sci 2013;932:134–43.

[50] Lee SH, An JH, Park HM, Jung BH. Investigation of endogenous metabolic changes in the urine of pseudo germ-free rats using a metabolomic approach. J Chromatogr B Analyt Technol Biomed Life Sci. 2012;887-888:8–18.

[51] Swann JR, Tuohy KM, Lindfors P, Brown DT, Gibson GR, Wilson ID, et al. Variation in antibiotic-induced microbial recolonization impacts on the host metabolic phenotypes of rats. J Proteome Res 2011;10(8):3590–603.

[52] Swann JR, Want EJ, Geier FM, Spagou K, Wilson ID, Sidaway JE, et al. Systemic gut microbial modulation of bile acid metabolism in host tissue compartments. Proc Natl Acad Sci USA 2011;108(Suppl. 1):4523–30.

[53] Witkin SS, Linhares IM, Giraldo P. Bacterial flora of the female genital tract: function and immune regulation. Best Pract Res Clin Obstet Gynaecol. 2007;21(3):347–54.

[54] Ravel J, Gajer P, Abdo Z, Schneider GM, Koenig SS, McCulle SL, et al. Vaginal microbiome of reproductive-age women. Proc Natl Acad Sci USA 2011;108(Suppl. 1):4680–7.

[55] Gajer P, Brotman RM, Bai G, Sakamoto J, Schütte UM, Zhong X, et al. Temporal dynamics of the human vaginal microbiota. Sci Transl Med 2012;4(132):132ra52.

[56] Verstraelen H, Verhelst R, Claeys G, De Backer E, Temmerman M, Vaneechoutte M. Longitudinal analysis of the vaginal microflora in pregnancy suggests that L. crispatus promotes the stability of the normal vaginal microflora and that L. gasseri and/or L. iners are more conducive to the occurrence of abnormal vaginal microflora. BMC Microbiol 2009;9:116.

[57] Aagaard K, Riehle K, Ma J, Segata N, Mistretta TA, Coarfa C, et al. A metagenomic approach to characterization of the vaginal microbiome signature in pregnancy. PLoS One. 2012;7(6):e36466.

[58] Romero R, Hassan SS, Gajer P, Tarca AL, Fadrosh DW, Nikita L, et al. The composition and stability of the vaginal microbiota of normal pregnant women is different from that of non-pregnant women. Microbiome 2014;2(1):4.

[59] MacIntyre DA, Chandiramani M, Lee YS, Kindinger L, Smith A, Angelopoulos N, et al. The vaginal microbiome during pregnancy and the postpartum period in a European population. Sci Rep. 2015;5:8988.

[60] Mossop H, Linhares IM, Bongiovanni AM, Ledger WJ, Witkin SS. Influence of lactic acid on endogenous and viral RNA-induced immune mediator production by vaginal epithelial cells. Obstet Gynecol 2011;118(4):840–6.

[61] Witkin SS, Mendes-Soares H, Linhares IM, Jayaram A, Ledger WJ, Forney LJ. Influence of vaginal bacteria and D- and L-lactic acid isomers on vaginal extracellular matrix metalloproteinase inducer: implications for protection against upper genital tract infections. MBio 2013;4(4).

[62] Koren O, Goodrich JK, Cullender TC, Spor A, Laitinen K, Bäckhed HK, et al. Host remodeling of the gut microbiome and metabolic changes during pregnancy. Cell. 2012;150(3):470–80.

[63] Zhang D, Huang Y, Ye D. Intestinal dysbiosis: an emerging cause of pregnancy complications? Med Hypotheses. 2015;84(3):223–6.

[64] Gomez Arango LF, Barrett HL, Callaway LK, Nitert MD. Probiotics and pregnancy. Curr Diab Rep. 2015;15(1):567.

[65] Priyadarshini M, Thomas A, Reisetter AC, Scholtens DM, Wolever TM, Josefson JL, et al. Maternal short-chain fatty acids are associated with metabolic parameters in mothers and newborns. Transl Res 2014;164(2):153–7.

[66] Pinto J, Barros AS, Domingues MR, Goodfellow BJ, Galhano E, Pita C, et al. Following healthy pregnancy by NMR metabolomics of plasma and correlation to urine. J Proteome Res 2015;14(2):1263–74.

[67] Maitre L, Fthenou E, Athersuch T, Coen M, Toledano MB, Holmes E, et al. Urinary metabolic profiles in early pregnancy are associated with preterm birth and fetal growth restriction in the Rhea mother-child cohort study. BMC Med 2014;12:110.

[68] Walsh JM, Wallace M, Brennan L, McAuliffe FM. Early pregnancy maternal urinary metabolomic profile and later insulin resistance and fetal adiposity. J Matern Fetal Neonatal Med 2014:1–4. [Epub ahead of print].

[69] Kuc S, Koster MP, Pennings JL, Hankemeier T, Berger R, Harms AC, et al. Metabolomics profiling for identification of novel potential markers in early prediction of preeclampsia. PLoS One 2014;9(5):e98540.

[70] Iacovidou N, Varsami M, Syggellou A. Neonatal outcome of preterm delivery. Ann NY Acad Sci 2010;1205:130–4.

[71] Lamont RF. Infection in the prediction and antibiotics in the prevention of spontaneous preterm labour and preterm birth. BJOG 2003;110(Suppl. 20):71–5.

[72] Watts DH, Krohn MA, Hillier SL, Eschenbach DA. The association of occult amniotic fluid infection with gestational age and neonatal outcome among women in preterm labor. Obstet Gynecol 1992;79(3):351–7.

[73] Romero R, Salafia CM, Athanassiadis AP, Hanaoka S, Mazor M, Sepulveda W, et al. The relationship between acute inflammatory lesions of the preterm placenta and amniotic fluid microbiology. Am J Obstet Gynecol 1992;166(5):1382–8.

[74] MacIntyre DA, Lee YS, Migale R, Herbert BR, Waddington SN, Peebles D, et al. Activator protein 1 is a key terminal mediator of inflammation-induced preterm labor in mice. FASEB J 2014;28(5):2358–68.

[75] Estrada-Gutierrez G, Gomez-Lopez N, Zaga-Clavellina V, Giono-Cerezo S, Espejel-Nuñez A, Gonzalez-Jimenez MA, et al. Interaction between pathogenic bacteria and intrauterine leukocytes triggers alternative molecular signaling cascades leading to labor in women. Infect Immun. 2010;78(11):4792–9.

[76] Donders G, Bellen G, Rezeberga D. Aerobic vaginitis in pregnancy. BJOG 2011; 118(10):1163–70.

[77] Leitich H, Bodner-Adler B, Brunbauer M, Kaider A, Egarter C, Husslein P. Bacterial vaginosis as a risk factor for preterm delivery: a meta-analysis. Am J Obstet Gynecol 2003;189(1):139–47.

[78] Laxmi U, Agrawal S, Raghunandan C, Randhawa VS, Saili A. Association of bacterial vaginosis with adverse fetomaternal outcome in women with spontaneous preterm labor: a prospective cohort study. J Matern Fetal Neonatal Med. 2012;25(1):64–7.

[79] Breugelmans M, Vancutsem E, Naessens A, Laubach M, Foulon W. Association of abnormal vaginal flora and Ureaplasma species as risk factors for preterm birth: a cohort study. Acta Obstet Gynecol Scand 2010;89(2):256–60.

[80] Kafetzis DA, Skevaki CL, Skouteri V, Gavrili S, Peppa K, Kostalos C, et al. Maternal genital colonization with Ureaplasma urealyticum promotes preterm delivery: association of the respiratory colonization of premature infants with chronic lung disease and increased mortality. Clin Infect Dis 2004;39(8):1113–22.

[81] Thorsen P, Jensen IP, Jeune B, Ebbesen N, Arpi M, Bremmelgaard A, et al. Few microorganisms associated with bacterial vaginosis may constitute the pathologic core: a population-based microbiologic study among 3596 pregnant women. Am J Obstet Gynecol 1998;178(3):580–7.

[82] Goldenberg RL, Culhane JF, Iams JD, Romero R. Epidemiology and causes of preterm birth. Lancet. 2008;371(9606):75–84.

[83] Asindi AA, Archibong EI, Mannan NB. Mother-infant colonization and neonatal sepsis in prelabor rupture of membranes. Saudi Med J 2002;23(10):1270–4.

[84] Simcox R, Sin WT, Seed PT, Briley A, Shennan AH. Prophylactic antibiotics for the prevention of preterm birth in women at risk: a meta-analysis. Aust NZJ Obstet Gynaecol 2007;47(5):368–77.

[85] Kenyon S, Boulvain M, Neilson JP. Antibiotics for preterm rupture of membranes. Cochrane Database Syst Rev 2013;12:CD001058.

[86] Tan YT, Tillett DJ, McKay IA. Molecular strategies for overcoming antibiotic resistance in bacteria. Mol Med Today 2000;6(8):309–14.

[87] Nicholson JK, Holmes E, Kinross J, Burcelin R, Gibson G, Jia W, et al. Host-gut microbiota metabolic interactions. Science. 2012;336(6086):1262–7.

[88] Scheline RR. Metabolism of phenolic acids by the rat intestinal microflora. Acta Pharmacol Toxicol (Copenh) 1968;26(2):189–205.

[89] Jain V, Das V, Agarwal A, Pandey A. Asymptomatic bacteriuria & obstetric outcome following treatment in early versus late pregnancy in north Indian women. Indian J Med Res 2013;137(4):753–8.

[90] Gupta A, Dwivedi M, Mahdi AA, Khetrapal CL, Bhandari M. Broad identification of bacterial type in urinary tract infection using (1)h NMR spectroscopy. J Proteome Res 2012;11(3):1844–54.

[91] Burillo A, Rodríguez-Sánchez B, Ramiro A, Cercenado E, Rodríguez-Créixems M, Bouza E. Gram-Stain Plus MALDI-TOF MS (Matrix-Assisted Laser Desorption Ionization-Time of Flight Mass Spectrometry) for a rapid diagnosis of urinary tract infection. PLoS One. 2014;9(1):e86915.

[92] Tadesse E, Teshome M, Merid Y, Kibret B, Shimelis T. Asymptomatic urinary tract infection among pregnant women attending the antenatal clinic of Hawassa Referral Hospital, Southern Ethiopia. BMC Res Notes 2014;7:155.

[93] Strittmatter N, Rebec M, Jones EA, Golf O, Abdolrasouli A, Balog J, et al. Characterization and identification of clinically relevant microorganisms using rapid evaporative ionization mass spectrometry. Anal Chem 2014;86(13):6555–62.

[94] Ozkan S, Ceylan Y, Ozkan OV, Yildirim S. Review of a challenging clinical issue: intrahepatic cholestasis of pregnancy. World J Gastroenterol 2015;21(23):7134–41.

[95] Milona A, Owen BM, Cobbold JF, Willemsen EC, Cox IJ, Boudjelal M, et al. Raised hepatic bile acid concentrations during pregnancy in mice are associated with reduced farnesoid X receptor function. Hepatology 2010;52:1341–9.

[96] Geenes V, Williamson C. Intrahepatic cholestasis of pregnancy. World J Gastroenterol 2009;15:2049–66.

[97] Wojcicka-Jagodzinska J, Kuczynska-Sicinska L, Czajkowski K, Smolarczyk R. Carbohydrate metabolism in the course of intrahepatic cholestasis in pregnancy. Am J Obstet Gynecol 1989;161:959–64.

[98] Tribe RM, Dann AT, Kenyon AP, Seed P, Shennan AH, Mallet A. Longitudinal profiles of 15 serum bile acids in patients with intrahepatic cholestasis of pregnancy. Am J Gastroenterol 2010;105(3):585–95.

[99] Reyes H, Zapata R, Hernández I, Gotteland M, Sandoval L, Jirón MI, et al. Is a leaky gut involved in the pathogenesis of intrahepatic cholestasis of pregnancy? Hepatology. 2006;43(4):715–22.

[100] Larsson E, Tremaroli V, Lee YS, Koren O, Nookaew I, Fricker A, et al. Analysis of gut microbial regulation of host gene expression along the length of the gut and regulation of gut microbial ecology through MyD88. Gut 2012;61:1124–31.

[101] Backhed F, Ding H, Wang T, Hooper LV, Koh GY, Nagy A, et al. The gut microbiota as an environmental factor that regulates fat storage. Proc Natl Acad Sci USA 2004;101: 15718–23.

[102] Van Mil SW, Milona A, Dixon PH, Mullenbach R, Geenes VL, Chambers J, et al. Functional variants of the central bile acid sensor FXR identified in intrahepatic cholestasis of pregnancy. Gastroenterology. 2007;133(2):507–16.

[103] Lamminpää R, Vehviläinen-Julkunen K, Gissler M, Selander T, Heinonen S. Pregnancy outcomes of overweight and obese women aged 35 years or older – a registry based study in Finland. Obes Res Clin Pract 2015. Available from: doi: http://dx.doi.org/10.1016/j.orcp.2015.05.008 (in press).

[104] Lutsiv O, Mah J, Beyene J, McDonald SD. The effects of morbid obesity on maternal and neonatal health outcomes: a systematic review and meta-analyses. Obes Rev 2015;16(7): 531–46.

[105] Khalak R, Cummings J, Dexter S. Maternal obesity: significance on the preterm neonate. Int J Obes (Lond) 2015;39:1433–6.

[106] Sharp GC, Lawlor DA, Richmond RC, Fraser A, Simpkin A, Suderman M, et al. Maternal pre-pregnancy BMI and gestational weight gain, offspring DNA methylation and later offspring adiposity: findings from the Avon Longitudinal Study of Parents and Children. Int J Epidemiol 2015;44:1288–304.

[107] Yaniv-Salem S, Shoham-Vardi I, Kessous R, Pariente G, Sergienko R, Sheiner E. Obesity in pregnancy: what's next? Long-term cardiovascular morbidity in a follow-up period of more than a decade. J Matern Fetal Neonatal Med 2015:1–5.

[108] Johansson K, Cnattingius S, Näslund I, Roos N, Trolle Lagerros Y, Granath F, et al. Outcomes of pregnancy after bariatric surgery. N Engl J Med 2015;372(9):814–24.

[109] Santangeli L, Sattar N, Huda SS. Impact of maternal obesity on perinatal and childhood outcomes. Best Pract Res Clin Obstet Gynaecol 2015;29(3):438–48.

[110] Roberts VH, Frias AE, Grove KL. Impact of maternal obesity on fetal programming of cardiovascular disease. Physiology (Bethesda) 2015;30(3):224–31.

[111] Ingvorsen C, Brix S, Ozanne SE, Hellgren LI. The effect of maternal Inflammation on foetal programming of metabolic disease. Acta Physiol (Oxf) 2015 [Epub ahead of print].

[112] Leon-Cabrera S, Solís-Lozano L, Suárez-Álvarez K, González-Chávez A, Béjar YL, Robles-Díaz G, et al. Hyperleptinemia is associated with parameters of low-grade systemic inflammation and metabolic dysfunction in obese human beings. Front Integr Neurosci 2013;7:62.

[113] Pantham P, Aye IL, Powell TL. Inflammation in maternal obesity and gestational diabetes mellitus. Placenta. 2015;36(7):709–15.

[114] Kozyrskyj AL, Kalu R, Koleva PT, Bridgman SL. Fetal programming of overweight through the microbiome: boys are disproportionately affected. J Dev Orig Health Dis 2015:1–10. [Epub ahead of print].

[115] Gohir W, Ratcliffe EM, Sloboda DM. Of the bugs that shape us: maternal obesity, the gut microbiome, and long-term disease risk. Pediatr Res 2015;77(1–2):196–204.

[116] Sacks DA, Hadden DR, Maresh M, Deerochanawong C, Dyer AR, Metzger. Frequency of gestational diabetes mellitus at collaborating centers based on iadpsg consensus panel-recommended criteria: the hyperglycemia and adverse pregnancy outcome (HAPO) study. Diabetes Care 2012;35:526–8.

[117] Zhang C, Hu FB, Olsen SF, Vaag A, Gore-Langton R, Chavarro JE, et al. DWH study team. Rationale, design, and method of the Diabetes & Women's Health study – a study of long-term health implications of glucose intolerance in pregnancy and their determinants. Acta Obstet Gynecol Scand 2014;93(11):1123–30.

[118] Gopalakrishnan V, Singh R, Pradeep Y, Kapoor D, Rani AK, Pradhan S, et al. Evaluation of the prevalence of gestational diabetes mellitus in North Indians using the International Association of Diabetes and Pregnancy Study groups (IADPSG) criteria. J Postgrad Med. 2015;61(3):155–8.

[119] Kc K, Shakya S, Zhang H. Gestational diabetes mellitus and macrosomia: a literature review. Ann Nutr Metab 2015;66(Suppl. 2):14–20.

[120] Hawryluk J, Grafka A, Gęca T, Łopucki M. Gestational diabetes in the light of current literature. Pol Merkur Lekarski 2015;38(228):344–7.

[121] Abell SK, De Courten B, Boyle JA, Teede HJ. Inflammatory and other biomarkers: role in pathophysiology and prediction of gestational diabetes mellitus. Int J Mol Sci 2015;16(6):13442–73.

[122] Huynh J, Xiong G, Bentley-Lewis R. A systematic review of metabolite profiling in gestational diabetes mellitus. Diabetologia. 2014;57(12):2453–64.

[123] Pinto J, Almeida LM, Martins AS, Duarte D, Barros AS, Galhano E, et al. Prediction of gestational diabetes through NMR metabolomics of maternal blood. J Proteome Res 2015;14(6):2696–706.

[124] Dudzik D, Zorawski M, Skotnicki M, Zarzycki W, Kozlowska G, Bibik-Malinowska K, et al. Metabolic fingerprint of gestational diabetes mellitus. J Proteomics. 2014;103:57–71.

[125] Isolauri E, Rautava S, Collado MC, Salminen S. Role of probiotics in reducing the risk of gestational diabetes. Diabetes Obes Metab 2015;17:713–9.

[126] Barrett HL, Dekker Nitert M, Conwell LS, Callaway LK. Probiotics for preventing gestational diabetes. Cochrane Database Syst Rev. 2014;2:CD009951.

[127] Xiang AH, Wang X, Martinez MP, Walthall JC, Curry ES, Page K, et al. Association of maternal diabetes with autism in offspring. J Am Med Assoc. 2015;313(14):1425–34.

[128] Yap IK, Angley M, Veselkov KA, Holmes E, Lindon JC, Nicholson JK. Urinary metabolic phenotyping differentiates children with autism from their unaffected siblings and age-matched controls. J Proteome Res 2010;9(6):2996–3004.

[129] Grapov D, Lemay DG, Weber D, Phinney BS, Azulay Chertok IR, Gho DS, et al. The human colostrum whey proteome is altered in gestational diabetes mellitus. J Proteome Res 2015;14(1):512–20.

[130] Smilowitz JT, Totten SM, Huang J, Grapov D, Durham HA, Lammi-Keefe CJ, et al. Human milk secretory immunoglobulin a and lactoferrin N-glycans are altered in women with gestational diabetes mellitus. J Nutr 2013;143(12):1906–12.

[131] Dani C, Bresci C, Berti E, Ottanelli S, Mello G, Mecacci F, et al. Metabolomic profile of term infants of gestational diabetic mothers. J Matern Fetal Neonatal Med 2014;27(6):537–42.

[132] Seely E, Tsigas E, Rich-Edwards JW. Preeclampsia and future cardiovascular disease in women: how good are the data and how can we manage our patients? Semin Perinatol. 2015;39:276–283.

[133] NHS Choices. *Pre-eclampsia*. Available from: <http://www.nhs.uk/conditions/Pre-eclampsia/Pages/Introduction.aspx>; 2013.

[134] Kenny LC, Broadhurst DI, Dunn W, Brown M, North RA, McCowan L, et al. Screening for pregnancy endpoints consortium. Robust early pregnancy prediction of later preeclampsia using metabolomic biomarkers. Hypertension 2010;56(4):741–9.

[135] Dawson Jr R, Liu S, Jung B, Messina S, Eppler B. Effects of high salt diets and taurine on the development of hypertension in the stroke-prone spontaneously hypertensive rat. Amino Acids 2000;19(3–4):643–65.

[136] Odibo AO, Goetzinger KR, Odibo L, Cahill AG, Macones GA, Nelson DM, et al. First-trimester prediction of preeclampsia using metabolomic biomarkers: a discovery phase study. Prenat Diagn 2011;31(10):990–4.

[137] Bahado-Singh RO, Akolekar R, Mandal R, Dong E, Xia J, Kruger M, et al. Metabolomics and first-trimester prediction of early-onset preeclampsia. J Matern Fetal Neonatal Med 2012;25(10):1840–7.

[138] Bahado-Singh RO, Akolekar R, Mandal R, Dong E, Xia J, Kruger M, et al. First-trimester metabolomic detection of late-onset preeclampsia. Am J Obstet Gynecol 2013;208(1) 58.e1–7.

[139] Heazell AE, Brown M, Dunn WB, Worton SA, Crocker IP, Baker PN, et al. Analysis of the metabolic footprint and tissue metabolome of placental villous explants cultured at different oxygen tensions reveals novel redox biomarkers. Placenta. 2008;29(8):691–8.

[140] Dunn WB, Brown M, Worton SA, Crocker IP, Broadhurst D, Horgan R, et al. Changes in the metabolic footprint of placental explant-conditioned culture medium identifies metabolic disturbances related to hypoxia and pre-eclampsia. Placenta. 2009;30(11):974–80.

[141] Austdal M, Skråstad RB, Gundersen AS, Austgulen R, Iversen AC, Bathen TF. Metabolomic biomarkers in serum and urine in women with preeclampsia. PLoS One 2014;9(3):e91923.

[142] Zeiger BB, Bergamo AC, Vidal DH, Aires FT, Ferreira DG, Scarpelini M, et al. [170-PO]: Preeclampsia in a reference hospital: possible association with lipotoxicity. Pregnancy Hypertens 2015;5(1):87–8.

[143] Amarasekara R, Jayasekara RW, Senanayake H, Dissanayake VH. Microbiome of the placenta in pre-eclampsia supports the role of bacteria in the multifactorial cause of pre-eclampsia. J Obstet Gynaecol Res 2015;41(5):662–9.

[144] Contreras A, Herrera JA, Soto JE, Arce RM, Jaramillo A, Botero JE. Periodontitis is associated with preeclampsia in pregnant women. J Periodontol. 2006;77(2):182–8.

[145] Kramer MS, Kakuma R. Optimal duration of exclusive breastfeeding. Cochrane Database Syst Rev 2012;8:CD003517.

[146] <https://www.gov.uk/government/uploads/system/uploads/attachment_data/file/139591/Annexe_D._Composition_of_breast_milk_review-v3.0.pdf>.

[147] Bode L. Human milk oligosaccharides: every baby needs a sugar mama. Glycobiology 2012;22(9):1147–62.

[148] Kobata A. Structures and application of oligosaccharides in human milk. Proc Jpn Acad Ser B Phys Biol Sci 2010;86:731–47.

[149] Ruiz-Palacios GM, Cervantes LE, Ramos P, Chavez-Munguia B, Newburg DS. Campylobacter jejuni binds intestinal H (O) antigen (Fucα1, 2Galβ1, 4GlcNAc), and fucosyloligosaccharides of human milk inhibit its binding and infection. J Biol Chem 2003;278:14112–20.

[150] Angeloni S, Ridet JL, Kusy N, Gao H, Crevoiser F, Guinchard S, et al. Glycoprofiling with micro-arrays of glycoconjugates and lectins. Glycobiology 2005;15:31–41.

[151] Eiwegger T, Stahl B, Haidl P, Schmitt J, Boehm G, Dehlink E, et al. Prebiotic oligosaccharides: in vitro evidence for gastrointestinal epithelial transfer and immunomodulatory properties. Pediatr Allergy Immunol 2010;21:1179–88.

[152] Jantscher-Krenn E, Zherebtsov M, Nissan C, Goth K, Guner YS, Naidu N, et al. The human milk oligosaccharide disialyllacto-N-tetraose prevents necrotising enterocolitis in neonatal rats. Gut 2011;61:1417–25.

[153] Wang B, Yu B, Karim M, Hu H, Sun Y, McGreevy P, et al. Dietary sialic acid supplementation improves learning and memory in piglets. Am J Clin Nutr 2007;85:561–9.

[154] Birch EE, Garfield S, Hoffman DR, Uauy R, Birch DG. A randomized controlled trial of early dietary supply of long-chain polyunsaturated fatty acids and mental development in term infants. Dev Med Child Neurol 2000;42(3):174–81.

[155] Fujita M, Roth E, Lo YJ, Hurst C, Voliner J, Kendell A. In poor families, mothers' milk is richer for daughters than sons: a test of Trivers–Willard hypothesis in agropastoral settlements in Northern Kenya. Am J Phys Athropol 2012;149:42–59.

[156] Vázquez-Fresno R, Llorach R, Urpi-Sarda M, Lupianez-Barbero A, Estruch R, Corella D, et al. Metabolomic pattern analysis after mediterranean diet intervention in a nondiabetic population: a 1- and 3-year follow-up in the PREDIMED study. J Proteome Res. 2015;14(1):531–40.

[157] O'Gorman A, Brennan L. Metabolomic applications in nutritional research: a perspective. J Sci Food Agric. 2015;95:2567–2570.

[158] Swann JR, Claus SP. Nutrimetabonomics: nutritional applications of metabolic profiling. Sci Prog. 2014;97(Pt 1):41–7.

[159] LeMieux MJ, Aljawadi A, Moustaid-Moussa N. Nutrimetabolomics. Adv Nutr 2014;5(6):792–4.

[160] Jaeggi T, Kortman GA, Moretti D, Chassard C, Holding P, Dostal A, et al. Iron fortification adversely affects the gut microbiome, increases pathogen abundance and induces intestinal inflammation in Kenyan infants. Gut. 2015;64(5):731–42.

[161] Bertram HC, Hoppe C, Petersen BO, Duus JØ, Mølgaard C, Michaelsen KF. An NMR-based metabonomic investigation on effects of milk and meat protein diets given to 8-year-old boys. Br J Nutr. 2007;97(4):758–63.

[162] Metzler-Zebeli BU, Mann E, Ertl R, Schmitz-Esser S, Wagner M, Klein D, et al. Dietary calcium concentration and cereals differentially affect mineral balance and tight junction proteins expression in jejunum of weaned pigs. Br J Nutr. 2015;113(7):1019–31.

[163] Merrifield CA, Lewis MC, Claus SP, Pearce JT, Cloarec O, Duncker S, et al. Weaning diet induces sustained metabolic phenotype shift in the pig and influences host response to *Bifidobacterium lactis* NCC2818. Gut. 2013;62(6):842–51.

[164] Suda Y, Villena J, Takahashi Y, Hosoya S, Tomosada Y, Tsukida K, et al. Immunobiotic *Lactobacillus jensenii* as immune-health promoting factor to improve growth performance and productivity in post-weaning pigs. BMC Immunol 2014;15:24.

[165] Deshpande G, Rao S, Patole S. Probiotics in neonatal intensive care – back to the future. Aust NZJ Obstet Gynaecol 2015;55(3):210–7.

[166] Laitinen K, Collado MC, Isolauri E. Early nutritional environment: focus on health effects of microbiota and probiotics. Benef Microbes 2010;1(4):383–90.

[167] Koleva PT, Bridgman SL, Kozyrskyj AL. The infant gut microbiome: evidence for obesity risk and dietary intervention. Nutrients 2015;7(4):2237–60.

[168] Rzehak P, Hellmuth C, Uhl O, Kirchberg FF, Peissner W, Harder U, European Childhood Obesity Trial Study Group. Rapid growth and childhood obesity are strongly associated with lysoPC(14:0). Ann Nutr Metab. 2014;64(3–4):294–303.

[169] Karlsson CL, Onnerfalt J, Xu J, Molin G, Ahrne S, Thorngren-Jerneck K. The microbiota of the gut in preschool children with normal and excessive body weight. Obesity 2012;20:2257–61.

[170] Azad MB, Bridgman SL, Becker AB, Kozyrskyj AL. Infant antibiotic exposure and the development of childhood overweight and central adiposity. Int J Obes (Lond) 2014;38(10):1290–8.

[171] Thomas EL, Parkinson JR, Hyde MJ, Yap IK, Holmes E, Doré CJ, et al. Aberrant adiposity and ectopic lipid deposition characterize the adult phenotype of the preterm infant. Pediatr Res 2011;70(5):507–12.

[172] Black RE, Victora CG, Walker SP, Bhutta ZA, Christian P, de Onis M, Maternal Child Nutrition Study G. Maternal and child undernutrition and overweight in low-income and middle-income countries. Lancet. 2013;382:427–51.

[173] Smith MI, Yatsunenko T, Manary MJ, Trehan I, Mkakosya R, Cheng J, et al. Gut microbiomes of Malawian twin pairs discordant for kwashiorkor. Science 2013;339(6119): 548–54.

[174] Wang M, Yang X, Ren L, Li S, He X, Wu X, et al. Biomarkers identified by urinary metabonomics for noninvasive diagnosis of nutritional rickets. J Proteome Res. 2014;13(9): 4131–42.

[175] Dessì A, Puddu M, Ottonello G, Fanos V. Metabolomics and fetal-neonatal nutrition: between "not enough" and "too much". Molecules. 2013;18(10):11724–32.

[176] Li Y, Han H, Chen S, Lu Y, Zhu L, Wen W, et al. Effects related to experiences of famine during early life on diabetes mellitus and impaired fasting glucose during adulthood. Zhonghua Liu Xing Bing Xue Za Zhi. 2014;35(7):852–5. Chinese.

[177] Klimek P, Leitner M, Kautzky-Willer A, Thurner S. Effect of fetal and infant malnutrition on metabolism in older age. Gerontology 2014;60(6):502–7.

[178] Canani RB, Costanzo MD, Leone L, Bedogni G, Brambilla P, Cianfarani S, et al. Epigenetic mechanisms elicited by nutrition in early life. Nutr Res Rev 2011;24(2):198–205.

[179] Wu Z, Li M, Zhao C, Zhou J, Chang Y, Li X, et al. Urinary metabonomics study in a rat model in response to protein-energy malnutrition by using gas chromatography-mass spectrometry and liquid chromatography-mass spectrometry. Mol Biosyst 2010;6(11):2157–63.

[180] Shen Q, Li X, Qiu Y, Su M, Liu Y, Li H, et al. Metabonomic and metallomic profiling in the amniotic fluid of malnourished pregnant rats. J Proteome Res 2008;7(5):2151–7.

[181] Sybesma W, Kort R, Lee YK. Locally sourced probiotics, the next opportunity for developing countries? Trends Biotechnol 2015;33(4):197–200.

[182] Kane AV, Dinh DM, Ward HD. Childhood malnutrition and the intestinal microbiome. Pediatr Res 2015;77(1–2):256–62.

[183] Subramanian S, Huq S, Yatsunenko T, Haque R, Mahfuz M, Alam MA, et al. Persistent gut microbiota immaturity in malnourished Bangladeshi children. Nature 2014;510(7505): 417–21.

[184] Guerrant RL, DeBoer MD, Moore SR, Scharf RJ, Lima AA. The impoverished gut – a triple burden of diarrhoea, stunting and chronic disease. Nat Rev Gastroenterol Hepatol 2013;10(4):220–9.

[185] Freemark M. Metabolomics in nutrition research: biomarkers predicting mortality in children with severe acute malnutrition. Food Nutr Bull 2015;36(1 Suppl.):S88–92.

[186] Solano-Aguilar G, Fernandez KP, Ets H, Molokin A, Vinyard B, Urban JF, et al. Characterization of fecal microbiota of children with diarrhea in 2 locations in Colombia. J Pediatr Gastroenterol Nutr 2013;56(5):503–11.

[187] Pop M, Walker AW, Paulson J, Lindsay B, Antonio M, Hossain MA, et al. Diarrhea in young children from low-income countries leads to large-scale alterations in intestinal microbiota composition. Genome Biol 2014;15(6):R76.

[188] Guard BC, Barr JW, Reddivari L, Klemashevich C, Jayaraman A, Steiner JM, et al. Characterization of microbial dysbiosis and metabolomic changes in dogs with acute diarrhea. PLoS One 2015;10(5):e0127259.

[189] Turin CG, Ochoa TJ. The role of maternal breast milk in preventing infantile diarrhea in the developing world. Curr Trop Med Rep 2014;1(2):97–105.

[190] Dylag K, Hubalewska-Mazgaj M, Surmiak M, Szmyd J, Brzozowski T. Probiotics in the mechanism of protection against gut inflammation and therapy of gastrointestinal disorders. Curr Pharm Des 2014;20(7):1149–55.

[191] Aroniadis OC, Brandt LJ, Greenberg A, Borody T, Kelly CR, Mellow M, et al. Long-term follow-up study of fecal microbiota transplantation for severe and/or complicated clostridium difficile infection: a multicenter experience. J Clin Gastroenterol 2015 [Epub ahead of print].

[192] Wang J, Xiao Y, Lin K, Song F, Ge T, Zhang T. Pediatric severe pseudomembranous enteritis treated with fecal microbiota transplantation in a 13-month-old infant. Biomed Rep 2015;3(2):173–5.

[193] Moore AM, Ahmadi S, Patel S, Gibson MK, Wang B, Ndao MI, et al. Gut resistome development in healthy twin pairs in the first year of life. Microbiome 2015;3:27. eCollection 2015.

[194] Hviid A, Svanström H, Frisch M. Antibiotic use and inflammatory bowel diseases in childhood. Gut. 2011;60(1):49–54.

[195] Kronman MP, Zaoutis TE, Haynes K, Feng R, Coffin SE. Antibiotic exposure and IBD development among children: a population-based cohort study. Pediatrics. 2012;130(4):e794–803.

[196] Mueller NT, Whyatt R, Hoepner L, Oberfield S, Dominguez-Bello MG, Widen EM, et al. Prenatal exposure to antibiotics, cesarean section and risk of childhood obesity. Int J Obes (Lond) 2015;39(4):665–70.

[197] Wang JY, Liu LF, Chen CY, Huang YW, Hsiung CA, Tsai HJ. Acetaminophen and/or antibiotic use in early life and the development of childhood allergic diseases. Int J Epidemiol 2013;42(4):1087–99.

[198] Vrieze A, Out C, Fuentes S, Jonker L, Reuling I, Kootte RS, et al. Impact of oral vancomycin on gut microbiota, bile acid metabolism, and insulin sensitivity. J Hepatol 2014;60(4):824–31.

[199] Riley LW, Raphael E, Faerstein E. Obesity in the United States – dysbiosis from exposure to low-dose antibiotics? Front Public Health 2013;1:69.

[200] Czekalski N, Sigdel R, Birtel J, Matthews B, Bürgmann H. Does human activity impact the natural antibiotic resistance background? Abundance of antibiotic resistance genes in 21 Swiss lakes. Environ Int. 2015;81:45–55.

[201] Narciso-da-Rocha C, Vaz-Moreira I, Svensson-Stadler L, Moore ER, Manaia CM. Diversity and antibiotic resistance of Acinetobacter spp. in water from the source to the tap. Appl Microbiol Biotechnol 2013;97(1):329–40.

[202] Murphy R, Stewart AW, Braithwaite I, Beasley R, Hancox RJ, Mitchell EA, ISAAC Phase Three Study Group. Antibiotic treatment during infancy and increased body mass index in boys: an international cross-sectional study. Int J Obes (Lond) 2014;38(8):1115–9.

[203] Trasande L, Blustein J, Liu M, Corwin E, Cox LM, Blaser MJ. Infant antibiotic exposures and early-life body mass. Int J Obes (Lond) 2013;37(1):16–23.

[204] Morel FB, Oosting A, Piloquet H, Oozeer R, Darmaun D, Michel C. Can antibiotic treatment in preweaning rats alter body composition in adulthood? Neonatology 2013;103(3):182–9.

[205] Vidal AC, Murphy SK, Murtha AP, Schildkraut JM, Soubry A, Huang Z, et al. Associations between antibiotic exposure during pregnancy, birth weight and aberrant methylation at imprinted genes among offspring. Int J Obes (Lond) 2013;37(7):907–13.

[206] Kanner L. Autistic disturbances of affective contact. Nervous Child 1943;2:217–50.

[207] Duchan E, Patel DR. Epidemiology of autism spectrum disorders. Pediatr Clin North Am 2012;59(1):27–43. ix–x.

[208] Guinchat V, Thorsen P, Laurent C, Cans C, Bodeau N, Cohen D. Pre-, peri- and neonatal risk factors for autism. Acta Obstet Gynecol Scand 2012;91(3):287–300.

[209] Faras H, Al Ateeqi N, Tidmarsh L. Autism spectrum disorders. Ann Saudi Med 2010; 30(4):295–300.

[210] Niehus R, Lord C. Early medical history of children with autism spectrum disorders. J Dev Behav Pediatr 2006;27(2 Suppl.):S120–7.

[211] Posey DJ, Kem DL, Swiezy NB, Sweeten TL, Wiegand RE, McDougle CJ. A pilot study of D-cycloserine in subjects with autistic disorder. Am J Psychiatry 2004;161(11):2115–7.

[212] de Magistris L, Familiari V, Pascotto A, Sapone A, Frolli A, Iardino P, et al. Alterations of the intestinal barrier in patients with autism spectrum disorders and in their first-degree relatives. J Pediatr Gastroenterol Nutr 2010;51(4):418–24.

[213] Parracho HM, Bingham MO, Gibson GR, McCartney AL. Differences between the gut microflora of children with autistic spectrum disorders and that of healthy children. J Med Microbiol 2005;54(Pt 10):987–91.

[214] Finegold SM, Dowd SE, Gontcharova V, Liu C, Henley KE, Wolcott RD, et al. Pyrosequencing study of fecal microflora of autistic and control children. Anaerobe. 2010;16(4):444–53.

[215] Shaw W. Increased urinary excretion of a 3-(3-hydroxyphenyl)-3-hydroxypropionic acid (HPHPA), an abnormal phenylalanine metabolite of Clostridia spp. in the gastrointestinal tract, in urine samples from patients with autism and schizophrenia. Nutr Neurosci 2010;13(3):135–43.

[216] Williams BL, Hornig M, Parekh T, Lipkin WI. Application of novel PCR-based methods for detection, quantitation, and phylogenetic characterization of Sutterella species in intestinal biopsy samples from children with autism and gastrointestinal disturbances. MBio 2012;3(1).

[217] Ek CJ, Dziegielewska KM, Habgood MD, Saunders NR. Barriers in the developing brain and neurotoxicology. Neurotoxicology 2012;33(3):586–604.

[218] Lis AW, Mclaughlin I, Mpclaughlin RK, Lis EW, Stubbs EG. Profiles of ultraviolet-absorbing components of urine from autistic children, as obtained by high-resolution ion-exchange chromatography. Clin Chem. 1976;22(9):1528–32.

[219] Schultz ST. Can autism be triggered by acetaminophen activation of the endocannabinoid system? Acta Neurobiol Exp (Wars) 2010;70(2):227–31.

[220] Alberti A, Pirrone P, Elia M, Waring RH, Romano C. Sulphation deficit in "low-functioning" autistic children: a pilot study. Biol Psychiatry 1999;46(3):420–4.

[221] Kern JK, Geier DA, Adams JB, Garver CR, Audhya T, Geier MR. A clinical trial of glutathione supplementation in autism spectrum disorders. Med Sci Monit 2011;17(12):CR677–82.

[222] Werling DM, Geschwind DH. Recurrence rates provide evidence for sex-differential, familial genetic liability for autism spectrum disorders in multiplex families and twins. Mol Autism 2015;6:27.

[223] Hofheimer JA, Sheinkopf SJ, Eyler LT. Autism risk in very preterm infants—new answers, more questions. J Pediatr 2014;164(1):6–8.

Chapter 9

The Aging Superorganism

James Kinross[1] and Julian R. Marchesi[2]

[1]*Section of Biosurgery and Surgical Technology, Department of Surgery and Cancer, Faculty of Medicine, Imperial College London, London, UK* [2]*Institute of Reproductive and Developmental Biology, Department of Surgery and Cancer, Imperial College London, The Hammersmith Hospital, London, UK; Centre for Digestive and Gut Health, Imperial College London, London, UK; School of Biosciences, Cardiff University, Cardiff, Wales, UK*

Chapter Outline

9.1 INTRODUCTION

Aging is a complex process and is defined as progressive functional deterioration associated with frailty, disease, and death [1]. The phenomenon of the aging population is both pronounced and historically unprecedented. In the next 4 decades, approximately 22% of the global population will be over the age of 60 years—a jump from 800 million to 2 billion people. During the next 5 years, for the first time in history, people aged 65 years and older in the world will outnumber children aged younger than 5 years [2–4]. Although health data suggest that older adults are healthier than ever, wide gaps in life expectancy between the worst (Sierra Leone) and the best (Japan) performing countries remain, with a range of 0–36 years for life expectancy at birth [5]. It is also not clear if the quality of life from these additional years is improving.

E. Holmes, J.K. Nicholson, A.W. Darzi & J.C. Lindon (Eds): Metabolic Phenotyping in Personalized and Public Healthcare. DOI: http://dx.doi.org/10.1016/B978-0-12-800344-2.00009-4

As expected, the biological changes that constitute and affect aging are complex, and molecular alterations only loosely correspond to chronologic age, which changes at a steady rate. As a consequence, great interindividual functional variability is a hallmark of older populations. Roughly 25% of the heterogeneity in health and function in older age is genetically determined [6], with the remainder strongly affected by the cumulative effect of health behaviors and environmental inequities across the life course. This difference suggests that age-related mortality is more heavily influenced by the molecular changes over time that result from a combination of environmental, epigenetic, post-translational, microbial, and lifestyle factors than genetics alone. What is known is that noncommunicable diseases are the major causes of death and disability in older age, with an increasing prevalence of chronic conditions such as dementia, stroke, cancer, chronic obstructive pulmonary disease, and diabetes. Much of this can be prevented or delayed, and increasing emphasis is now being placed on early-life preventive strategies that enable healthy behaviors and minimize metabolic risk factors. In reality, however, older people have multiple co-morbidities commonly manifesting as loss of physiologic functions as well as broad geriatric syndromes such as frailty and impaired cognition, continence, gait, and balance [7].

The goals of aging research, therefore, are to determine the biological basis of physiologic aging and to define the risk factors that lead to diversity in morbidity and lifespan. Current evidence suggests that the gut microbiome plays a critical component in defining the physiologic response to aging and, more importantly, in determining the efficacy of preventive strategies or drug therapies. Given the complex, multiple biological pathways inherent in the aging process, systems medicine based on an integrated "-omics" analysis of aging provides a compelling methodology not only for delineating the mammalian mechanics of aging but also for providing a more comprehensive analysis of the prokaryotic and nonmammalian biology that defines how we age. This approach is allowing us to answer some fundamental questions in aging research; for example, do microbiome dynamics severely alter the effects of single gene interventions? Could researchers be targeting the wrong genome in the superorganism?

9.2 SYSTEMS BIOLOGY OF AGING

9.2.1 Considerations in Experimental Design

Several evolutionarily conserved pathways that modulate lifespan have been identified in organisms ranging from yeasts to primates. These are pertinent, since even the fruit fly *Drosophila melanogaster* and the nematode worm *Caenorhabditis elegans* (*C. elegans*) live in the presence of microbiota, both in nature and in the laboratory, although the compositions of their microbiomes vary from that of the human. The complexity of the microbiota component, therefore, should not be ignored when interpreting the variations in data from different laboratories studying the effects of single genes on aging. Because of their genetic tractability and inexpensive husbandry, invertebrate studies allow for assigning direct causality to the presence or absence of microbiota. Because of their short lifespans, *Drosophila* and *C. elegans* are also powerful models for studying established genetic pathways that modulate the aging process, including the insulin/insulin-like growth factor (IGF-1) signaling (IIS) pathway, the target of rapamycin (TOR), and adenosine monophosphate –activated protein kinase.

Good experimental design is, as always, essential for providing deep insight into the aging process. In the laboratory, longitudinal studies of longevity permits the analysis of target mechanisms as an organism ages under controlled conditions, and in human studies, cross-sectional cohort studies are more practical. Intervention studies that re-create the conditions of aging have also been applied; for example, D-galactose injection has been shown to induce many changes in mice that represent accelerated aging. This mouse model has been widely used for pharmacologic studies of antiaging agents. The effects of D-galactose on aging have been attributed to glucose and lipid metabolic disorders, oxidative damage, accumulation of advanced glycation end products, reduction in abnormal substance elimination, cell apoptosis, and insulin resistance [8]. It also induces senescence characteristics in cultured cortical astrocytes [9].

There are, however, specific challenges in studying age-related changes that must be considered when interpreting aging data. For example, gene expression changes with age are usually tissue specific [10]. Although well-established challenges, eg, contamination, exist in the analysis of bacteria, the difficulties of studying anaerobic organisms in ambient conditions and animal husbandry techniques will also influence microbial ecology. But the analysis of the microbiome in aging presents some more specific problems; for example, with aging, the composition of the human microbiome changes; how do we replicate this in the laboratory? The use of germ-free (GF) animals has provided us with a useful methodology for studying how bacteria influence the developing gut and the aging process; indeed, these experiments have clearly described the importance of the microbiome to the development of a healthy gut. It is also worth noting that GF rodents live longer compared with their age- and sex-matched conventional cousins. But it must be remembered that these animals often manifest behavioral deficits, morphologic and immunologic abnormalities as a result of altered metabolism, and they have abnormal organ development. They cannot, therefore, be considered appropriate representations of a "normal" animal. Humanized animal models developed through fecal transplantation have served as a useful work-around for this issue. These experiments are described in more detail in this chapter, and they have provided invaluable data on the role of the microbiome in human disease. However, it should be kept in mind that a "humanized" microbiome does not imply "humanized" biology. For example, it is not possible to humanize a mouse metabonome simply by transplanting neonatal feces.

9.2.2 "-omics" Analysis of Aging

Established pathways in aging are outlined in Fig. 9.1. Before one can consider how the microbiome influences the aging process, it is useful to consider how the "-omics" technologies are transforming our understanding of the aging process and identify common targets of microbiome–host interactions. Recently, an attempt has been made to create a central repository for system-level studies of aging through the development of the Digital Ageing Atlas, where human

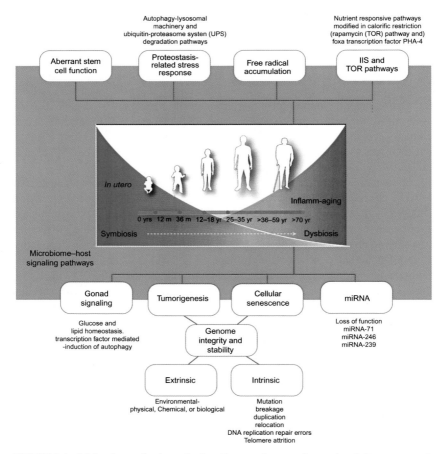

FIGURE 9.1 Molecular mechanisms of aging. Gene–environment interactions influence genomic integrity throughout life, although stochastic genetic mutations will also play a significant role in influencing the risk of cellular senescence and tumorigenesis. Many of these pathways are interconnected, although central pathways such as the insulin-like signaling (ILS) and target of rapamycin (TOR) pathways are critical. During the aging process, microbiome–host interactions lead to a state of dysbiosis and persistent inflammation in the bowel, which also has a significant impact; however, the mechanisms through which the microbiome modulates these pathways are poorly understood.

age-related data covering different biological levels (molecular, cellular, physiologic, psychological, and pathologic) is freely available online (http://ageing-map.org/) [11].

9.2.2.1 Genomics

Somatic mutations in both the nuclear and the mitochondrial genomes are well established in human models of disease [12], but genome-wide association studies (GWASs) are now permitting these to be studied on an unprecedented scale. For example, data from over 100,000 participants have demonstrated that clonal mosaic karyotype anomalies become more frequent both with increasing age and in cancer [13,14], and there is evidence of the accumulation of somatic mitochondrial DNA (mtDNA) mutations with increasing age [15] and age-related diseases. All regions of the mitochondrial genome are affected, but these age-related changes are seen only in certain tissues and not across all age ranges [16]. A recent meta-analysis of age-related gene expression profiles from mice, rats, and humans identified 73 genes that were consistently expressed with increasing age [17]. The gene with the greatest increase in expression was apolipoprotein D (*APOD*), which is known to be associated with neurodegenerative diseases; however, the most common genetic categories for altered gene expression with aging were related to the immune response, including complement activation, antigen processing by the lysosome, apoptosis, and anti-apoptosis pathways [17]. Many of the expression changes that have so far been found in studies of human blood have also been found in genes related to lymphocytes and the immune system [18]. This is interesting because this is a major point of microbiome–host interaction. Some studies of gene expression in blood from long-lived families have also implicated the mammalian target of rapamycin (mTOR) pathway as relevant to longevity in humans, which is consistent with findings in model organisms [19]. Other GWAS analyses are identifying novel loci associated with longevity. For example, rs2149954, on chromosome 5q33.3 (rs4420638 on chromosome 19q13.32, representing the TOMM40/APOE/APOC1 locus), was identified in an analysis of over 7000 Europeans. This allele has previously been reported to be associated with low blood pressure in middle age, and the minor allele was also associated with decreased cardiovascular mortality risk, independent of blood pressure. However, the pleiotropic mechanisms by which this intragenic variation contributes to lifespan regulation have yet to be elucidated [20].

9.2.2.2 Transcriptomics and Epigenetics

Changes in DNA methylation have been shown to occur during both cellular senescence and in vivo aging [21]. DNA methylation changes result in gene expression patterns, which play an important role in cellular differentiation and development. However, random DNA methylation changes can arise over time or as a consequence of environmental damage, and this observation correlates with age. An analysis of cord blood from newborns and blood taken peripherally from centenarians found a marked reduction in global methylation and CpG island promoter hypermethylation with extreme age [22]. The main sites with altered age-related methylation identified in this analysis were the promoters of genes with roles in development or established functions in cellular aging in the brain and blood. However, similar findings have been described elsewhere. A cross-sectional study that looked at DNA methylation in boys aged 3–17 years and adult populations found age-associated changes at 2078 of 27,000 loci [23]. Meta-analysis of age-associated DNA methylation changes in adult populations, and 74% of age-associated loci in the pediatric group demonstrated a threefold to fourfold increased rate of DNA methylation in children versus adults, suggesting that some of the key methylation changes are, in fact, age programmed and take place during early life stages; by contrast, changes in methylation during adulthood may be the result of damage accumulation and the corresponding transcriptional responses. Indeed, many of these DNA methylation changes were found to be enriched on genes with developmental and immune system functions.

9.2.2.3 Metabonomics

In the absence of disease or pathology, an individual's metabolic phenotype changes steadily over time. As a result, age-related changes in the urinary and fecal metabonomes of nematode [24], animal, and human models have been established. The metabolic signature of aging consists of changes in the levels of metabolites associated with amino acid metabolism, tricarboxylic acid (TCA) cycle, the tryptophan–nicotinamide adenine dinucleotide pathway, and the host–microbiota metabolic axis; specific examples of how each of these metabolic pathways influences aging are now being defined [25]. For example, a secondary metabolite of tryptophan, C-glycosyl tryptophan, strongly correlates with age and lung function and is associated with bone mineral density and birth weight [26]. Adenosine is an established neuroprotective agent, and adenosine receptors are deregulated in Alzheimer's disease (AD), and purine metabolism undergoes stage- and region-dependent deregulation of purine metabolism in AD [27]. Lipid species and their metabolism are closely linked to the aging process. Vitamin D_2-related compounds phosphoserine (40:5), monoacylglyceride (22:1), diacylglyceride (33:2), and resolvin D6 all decrease with the aging process. The hydroxyl fatty acid (25-hydroxy-hexacosanoic), a polyunsaturated fatty acid (eicosapentaenoic acid), two phospholipids (phosphatidylcholine [42:9] and phosphatidylserine [42:3]), and a prostaglandin (15-keto-prostaglandin F2alpha) have also been found to decrease with aging [28].

Metabonomic analyses are able to describe exposome, microbiome, and environmental influences on aging and are specific to cultures or subpopulations. For example, ^1H nuclear magnetic resonance (NMR) spectral profiles from urine specimens collected from the Taiwanese Social Environment and Biomarkers of Aging Study (SEBAS; $n = 857$; age 54–91 years) and the Mid-Life in the US study (MIDUS II; $n = 1148$; age 35–86 years) revealed common age-related characteristics in urinary metabolite profiles in the Taiwanese and American populations, as well as distinctive features [29]. In both cases, the gut microbial co-metabolites 4-cresyl sulfate (4CS) and phenylacetylglutamine (PAG) were positively associated with age. In addition, creatine and β-hydroxy-β-methylbutyrate (HMB) were negatively correlated with age in both populations probably reflecting the decreasing muscle mass associated with older age. The further importance of spectral analysis in this area is demonstrated by the studies explored in the sections later.

9.3   NUTRITION, DIET, AND CALORIFIC RESTRICTION

Dietary restriction (DR) or calorific restriction (CR) is a robust means of extending the adult lifespan and postponing age-related diseases in many species, including yeasts, nematode worms, flies, and rodents. The genetic requirements for lifespan extension by DR in the nematode *C. elegans* have implicated a number of key molecules in this process, including the nutrient-sensing TOR pathway and the Foxa transcription factor PHA-4 [30]. However, multiple molecular pathways are responsible for determining the increase in lifespan. For example, CR induces a specific lipidome and metabolome reprogramming event in mouse liver, which is associated with lower protein oxidative damage [31].

In a study of 48 Labrador retrievers paired at age 6 weeks by sex and weight assigned randomly to control feeding (CR) or DR groups, the dogs assigned to 25% DR had a 1.8 year longer median lifespan and delayed onset of late-life diseases, especially osteoarthritis. Long-term DR did not negatively affect skeletal maturation, structure, or metabolism. Fat mass above 25% was associated with increasing insulin resistance, which independently predicted lifespan and chronic diseases. DR dogs also required 17% less energy to maintain each lean kilogram. Metabonomics-based urine metabolite trajectories suggest that signals from gut microbiota may be involved in DR longevity and health responses [32]. These findings have been replicated in other dog and monkey experimental studies, suggesting a potential connection among the gut microbiota, CR, and aging [33,34]. This, in itself, is not surprising, as a major function of the microbiome is nutritional.

DR alters host gene expression in mice with lifelong dietary and exercise interventions. Indeed, it is not simply volumetric restriction, but a low-fat diet also infers a greater life expectancy in animal models [35] by stabilizing glucose homoeostasis and lipid metabolism constantly throughout the lifespan. A restricted high-fat diet intake (HFD + CR) results in dramatic extensions by up to 36% of the median and maximum lifespans in mice models compared with animals subjected to a high-fat diet. In general, experimental data now confirm what common sense dictates, that is, obesity-related metabolic syndrome is highly associated with accelerated aging and reduced lifespan. Interestingly, it has now been recently shown that lifelong CR on both high-fat and low-fat diets, but not voluntary exercise, significantly changes the overall structure of the gut microbiota in C57BL/6 J mice. CR enriches phylotypes positively correlated with lifespan, such as the genus *Lactobacillus* on a low-fat diet, and reduces phylotypes negatively correlated with lifespan. These CR-induced changes in the gut microbiota are concomitant with significantly reduced serum levels of lipopolysaccharide-binding protein, suggesting that animals under CR can establish a structurally balanced architecture of the gut microbiota, which may exert a health benefit to the host via reduction of antigen load from the gut [36].

The free radical theory of aging proposes that accumulating macromolecular damage from increased reactive oxygen species (ROS) over time causes aging. However, CR in mammals or flies enhances mitochondrial biogenesis and function, effectively countering what is known as the "Warburg effect" or anaerobic glycolysis. This metabolism is, in some ways, counterintuitive to the ROS theory, since mitochondria are the primary source of ROS. However, the enhancement of mitochondrial function helps the organism switch from glycolysis to oxidative metabolism, which, in turn, mediates the protective effects of DR. Mitochondrial TCA cycle metabolites such as pyruvate, fumarate, malate, and oxaloacetate have been previously shown in *C. elegans* to extend lifespan upon feeding [37]. The exact mechanistic effects have yet to be elucidated; however, these mitochondrial metabolites (or "mitobolites") such as α-ketoglutarate can extend lifespan by inhibition of mitochondrial adenosine triphosphatase in *C. elegans* models through a TOR-dependent mechanism [38].

Aging mice harbor a distinct microbiota, which can be modulated by resistant starch and enriched for bacteria that are associated with improved health. For example, the cecal microbiota of mice fed a diet depleted in resistant starch and containing the readily digestible carbohydrate amylopectin were dominated by bacteria in the *Firmicutes* phylum and contained low levels of *Bacteroidetes* and *Actinobacteria*. The proportions of *Bifidobacterium* and *Akkermansia* positively correlate with mouse feeding responses, gut weight, and expression levels of proglucagon, the precursor of the gut antiobesity/antidiabetic hormone GLP-1 [39].

Healthy aging depends on removal of damaged cellular material that is, in part, mediated by autophagy. The nutritional status of cells affects both aging and autophagy through metabolic circuits that have yet to be fully defined. For example, Nucleocytosolic acetyl-coenzyme A (AcCoA) production is a metabolic repressor of autophagy during aging in yeasts [40]. This control is achieved through hyperactivation of the nucleocytosolic AcCoA synthetase Acs2p, which, in turn, triggers histone acetylation, repression of autophagy genes, and an age-dependent defect in autophagic flux, culminating in reduced lifespan. Brain-specific knockdown of *Drosophila* AcCoA synthetase is sufficient to enhance autophagic protein clearance and prolong lifespan. AcCoA integrates various nutrition pathways, and this may explain diet-dependent lifespan and autophagy regulation. Autophagy promotes lifespan extension upon MetR and requires the subsequent stimulation of vacuolar acidification, whereas it is epistatic to the equally autophagy-dependent antiaging pathway triggered by TOR1 inhibition or deletion [41].

9.4 THE AGING SUPERORGANISM

9.4.1 The Infant Microbiome

The microbial colonization of the infant gut is of vital importance to the human lifecycle, and the early establishment of host–microbe interactions has consequences for human health and disease later in life. For example, exposure to microbes during early childhood is associated with protection from immune-mediated diseases such as inflammatory bowel disease (IBD), asthma, and eczema. Therefore, understanding how the neonatal gut is colonized and its development in early life is essential for understanding what role the microbiome plays in aging in general. Until recently, it was thought that bacteria cannot cross the placental barrier; however, it is now understood that the maternal *E. faecium* strain has been detected in the meconium of mouse pups after aseptic cesarean section (C-section), although it is has been suggested that these data may be prone to sampling errors and contamination. In the first 3 months, the relative abundance of *Proteobacteria* declines, and the abundance of *Actinobacteria* increases. After this, the *Actinobacteria* decrease, and the relative abundance of the phylum *Firmicutes* starts to increase. These trends continue through to age 6 months, with the emergence of a low level of *Verrucomicrobia*, and over the course of the first 2 years of life, the progression of the gut microbiota toward an adult-like microbiota is quite clear at the phylum level. However, the route of delivery may play a significant role in defining how the microbiome matures. For example, after a C-section delivery, the microbiota of newborns resembles the skin microbiota of the mother, at all body sites, whereas naturally delivered infants have a microbiota more similar to that of the vaginal canal. Vaginally delivered infants have a greater relative abundance of phylum *Bacteroidetes* and a lower abundance of *Firmicutes* compared with children delivered via C-section. It then takes a year until the levels of *Bacteroidetes* begin to rise in C-section-delivered children to levels similar to those in vaginally delivered children. Even at the age of 7 years, levels of clostridia are still significantly higher in vaginally delivered children. Over 20% of children in the United Kingdom are now delivered by C-section. The genus *Bifidobacterium* is numerically dominant throughout the first year at least, regardless of the delivery mode, a finding that recurs throughout the literature. High levels of *Bifidobacterium* are driven by breastfeeding, highlighted by a much slower colonization in bottlefed infants. Other differences are caused by breastfeeding over formula feeding, including lower clostridia levels, and different colonization times. This may be clinically important. For example, abnormal intestinal colonization during the first weeks of life may alter the barrier and the nutritional and immunologic functions of the host microbiota and, as a consequence, susceptibility to disease increases, although this is poorly defined. Although large interindividual differences are detected in fecal microbiota profiles, the meconium microbiota is distinct from that of fecal samples [42].

It is clear, however, that the phylogenetic diversity of the microbiome increases gradually over time, and changes in community composition follow a smooth temporal gradient. However, major taxonomic groups are susceptible to abrupt shifts in abundance, corresponding to changes in diet or health. Community assembly is not random, and data suggest that discrete steps of bacterial succession are modified by early-life events. The infant microbiome is enriched in genes facilitating lactate utilization, and intriguingly, functional genes responsible for plant polysaccharide metabolism are present before the introduction of solid food; that is, the neonatal gut is primed for an adult diet. During the initial few months of a milk diet, bacteria such as *Bifidobacteria*, highly adapted to process milk oligosaccharides are abundant. Following the ingestion of table foods, there is a sustained increase in the abundance of *Bacteroidetes*, elevated fecal short-chain fatty acid (SCFA) levels, enrichment of genes associated with carbohydrate utilization; vitamin biosynthesis; and xenobiotic degradation; and a more stable community composition, all of which are characteristic of the adult microbiome [43] (Table 9.1).

9.4.2 The Aging Microbiome

It is important to understand the relationship between the aging gut and the aging luminal and mucosal ecosystem it supports. Aging brings with it social and physiologic changes that have a profound effect on the gut. For example, with increasing age comes poor dentition, masticatory dysfunction, gastric atrophy, increased risk of *Helicobacter pylori* infection, reduced expression of gut hormones such as leptin and ghrelin, increased prevalence of chronic diseases of the intestine (eg, diverticulosis), and the increased risks of social isolation, reduced mobility, and malnutrition [44]. These factors, in turn, exert significant modifying forces on the microbiome.

At present, it is not feasible to define precisely the "age threshold" for the starting of the aging of the gut microbiota, and current studies have incorporated subjects with a wide range of age (60–80 years). However, it is generally accepted to be somewhere between 70 and 80 years [44]. This issue is critical as a lack of standardization makes robust interpretation of much of the microbiome aging data challenging. For example, primary outcomes (eg, frailty) vary and are often not standardized among trials, and the microbiome is subject to significant confounding through hospital and community environments.

The microbiota of older people displays greater interindividual variation than that of younger adults. Middle-aged people also have more distinct

TABLE 9.1 Summary Data of the Gut Microbiome in Aging

		Neonatal (3 weeks)	6 months	Middle Age	Old Age
Mouse	Feces		Lactobacillaceae, Prevotellaceae, and Porphyromonadaceae	Rikenellaceae, Lactobacillaceae, Mycoplasmataceae, and Erysipelotrichaceae, and less Lachnospiraceae, Clostridiaceae, Ruminococcaceae, Prevotellaceae, and Porphyromonadaceae	Rikenellaceae, Lachnospiraceae, Ruminococcaceae, and Clostridiaceae
Human	Feces	Proteobacteria (Escherichia coli, E. fergusonii, Klebsiella pneumoniae, and Serratia marcescens, Enterococcus, and Yersinia	Bacteroides and Firmicutes predominate Common occurrence of Verrucomicrobia		Reduced Firmicutes, Clostridium cluster XIVa (Clostridium coccoides/Eubacterium rectale) Reduced Faecalibacterium prausnitzii and Bacteroidetes. Increase in facultative anaerobes, including streptococci, staphylococci, enterococci, and enterobacteria Decreased bifidobacteria
	Meconium	Bacilli:Firmicutes, Staphylococcus, Lactobacillus plantarum, and Streptococcus mitis			

taxonomic and functional compositions compared with other adult age groups. What is known is that as individuals age, there is an alteration in the relative proportions of the *Firmicutes* and the *Bacteroidetes*, with older adults having a higher proportion of *Bacteroidetes* and young adults having higher proportions of *Firmicutes* [45]. A significant reduction in bifidobacteria, *Bacteriodes*, and *Clostridium* cluster IV have also been reported [46]. Despite these findings, it should be remembered that the variability of these bacteria among individuals is considerable, and ranges from 3% to 92% for *Bacteroidetes* and between 7% and 94% for *Firmicutes*. Importantly, however, as described earlier, within individual subjects there is less temporal variability [47]. Broadly, these changes remain constant in geographic subgroups and at varying extremes of aging. However, there are some contradictions in the published data. For example, expression of *Firmicutes* and *Clostridium* cluster XIVa decreases in healthy Japanese subjects aged 74–94 years [48], in Italians over 60 years [49] and Italian centenarians [50], and in the Finnish over 70 years [51]. However, it increased in Germans over 60 years [49]. *Faecalibacterium prausnitzii* was also found to decline in populations of Italians over 60 years and in Italian centenarians [50]; in French, German, and Swedish subjects, it increased [49]. However, a reduced expression of *Clostridium* cluster XIVa and *F. prausnitzii* group members correlate with frailty, hospitalization, antibiotic treatment, and nonsteroidal anti-inflammatory drug (NSAID) therapy [44]. The *Bacteroidetes* population during aging is even more country dependent. The gut microbiota of the Irish seems to be characterized by a higher proportion of *Bacteroidetes* than that of Italians. In the case of the Irish, this proportion increases with the aging process (from 41% to 57%, in average), whereas in Italian older adults and centenarians, the proportion of *Bacteroidetes* remains approximately constant or even decreases. Evidence from studies using 16S rRNA gene-based molecular techniques suggests that a decrease in *Bacteroidetes* is more strictly related to frailty, antibiotic treatment, hospitalization, and *C. difficile*-associated diarrhea (CDAD) than to the aging process itself. Further discrepancies are also seen in *Bifidobacterium* expression in populations across countries.

However, although previous studies have focused on analyzing ecosystem-wide configurations in human intestinal microbiota, a recent analysis of a 1000 Western adults shows that there are distinct groups of bimodally distributed bacteria that tend to be either very abundant or nearly absent and exhibit reduced stability at the intermediate abundance range between these two contrasting states. The authors of this analysis proposed that the bistable groups represent "tipping elements," specific components of the intestinal microbiota that exhibit alternative stable states linked to the overall ecosystem state and the human physiology. For examples, the statistically significant association between groups of uncultured *Clostridiales* and age suggested that the resilience of the low-abundance state decreases with increasing age, and eventually, even a small perturbation may induce a shift to the high-abundance state. Moreover, the analysis shows that population-level variations may mask alternative states that are more pronounced in specific subpopulations, such as particular age groups. According to a logistic regression model, each age decade increased the odds of the uncultured clostridia high-abundance state by $2.6 \pm 1.2\%$ (95% confidence intervals), but the changes were particularly emphasized in the young and older populations [52].

The context of this relationship in the extremely old has yet to be defined. It is not yet clear if fluctuations in the gut microbiome occur as a consequence of the physiologic changes occurring in the human host with aging or if they drive them. Irrespective of this, the metabolic function of the microbiome remains critical for maintaining the health of an aging population. The age-related functional trajectory of the human gut microbiome is thus characterized by loss of genes for SCFA production and an overall reduction in the saccharolytic potential. Proteolytic functions are more dominant in the intestinal metagenome of younger adults and high-fat, high-protein diets lead to the enrichment of "pathobionts," that is, opportunistic proinflammatory bacteria generally present in the adult gut ecosystem in low numbers. On this basis, it is now possible to identify microbial genes that correlate strongly with mammalian aging [53]. In mice models, there is an under-representation of bacterial-encoded functions for cobalamin (B_{12}) and biotin (B_7) biosynthesis, and bacterial *SOS* genes associated with DNA repair in older mice. Conversely, creatine degradation, associated with muscle wasting, is over-represented within the gut microbiome as are bacterial-encoded β-glucuronidases, which play an important role in drug-induced epithelial cell toxicity. Older mice also show an overabundance of monosaccharide utilization genes relative to di-, oligo-, and polysaccharide utilization genes, which may have a substantial impact on gut homeostasis [54].

In their seminal study, Claesson et al. demonstrated that the fecal microbiota composition from 178 older subjects formed groups that were correlated with the residence location in the community, day hospitals, rehabilitation centers, or long-term residential care facilities. However, clustering of subjects by diet separated them by the same residence location and microbiota groupings. The separation of microbiota composition significantly correlated with measures of frailty, co-morbidity, nutritional status, markers of inflammation, and metabolites in fecal water. The individual microbiota of people in long-term care facilities was significantly less diverse than that of community dwellers. Loss of community-associated microbiota correlated with increased frailty. This provided further support for the concept of the relationship among diet, microbiota, and health status; they indicate a role for diet-driven microbiota alterations in varying rates of health decline with aging.

In the physiologically aging gut, age-related molecular changes accumulate in long-lived stem/progenitor cells, and tissue renewal and regeneration are compromised in the aging intestinal epithelium. Intestinal stem cells (ISCs) are the only dividing cells in the intestinal epithelium that can give rise to at least two differentiated intestinal cell types: enteroendocrine cells and enterocytes. This lineage is critical for normal tissue turnover as well as for epithelial recovery after damage or infection but is deregulated in the aging intestine, where ISC proliferation strongly increases, and polyploid, misdifferentiated cells accumulate in the epithelium. As flies age, hyperproliferation of ISCs leads to intestinal dysfunction, and interestingly, this does not occur in animals reared axenically. ISC proliferation rates are regulated by multiple stress and growth factor signaling pathways, including the JNK, Jak/Stat, p38 MAPK, and EGFR pathways, whose activity is influenced by the commensal microbiota and by infection with pathogenic bacteria. In the aging intestine of *Drosophila*, chronic activation of the transcription factor FOXO reduces expression of peptidoglycan recognition protein SC2 (PGRP-SC2), a negative regulator of IMD/Relish innate immune signaling, and a homolog of the anti-inflammatory molecules PGLYRP1–4 [55]. This repression causes deregulation of Rel/NFkB activity, resulting in commensal dysbiosis, stem cell hyperproliferation, and epithelial dysplasia. Restoring PGRP-SC2 expression in the enterocytes of the intestinal epithelium, in turn, prevents dysbiosis, promotes tissue homeostasis, and extends lifespan. This observation provides strong evidence for how host–bacteria interactions extend beyond intestinal homeostasis to influence systemic health and longevity.

It is becoming apparent that beyond their roles as nutrient sources and potential pathogens, gut bacteria are able to produce diffusible molecules, including metabolites and small RNAs that can directly impact aging [56]. A good example of this is *C. elegans*, which cannot produce its own nitric oxide (NO) because it lacks NO synthase. Therefore, these worms utilize the NO produced by the bacteria within their microenvironment. When *C. elegans* is co-cultured with NO-deficient *Bacillus subtilis* it shortens its lifespan, whereas exogenous supplementation of NO increases it. NO initiates specific transcriptional responses regulated by two transcription factors, DAF-16/FOXO and heat shock factor 1 (HSF-1), both known to mediate lifespan. Worms lacking DAF-16/FOXO or HSF-1 do not respond to the life-prolonging effects of bacterial NO; that is, bacterial NO modulates *C. elegans* lifespan via effects on host transcription [57]. Small noncoding RNAs (ncRNAs) expressed endogenously by *Escherichia coli* (*E. coli*) can modulate *C. elegans* behavior and longevity. Feeding worms *E. coli* deficient in ncRNA DsrA significantly increases their lifespan. In effect, *E. coli* are able to modulate longevity of the worm by hijacking the innate RNAi machinery. There are other examples of microbes promoting growth in low-nutrient conditions. For example, in larvae given no dietary yeast, a condition that usually arrests development, *Lactobacillus plantarum* promotes growth to adulthood. *L. plantarum* overrides the host's ability to match growth with nutrient availability and does so via upregulation of the TOR pathway. Flies overexpressing the inhibitor of TOR complex 1 are resistant to the effects of *L. plantarum* on growth.

9.5 MICROBIOME IMMUNE MODULATION AND INFLAMM-AGING

One of the most important effects of the aging process is a significant decline in the efficacy of both the adaptive and the innate immune systems; in particular, aging has a profound influence on the intestinal immune system, and it would appear that age-associated alterations arise in the mucosal immune system of the gastrointestinal (GI) tract earlier than in systemic immune compartments [58]. The GI tract represents the largest immunologic organ in terms of numbers of lymphocytes, and at any given time, the gut-associated lymphoid tissue (GALT) will harbor up to 20% of the total lymphocytes in the body. The intestinal immune system is exposed to a daily antigen load that surpasses that encountered by the systemic immune system during a person's entire lifetime. As a result, it has complex anatomic and physiologic architecture, which allows it to function and to educate the innate immune response (Fig. 9.2). The most well-understood signaling mechanisms involve innate pattern recognition receptors such as toll-like receptors (TLRs), nucleotide oligomerization domain-like receptors, and C-type lectin receptors. Binding of microbe-associated molecular patterns with these receptors can activate the antigen presenting cells and modulate their function through the expression of surface receptors, secreted cytokines, and chemokines. However, it is increasingly realized that the microbiome is able to modulate the immune response through a large repertoire of signaling mechanisms (Table 9.2); and a potentially important consequence of the age-related impairment of the balance between inflammatory and anti-inflammatory agents is a profound, systemic modification of cellular microenvironment(s), which, in turn, can determine a different rate of aging of organs and tissues, leading to the so-called mosaic of aging. This age-related inflammation-related change of the cell microenvironment may, in turn, also cause a remodeling in the epigenetic and genetic factors that influence lifespan. The dynamics of cross-talk signals between gut microbiota and the GALT immune system allows the host to tolerate and define the gut microbiota ecology through a constitutive low-grade inflammatory state.

The term *inflamm-aging* describes the consequences of the global reduction in the capability to cope with antigenic, chemical, physical, and nutritional stressors and of the concomitant progressive increase in proinflammatory markers found in older adults [59]. Inflamm-aging is a multifactorial and systemic process, characterized by complex interactions of a plethora of molecular mediators, as exemplified by the nuclear factor (NF)-κB interactome [60]. As populations have been living longer, the human immune system has had to face antigen exposure for several decades more than in our recent evolutionary past. Centenarians demonstrate a complex balance between proinflammatory and anti-inflammatory pathways, and the net result is a slower, more limited and balanced development of inflamm-aging. Disruption of this balance may be significant because it explains, to a certain extent, the continuing rise in noncommunicable diseases associated with inflammation, such as atherosclerosis, cardiovascular diseases, type 2 diabetes, metabolic syndrome, sarcopenia, osteoporosis, cognitive decline, and frailty.

Thus, a continuous low-level physiologic inflammatory tone preserves the symbiotic nature of the microbiota–host relationship, which, in turn, is important for the development and homeostasis of the human immune system. *Meta-flammation*

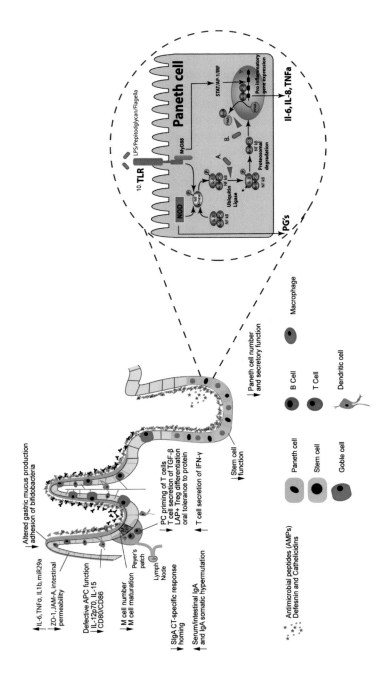

FIGURE 9.2 Summary of ageing on the immune function of the gut. *APC*, antigen presenting cell. In the aging gut, complex changes occur in microbial ecology a reduction in biodiversity, which is a vital part of the intestinal defense mechanism; once luminal populations of commensal microbiota are depleted, the intestine is vulnerable to the detrimental effects of pathologic bacteria. For example, certain bacteria have been shown to specifically modulate the TLR pathway for the benefit of the host, and they are able to directly interfere with inflammatory response transcription factors such as NF-κB. (A) Nonpathogenic *Salmonella* antagonizes IκB ubiquitination; (B) *Bacteroides thetaiotaomicron* upregulates the PPARγ-dependent export of RelA.

(or metabolic inflammation) describes the orchestrated response of cells to excess nutrients, as seen in obese states, which induce stress in insulin-sensitive organs and tissues such as the pancreas, adipose tissue, liver, and muscle. This response, in turn, leads to an upregulation and release of cytokines, chemokines, and adipokines. Visceral adipose tissue, which increases quantitatively with age, is also invaded by activated macrophages and T cells, which, in turn, are able to induce the activation of multiple signaling and further exacerbate the inflammatory state.

For example, defensins (Fig. 9.2) have been shown to alter the commensal bacterial population in the gut of mice, decreasing the *Firmicutes/Bacteroidetes* ratio. Commensal bacteria in the human gut are resistant to the action of secretory immunoglobulin A and are found to be coated with immunoglobulins. Experiments in GF mice have also demonstrated that invariant natural killer T (iNKT) cells accumulate in the colonic lamina propria and the lungs, resulting in increased morbidity in models of IBD and allergic asthma compared with that of specific GF mice. This accumulation is associated with increased intestinal and pulmonary expression of the chemokine ligand CXCL16, which is not associated with increased mucosal iNKT cells. Colonization of neonatal, but not adult, GF mice with the conventional microbiota is protective of mucosal iNKT accumulation and related pathology. These results suggest that age-sensitive contact with commensal microbes is critical for establishing mucosal iNKT cell tolerance to later environmental exposures [64].

Another source of recent interest has been the description of cellular senescence. During life, cells are continually exposed to a variety of damaging agents (eg, ionizing radiation and oxygen free radicals, among others) as well as to proinflammatory cytokines. Proliferating cells can initiate an additional response by adopting a state of permanent cell cycle arrest, termed "cellular senescence"; it is proposed to be a tumor suppression mechanism that possibly contributes to aging. The senescent phenotype is accompanied by upregulation of the DNA damage response system and by the robust secretion of numerous growth factors, proinflammatory cytokines (eg, interleukin-6 [IL-6], IL-8), proteases, and other proteins, globally named *senescence-associated secretory phenotype* [10]. Senescent cells, over time, develop a phenotype that becomes increasingly complex, with both beneficial effects (such as tumor suppression and tissue repair) and deleterious effects (tumor promotion and ageing) on the health of the organism. A large number of environmental stressors may trigger the senescence state, but it seems that it this susceptible to dietary manipulation and microbiome–host interactions; regulation of colonic luminal polyamines produced by intestinal microbiota delays senescence in mice [68]

9.6 DISEASES OF THE AGING POPULATION: ALZHEIMER'S DISEASE AND DEMENTIA

Although the microbiome is linked to numerous disease of old age, it is pertinent to focus on one of the most important: dementia. Globally, 44 million

TABLE 9.2 Summary Data of Studies that Have Examined How the Microbiome Modulates the Immune Response

Author	Year	Bacteria	Mechanism	Model	Disease Implications
Atarashi et al. [61]	2011	Clostridia class	T-cell differentiation	Mouse	Colitis
Gaboriau-Routhiau et al. [62]	2009	Segmented filamentous bacteria (SFB)	T-helper cell response	Gnotobiotic mice	Postnatal microbiome development
Atarashi et al. [63]	2014	Clusters IV, XIVa, and XVIII of Clostridia	Treg and interleukin (IL)-10 induction	Mice	Colitis
Olszak et al. [64]	2012	Commensal colonization	CXCL16 regulated modulation of invariant natural killer cells (iNKT)	Mouse (germ-free)	Colitis, asthma
Habil et al. [65]	2014	*Lactobacillus casei* strain *Shirota* (LcS), *Lactobacillus fermentum* strain MS15	Cathelicidin-related peptides and defensin production	Caco-2	Colitis
Chinen (Nature Comms) [66]	2011	–	Suppressor of cytokine signaling 1 (SOCS1) PGE2	Mice	Colitis
Yoshimoto [67]	2013	–	Senescence-associated secretory phenotype (SASP)	Mice	Hepatocellular carcinoma, obesity

people are living with dementia, and this number is predicted to nearly double every 20 years. With one case diagnosed every 4 s, and a global economic cost of US$604 billion, which is 1% of global gross domestic product (GDP), the challenge posed by the social and economic burden of dementia is unequivocal [69]. A major secretory product of microbes is amyloid, and the contribution of microbial amyloid to the pathophysiology of the human central nervous system (CNS) and to the cause of dementia is potentially substantial. Early work suggested that amyloids of microbial origins may serve as part of an immune-evasive strategy. However, it has now been determined that humans have a very large systemic burden of amyloid that could contribute to the pathology of progressive neurologic diseases with an amyloidogenic component, such as AD. Amyloidogenic fungal proteins and diffuse mycoses have been found in the blood of patients with AD, which suggests that chronic fungal infection may increase AD risk [70]. *E. coli* also secretes extracellular amyloids, known as *curli fibers*, composed of the major structural subunit CsgA. These are common secretory components that facilitate surface attachment and adhesion, biofilm development, and protection against host defenses [71].

Biofilms are a matrix of extracellular polymeric amyloids and other lipoproteins in a variation of structural forms. The extracellular 17.7 kDa CsgA amyloid precursor contains a pathogen-associated molecular pattern, which, like the Aβ42 peptide, is recognized by the human immune system TLR2. Of the 13 currently identified TLRs (TLR1–TLR13), the microglial TLR2s are activated by amyloid, lipoproteins, and other microbial triggers, which subsequently induce cytokine production, inflammation, phagocytosis, and innate immune defense responses that directly impact CNS homeostasis and drive neuropathology. More specifically, the TLR2/TLR1 complex can recognize biofilm-associated amyloids produced by *Firmicutes*, *Bacteroidetes*, and *Proteobacteria* [72]. An expanding list of bacterial amyloid systems include those associated with gram-negative species such as *Pseudomonas* and others, suggesting that functional amyloids are a widespread phenomenon utilized by a wide diversity of microbiome bacteria. Indeed, the extremely large number and variety of microbiome bacteria and their capability to produce vast quantities of amyloids indicates that human physiology may be potentially exposed to a tremendous systemic amyloid burden, especially during aging when the GI tract epithelial and blood–brain barriers become significantly more restructured and permeable.

A major implication of this work is that it may be possible to identify a dementia-associated microbiome. Critically, this could be modified through probiotic, nutritional, or even antibiotic strategies to reduce the burden of this epidemic. However, at present longitudinal data are missing, and further mechanistic evidence is required.

9.7   PERSONALIZED HEALTH CARE IN AN AGING POPULATION

In the past 10 years, the average number of items prescribed for each person per year in the United Kingdom has increased by 53.8%, from 11.9 in 2001 to 18.3 in 2011 [73]. This is largely being driven by the rise in the number of frail older adults. Polypharmacy not only represents a major economic burden, but it carries significant health risks. The gut microbiome plays a significant role in the metabolism of many drugs commonly used by older people. Adverse drug reactions are at least twice as common in older adults compared with younger adults, but the side effects in this age group are not well documented, since most clinical trials exclude patients older than 75–80 years.

For example, the commonly used cardiac drug digoxin is inactivated in the gut by the actinobacterium *Eggerthella lenta*. The mechanism for this inactivation is based on a cytochrome-encoding operon, which is upregulated by digoxin but is inhibited by arginine, which is absent in nonmetabolizing *Eg. lenta* strains, and is predictive of digoxin inactivation by the human gut microbiome. Pharmacokinetic studies using gnotobiotic mice have revealed that dietary protein reduces the in vivo microbial metabolism of digoxin, with significant changes to drug concentration in serum and urine [74]. Further examples of the importance of the gut microbiome in drug metabolism can be found in paracetamol [75] and NSAIDs. Up to 70% of older adults use analgesics, and NSAIDs are the most widely used analgesics at a rate of 50%. NSAIDs, however, pose significant risks to older patients, including the risk of GI bleed and renal failure. Recent findings have also suggested that there are differences in microbial markers and microbiota composition between older NSAID users and nonusers. The total number of microbes is higher in older adults who use NSAIDs than in those who do not. At the genus level, reductions were demonstrated in the proportion of known butyrate producers belonging to *Clostridium* cluster XIVa, such as *Roseburia* and *Ruminococcus*. Moreover, in the *Actinobacteria* group, lower numbers of *Collinsella* spp. were evident in older subjects using NSAIDs compared both with young adults and older adults not using NSAIDs, which suggests that the use of NSAIDs along with an advanced age may also influence the composition of the intestinal microbiota. Furthermore, relatively high numbers of *Lactobacillus* appeared only in the older subjects not using NSAIDs. In general, the number of microbial members in the major phyla was lower in these older adults. Some evidence suggests that it is possible to influence the microbiome through synbiotics to optimize the function NSAIDs, but further data are required [76].

The biguanide drug metformin is widely prescribed to treat type 2 diabetes and metabolic syndrome, but it also increases lifespan in *C. elegans* co-cultured with *E. coli*. This bacterium exerts complex nutritional and pathogenic effects on its nematode predator/host, and these effects impact health and aging. However, metformin is able to increase the lifespan of *C. elegans* by altering microbial folate and methionine metabolism. Alterations in metformin-induced longevity by mutation of worm methionine synthase and *S*-adenosylmethionine synthase point to metformin-induced methionine restriction in the host, consistent with the action of this drug as a DR mimetic. Metformin thus increases or decreases worm lifespan depending on the *E. coli* strain, metformin sensitivity, and glucose concentration. In mammals, the intestinal microbiome influences host metabolism, including development of metabolic disease. Thus, metformin-induced alteration of microbial metabolism could contribute to therapeutic efficacy and also to its side effects, which include folate deficiency and GI upset [77]. This finding is further supported by studies where *C. elegans* treated with the antibiotic trimethoprim, which inhibits dihydrofolate reductase in bacteria, or *E. coli* mutants treated with suppressed folate metabolism also demonstrate an increased lifespan [78]. Although reduced dietary methionine is known to promote longevity in both mammals and *Drosophila*, these data add a new level of complexity to how pro-longevity drugs might mediate their effects on an organism indirectly by affecting the metabolites in the associated microbiota.

CDAD is a common problem in geriatric care, especially in hospitalized older patients with co-morbid illnesses and ongoing exposure to antibiotics. Antibiotic use has been shown to reduce both the richness and the diversity of the gut microbiome in older adults, with reductions in populations of bifidobacteria and *Bacteroides-Prevotella* and increases in *Lactobacillus* and clostridial diversity [79]. This observation implies that antibiotic therapy may have a long-term effect on the intestinal microbiota in older adults. Manipulation of the fecal microbiota using fecal transplantation from healthy donors is a plausible approach to treating recurrent diseases refractory to repeated antibiotic treatment. Numerous randomized controlled trials have now shown fecal microbiota transplantation (FMT) to be effective in the treatment of recurrent *C. difficile* infection. Forty-three individuals were randomized to receive FMT plus oral vancomycin treatment for 4–5 days, oral vancomycin treatment for 14 days alone, or oral vancomycin treatment for 14 days plus bowel lavage on day 4 or 5. Enrolled subjects were individuals with a mean age greater than 65 in all three treatment arms. The primary endpoint was cure without evidence of relapse within 10 weeks of treatment initiation. However, this is still an ineffective instrument in the treatment of older patients, who are often frail, and its wider deployment requires more data. More importantly, there is no evidence that fecal transplantation influences aging.

9.8   WHO WANTS TO LIVE FOREVER?

The immunomodulatory and nutritional properties of the gut microbiota must now be considered and exploited as powerful tools to promote healthy aging and to extend lifespan. However, some fundamental questions have still to be answered. For example, can transplanting a "DR microbiome" promote healthy aging? More precisely, can we identify specific bacteria and their functions that can promote health in old age and prevent disease? Other targets include dietary interventions that mimic chronic DR (periodic fasting mimicking diets, protein restriction, etc.).

Systems biology approaches are providing us with novel mechanistic insights that can be mined to develop personalized therapeutic strategies, biomarkers, and preventive nutritional approaches. Metabolic phenotyping, therefore, provides a compelling platform from which to study the aging process, since it allows us to study critical gene–environment interactions that influence the biology of aging in a dynamic and time-dependent manner. Moreover, metabolic phenotyping also provides us with unique functional insights into the role of the microbiome in aging; metagenomic cataloguing exercises are unlikely to deliver these alone. As has been discussed here, aging is linked to polypharmacy and complex drug–host–microbiome interactions that influence both drug efficacy and toxicity. Metabolic phenotyping and age-specific pharmaco-metabonomic approaches will provide personalized strategies for reducing the economic and health burdens imposed by polypharmacy and for improving the safety of drugs in this cohort. Translational studies will now determine whether systems medicine technologies are able to deliver next-generation biomarkers and therapeutic targets for the growing epidemic of age-related diseases such as dementia, as we enter an unprecedented chapter in the history of human aging.

REFERENCES

[1] Fontana L, Partridge L, Longo VD. Extending healthy life span—from yeast to humans. Science 2010;328:321–6.

[2] Beard JR, Bloom DE. Towards a comprehensive public health response to population ageing. Lancet 2015;385:658–61.

[3] Bloom DE, Chatterji S, Kowal P, Lloyd-Sherlock P, McKee M, Rechel B, et al. Macroeconomic implications of population ageing and selected policy responses. Lancet 2015;385:649–57.

[4] Mortality GBD, Causes of Death C Global, regional, and national age-sex specific all-cause and cause-specific mortality for 240 causes of death, 1990–2013: a systematic analysis for the Global Burden of Disease Study 2013. Lancet 2015;385:117–71.

[5] Mathers CD, Stevens GA, Boerma T, White RA, Tobias MI. Causes of international increases in older age life expectancy. Lancet 2015;385:540–8.

[6] Deelen J, Beekman M, Capri M, Franceschi C, Slagboom PE. Identifying the genomic determinants of aging and longevity in human population studies: progress and challenges. Bioessays 2013;35:386–96.

[7] Lee PG, Cigolle C, Blaum C. The co-occurrence of chronic diseases and geriatric syndromes: the health and retirement study. J Am Geriatr Soc 2009;57:511–16.

[8] Zhou YY, Ji XF, Fu JP, Zhu XJ, Li RH, Mu CK, et al. Gene transcriptional and metabolic profile changes in mimetic aging mice induced by D-galactose. PLoS One 2015;10:e0132088.

[9] Shen Y, Gao H, Shi X, Wang N, Ai D, Li J, et al. Glutamine synthetase plays a role in D-galactose-induced astrocyte aging *in vitro* and *in vivo*. Exp Gerontol 2014;58:166–73.

[10] Zahn JM, Sonu R, Vogel H, Crane E, Mazan-Mamczarz K, Rabkin R, et al. Transcriptional profiling of aging in human muscle reveals a common aging signature. PLoS Genet 2006;2:e115.

[11] Craig T, Smelick C, Tacutu R, Wuttke D, Wood SH, Stanley H, et al. The Digital ageing atlas: integrating the diversity of age-related changes into a unified resource. Nucleic Acids Res 2015;43:D873–8.

[12] Kennedy SR, Loeb LA, Herr AJ. Somatic mutations in aging, cancer and neurodegeneration. Mech Ageing Dev 2012;133:118–26.

[13] Laurie CC, Laurie CA, Rice K, Doheny KF, Zelnick LR, McHugh CP, et al. Detectable clonal mosaicism from birth to old age and its relationship to cancer. Nat Genet 2012;44:642–50.

[14] Jacobs KB, Yeager M, Zhou W, Wacholder S, Wang Z, Rodriguez-Santiago B, et al. Detectable clonal mosaicism and its relationship to aging and cancer. Nat Genet 2012;44:651–8.

[15] Sondheimer N, Glatz CE, Tirone JE, Deardorff MA, Krieger AM, Hakonarson H. Neutral mitochondrial heteroplasmy and the influence of aging. Hum Mol Genet 2011;20:1653–9.

[16] Andrew T, Calloway CD, Stuart S, Lee SH, Gill R, Clement G, et al. A twin study of mitochondrial DNA polymorphisms shows that heteroplasmy at multiple sites is associated with mtDNA variant 16093 but not with zygosity. PLoS One 2011;6:e22332.

[17] de Magalhaes JP, Curado J, Church GM. Meta-analysis of age-related gene expression profiles identifies common signatures of aging. Bioinformatics 2009;25:875–81.

[18] Hong MG, Myers AJ, Magnusson PK, Prince JA. Transcriptome-wide assessment of human brain and lymphocyte senescence. PLoS One 2008;3:e3024.

[19] Passtoors WM, Beekman M, Deelen J, van der Breggen R, Maier AB, Guigas B, et al. Gene expression analysis of mTOR pathway: association with human longevity. Aging Cell 2013;12:24–31.

[20] Deelen J, Beekman M, Uh HW, Broer L, Ayers KL, Tan Q, et al. Genome-wide association meta-analysis of human longevity identifies a novel locus conferring survival beyond 90 years of age. Hum Mol Genet 2014;23:4420–32.

[21] Horvath S, Zhang Y, Langfelder P, Kahn RS, Boks MP, van Eijk K, et al. Aging effects on DNA methylation modules in human brain and blood tissue. Genome Biol 2012;13:R97.

[22] Heyn H, Moran S, Esteller M. Aberrant DNA methylation profiles in the premature aging disorders Hutchinson–Gilford Progeria and Werner syndrome. Epigenetics 2013;8:28–33.

[23] Alisch RS, Barwick BG, Chopra P, Myrick LK, Satten GA, Conneely KN, et al. Age-associated DNA methylation in pediatric populations. Genome Res 2012;22:623–32.

[24] Patti GJ, Tautenhahn R, Johannsen D, Kalisiak E, Ravussin E, Bruning JC, et al. Meta-analysis of global metabolomic data identifies metabolites associated with life-span extension. Metabolomics 2014;10:737–43.

[25] Calvani R, Brasili E, Pratico G, Capuani G, Tomassini A, Marini F, et al. Fecal and urinary NMR-based metabolomics unveil an aging signature in mice. Exp Gerontol 2014;49:5–11.

[26] Menni C, Kastenmuller G, Petersen AK, Bell JT, Psatha M, Tsai PC, et al. Metabolomic markers reveal novel pathways of ageing and early development in human populations. Int J Epidemiol 2013;42:1111–9.

[27] Ansoleaga B, Jove M, Schluter A, Garcia-Esparcia P, Moreno J, Pujol A, et al. Deregulation of purine metabolism in Alzheimer's disease. Neurobiol Aging 2015;36:68–80.

[28] Jove M, Mate I, Naudi A, Mota-Martorell N, Portero-Otin M, De la Fuente M, et al. Human aging is a metabolome-related matter of gender. J Gerontol A Biol Sci Med Sci 2015. 2015 May 26. pii: glv074. [Epub ahead of print].

[29] Swann JR, Spagou K, Lewis M, Nicholson JK, Glei DA, Seeman TE, et al. Microbial-mammalian cometabolites dominate the age-associated urinary metabolic phenotype in Taiwanese and American populations. J Proteome Res 2013;12:3166–80.

[30] Lucanic M, Held JM, Vantipalli MC, Klang IM, Graham JB, Gibson BW, et al. N-acylethanolamine signalling mediates the effect of diet on lifespan in *Caenorhabditis elegans*. Nature 2011;473:226–9.

[31] Jove M, Naudi A, Ramirez-Nunez O, Portero-Otin M, Selman C, Withers DJ, et al. Caloric restriction reveals a metabolomic and lipidomic signature in liver of male mice. Aging Cell 2014;13:828–37.

[32] Lawler DF, Larson BT, Ballam JM, Smith GK, Biery DN, Evans RH, et al. Diet restriction and ageing in the dog: major observations over two decades. Br J Nutr 2008;99:793–805.

[33] Rezzi S, Martin FP, Shanmuganayagam D, Colman RJ, Nicholson JK, Weindruch R. Metabolic shifts due to long-term caloric restriction revealed in nonhuman primates. Exp Gerontol 2009;44:356–62.

[34] Richards SE, Wang Y, Claus SP, Lawler D, Kochhar S, Holmes E, et al. Metabolic phenotype modulation by caloric restriction in a lifelong dog study. J Proteome Res 2013;12:3117–27.

[35] Zhou B, Yang L, Li S, Huang J, Chen H, Hou L, et al. Midlife gene expressions identify modulators of aging through dietary interventions. Proc Natl Acad Sci USA 2012;109:E1201–9.

[36] Zhang C, Li S, Yang L, Huang P, Li W, Wang S, et al. Structural modulation of gut microbiota in life-long calorie-restricted mice. Nat Commun 2013;4:2163.

[37] Mouchiroud L, Molin L, Kasturi P, Triba MN, Dumas ME, Wilson MC, et al. Pyruvate imbalance mediates metabolic reprogramming and mimics lifespan extension by dietary restriction in *Caenorhabditis elegans*. Aging Cell 2011;10:39–54.

[38] Chin RM, Fu X, Pai MY, Vergnes L, Hwang H, Deng G, et al. The metabolite alpha-ketoglutarate extends lifespan by inhibiting ATP synthase and TOR. Nature 2014;510:397–401.

[39] Tachon S, Zhou J, Keenan M, Martin R, Marco ML. The intestinal microbiota in aged mice is modulated by dietary resistant starch and correlated with improvements in host responses. FEMS Microbiol Ecol 2013;83:299–309.

[40] Eisenberg T, Schroeder S, Andryushkova A, Pendl T, Kuttner V, Bhukel A, et al. Nucleocytosolic depletion of the energy metabolite acetyl-coenzyme a stimulates autophagy and prolongs lifespan. Cell Metab 2014;19:431–44.

[41] Ruckenstuhl C, Netzberger C, Entfellner I, Carmona-Gutierrez D, Kickenweiz T, Stekovic S, et al. Lifespan extension by methionine restriction requires autophagy-dependent vacuolar acidification. PLoS Genet 2014;10:e1004347.

[42] Moles L, Gomez M, Heilig H, Bustos G, Fuentes S, de Vos W, et al. Bacterial diversity in meconium of preterm neonates and evolution of their fecal microbiota during the first month of life. PLoS One 2013;8:e66986.

[43] Koenig JE, Spor A, Scalfone N, Fricker AD, Stombaugh J, Knight R, et al. Succession of microbial consortia in the developing infant gut microbiome. Proc Natl Acad Sci USA 2011;108(Suppl. 1):4578–85.

[44] Biagi E, Candela M, Fairweather-Tait S, Franceschi C, Brigidi P. Aging of the human metaorganism: the microbial counterpart. Age 2012;34:247–67.

[45] Mariat D, Firmesse O, Levenez F, Guimaraes V, Sokol H, Dore J, et al. The Firmicutes/Bacteroidetes ratio of the human microbiota changes with age. BMC Microbiol 2009;9:123.

[46] Zwielehner J, Liszt K, Handschur M, Lassl C, Lapin A, Haslberger AG. Combined PCR-DGGE fingerprinting and quantitative-PCR indicates shifts in fecal population sizes and diversity of Bacteroides, bifidobacteria and Clostridium cluster IV in institutionalized elderly. Exp Gerontol 2009;44:440–6.

[47] Claesson MJ, Jeffery IB, Conde S, Power SE, O'Connor EM, Cusack S, et al. Gut microbiota composition correlates with diet and health in the elderly. Nature 2012;488:178–84.

[48] Hayashi H, Sakamoto M, Kitahara M, Benno Y. Molecular analysis of fecal microbiota in elderly individuals using 16S rDNA library and T-RFLP. Microbiol Immunol 2003;47:557–70.

[49] Mueller S, Saunier K, Hanisch C, Norin E, Alm L, Midtvedt T, et al. Differences in fecal microbiota in different European study populations in relation to age, gender, and country: a cross-sectional study. Appl Environ Microbiol 2006;72:1027–33.

[50] Biagi E, Nylund L, Candela M, Ostan R, Bucci L, Pini E, et al. Through ageing, and beyond: gut microbiota and inflammatory status in seniors and centenarians. PLoS One 2010;5:e10667.

[51] Makivuokko H, Tiihonen K, Tynkkynen S, Paulin L, Rautonen N. The effect of age and nonsteroidal anti-inflammatory drugs on human intestinal microbiota composition. Br J Nutr 2010;103:227–34.

[52] Lahti L, Salojarvi J, Salonen A, Scheffer M, de Vos WM. Tipping elements in the human intestinal ecosystem. Nat Commun 2014;5:4344.

[53] Rampelli S, Candela M, Turroni S, Biagi E, Collino S, Franceschi C, et al. Functional metagenomic profiling of intestinal microbiome in extreme ageing. Aging 2013;5:902–12.

[54] Langille MG, Meehan CJ, Koenig JE, Dhanani AS, Rose RA, Howlett SE, et al. Microbial shifts in the aging mouse gut. Microbiome 2014;2:50.

[55] Guo L, Karpac J, Tran SL, Jasper H. PGRP-SC2 promotes gut immune homeostasis to limit commensal dysbiosis and extend lifespan. Cell 2014;156:109–22.

[56] Zhang R, Hou A. Host–microbe interactions in *Caenorhabditis elegans*. ISRN Microbiol 2013;2013:356451.

[57] Heintz C, Mair W. You are what you host: microbiome modulation of the aging process. Cell 2014;156:408–11.

[58] Man AL, Gicheva N, Nicoletti C. The impact of ageing on the intestinal epithelial barrier and immune system. Cell Immunol 2014;289:112–8.

[59] Franceschi C, Bonafe M, Valensin S, Olivieri F, De Luca M, Ottaviani E, et al. Inflamm-aging. An evolutionary perspective on immunosenescence. Ann N Y Acad Sci 2000;908:244–54.

[60] Cevenini E, Monti D, Franceschi C. Inflamm-ageing. Curr Opin Clin Nutr Metab Care 2013;16:14–20.

[61] Atarashi K, Tanoue T, Shima T, Imaoka A, Kuwahara T, Momose Y, et al. Induction of colonic regulatory T cells by indigenous Clostridium species. Science 2011;331:337–41.

[62] Gaboriau-Routhiau V, Rakotobe S, Lecuyer E, Mulder I, Lan A, Bridonneau C, et al. The key role of segmented filamentous bacteria in the coordinated maturation of gut helper T cell responses. Immunity 2009;31:677–89.

[63] Atarashi K, Tanoue T, Oshima K, Suda W, Nagano Y, Nishikawa H, et al. Treg induction by a rationally selected mixture of Clostridia strains from the human microbiota. Nature 2013;500:232–6.

[64] Olszak T, An D, Zeissig S, Vera MP, Richter J, Franke A, et al. Microbial exposure during early life has persistent effects on natural killer T cell function. Science 2012;336:489–93.

[65] Habil N, Abate W, Beal J, Foey AD. Heat-killed probiotic bacteria differentially regulate colonic epithelial cell production of human beta-defensin-2: dependence on inflammatory cytokines. Benefic Microbes 2014;5:483–95.

[66] Chinen T, Komai K, Muto G, Morita R, Inoue N, Yoshida H, et al. Prostaglandin E2 and SOCS1 have a role in intestinal immune tolerance. Nat Commun 2011;2:190.

[67] Yoshimoto S, Loo TM, Atarashi K, Kanda H, Sato S, Oyadomari S, et al. Obesity-induced gut microbial metabolite promotes liver cancer through senescence secretome. Nature 2013;499:97–101.

[68] Kibe R, Kurihara S, Sakai Y, Suzuki H, Ooga T, Sawaki E, et al. Upregulation of colonic luminal polyamines produced by intestinal microbiota delays senescence in mice. Sci Rep 2014;4:4548.

[69] [No authors listed] Addressing global dementia. Lancet 2014;383:2185.

[70] Alonso R, Pisa D, Marina AI, Morato E, Rabano A, Carrasco L. Fungal infection in patients with Alzheimer's disease. J Alzheimers Dis 2014;41:301–11.

[71] Asti A, Gioglio L. Can a bacterial endotoxin be a key factor in the kinetics of amyloid fibril formation? J Alzheimers Dis 2014;39:169–79.

[72] Hill JM, Lukiw WJ. Microbial-generated amyloids and Alzheimer's disease (AD). Front Aging Neurosci 2015;7:9.

[73] Wise J. Polypharmacy: a necessary evil. BMJ 2013;347:f7033.

[74] Haiser HJ, Gootenberg DB, Chatman K, Sirasani G, Balskus EP, Turnbaugh PJ. Predicting and manipulating cardiac drug inactivation by the human gut bacterium *Eggerthella lenta*. Science 2013;341:295–8.

[75] Clayton TA, Baker D, Lindon JC, Everett JR, Nicholson JK. Pharmacometabonomic identification of a significant host–microbiome metabolic interaction affecting human drug metabolism. Proc Natl Acad Sci USA 2009;106:14728–33.

[76] Bjorklund M, Ouwehand AC, Forssten SD, Nikkila J, Tiihonen K, Rautonen N, et al. Gut microbiota of healthy elderly NSAID users is selectively modified with the administration of *Lactobacillus acidophilus* NCFM and lactitol. Age 2012;34:987–99.

[77] Cabreiro F, Au C, Leung KY, Vergara-Irigaray N, Cocheme HM, Noori T, et al. Metformin retards aging in *C. elegans* by altering microbial folate and methionine metabolism. Cell 2013;153:228–39.

[78] Virk B, Correia G, Dixon DP, Feyst I, Jia J, Oberleitner N, et al. Excessive folate synthesis limits lifespan in the *C. elegans*: *E. coli* aging model. BMC Biol 2012;10:67.

[79] O'Sullivan O, Coakley M, Lakshminarayanan B, Conde S, Claesson MJ, Cusack S, et al. Alterations in intestinal microbiota of elderly Irish subjects post-antibiotic therapy. J Antimicrob Chemother 2013;68:214–21.

Chapter 10

Phenome Centers and Global Harmonization

Seth Chitayat[1,2] and John F. Rudan[2,3]

[1]*Department of Biomedical and Molecular Sciences, Queen's University, Kingston, ON, Canada* [2]*Department of Surgery, Queen's University, Kingston, ON, Canada* [3]*Human Mobility Research Centre, Queen's University and Kingston General Hospital, Kingston, ON, Canada*

Chapter Outline

10.1 PHENOMICS AND THE HUMAN CONDITION

Phenome is a term that describes the measurable physical and chemical outcomes of the interactions between genes and the environments experienced by individuals and organisms [1]. *Phenomics*, which is a branch of science that explores the basis of how our genes respond to environmental changes, is emerging as a powerful approach to unearth important human attributes at the

E. Holmes, J.K. Nicholson, A.W. Darzi & J.C. Lindon (Eds): Metabolic Phenotyping in Personalized and Public Healthcare. DOI: http://dx.doi.org/10.1016/B978-0-12-800344-2.00010-0

molecular level to explain how we adapt and how and why we are affected by disease [2–6]. Notionally, phenomics is very similar to Newton's Third Law of Motion [7], where, in general terms, the forces that we exert on our environment will result in the environment exerting "a force that is equal in magnitude and opposite in direction" on us. It is also similar to the concept of homeostasis, as previously described by Claude Bernard and Walter Cannon in that the internal conditions in cells and other living organisms undergo significant changes to maintain stability in the presence of changing external conditions.[1] Although framing phenomics in this way helps structure the relationship we share with our environment (eg, climate) at a macro-level, it also embraces the importance of paying attention to the micro-environmental impact (eg, diet, medicines) on our physiology. It recognizes that everything from the rays in sunshine we absorb, all the way through to the foods we eat, interacts with our genes in nondiscrete and sometimes unpredictable ways, and this gives rise to the key personalized nature of the human phenome.

The inclusion of these macro- and micro-environmental contributions in phenomics helps resolve the subtleties that explain the basis for why humans respond differently to similar environmental conditions. At the extreme, we certainly see this in the case of identical, monozygotic twins, whose genes are identical, but their phenomes are very different, and rarely do they interact with their environment in the same way. Phenomics captures our personalized experience with our surroundings. For example, what is the molecular basis for some smokers developing lung cancer whereas other smokes do not? Similarly, why do some individuals with a body mass index (BMI) less than 30 still have type II diabetes as opposed to some obese individuals who do not[2] [8]? Why is diabetes more prevalent in some populations than in others? What is the underlying molecular reason(s) for why patients respond to a therapy while on a clinical trial? These unresolved questions limit our understanding of the human condition, which is the gap that phenomics will bridge.

We are very much in the infancy of learning about the interaction between humans and their environments. The reason for this is because scientific initiatives have always focused inward, primarily on us, as a way to expand our knowledge about the molecular basis of our human physiology [9–13]. This is evidenced by large-scale investments in genomic and proteomic projects, which have helped define our genome and the structure, function, and interaction networks of proteins that comprise our proteome. We have used this accumulated wealth of knowledge to define disease on a genetic basis, with the expectation that we could develop a correspondence between such alterations in our genetic code and specific disruptions in protein structure and function. Such mutations that serve to destabilize the tertiary structure and function of proteins cause its cellular function to cease, which ultimately manifests in the onset and progression of disease. Although there are many examples, wherein genomic and proteomic analyses are sufficient to explain why disease processes are initiated, it is limited because these data are highly complex and lack the resolution required to capture how cellular processes react to disease. There are 25,000 human genes, 1,000,000 human proteins, and orders of magnitude fewer metabolites; in comparing the numbers alone, we can make more headway by investigating the role of our metabolome to understand homeostasis in physiology and pathobiology [14–17].

The limitation of genomic and proteomic analyses to study disease is also evidenced by a lack of standardized approaches to handle and manipulate these samples as well as nonstandardized ways to collect and analyze their data [18]. If our understanding of disease is limited by simply focusing on genomes and proteomes, we would be excluding important metabolic information that does reflect the true dynamism of our cellular activity, as well as the changing cellular circumstance that is pervasive in chronic disease. In this regard, the introversion of our basic and clinical science has limited this essential understanding of how our macro- and micro-environment help define us [19]. The saying "We are what we eat" cannot fully describe the long-term effects of poor eating habits; similarly, we cannot rely on genomics and proteomics alone to help us fully understand the natural and unnatural changes that occur in our physiology. Given that phenomics is really the sum of our genomic outputs and environmental inputs, the study of phenomes helps us realize the value of our scientific efforts by integrating our knowledge of human genomes and proteomes.

By no means is the aforementioned intended to discredit previous scientific efforts; rather, it is an attempt to hone in on the idea that the abstraction of human existence makes it difficult to characterize it at the outset. Metaphorically, it is very similar to the layers of an onion; we cannot get to its core unless we delicately and painstakingly go through the sequential process of removing every layer. Each layer represents a different level of scientific inquiry, with the expectation that inching our way closer to the core will improve our knowledge of the human condition. Scientists have tried to manage this complex process by orienting themselves to trends that they can capture and measure and have helped define hypotheses and rigorous experimentation. For example, diabetes was detected as early as 1500 BC, but we did not understand the role of insulin in diabetes until 1921 [20]. Despite the international recognition, in the form of a Nobel Prize, for Banting and Best for the discovery of insulin and its use as a strategy to treat patients with diabetes, it clearly would not have been an effective therapy for everyone, given the stratification of the populations with diabetes, as well as the notable differences between type I and type II diabetes [20]. The diabetes story clearly illustrates the idea of the human condition being layered, very similar to an onion, in that major accomplishments in understanding disease may require ongoing experimentation to develop more customized approaches that can help all patients.

However, even in this respect, the reason for our limited understanding of the human condition is not solely where we started from. We must also recognize the role and importance of infrastructure and technology, and combinations thereof, to effectively delineate disease processes. It speaks to the need to ensure that we maintain equilibrium between our innate curiosity and available technology. Sometimes, this leads to the repurposing of technologies such as the use of X-rays in the clinical setting to the use of X-rays for diffraction experiments of protein crystals [21]. Another example is that of nuclear magnetic resonance (NMR), which was discovered in condensed matter in 1945 and then employed, from the late 1950s, as a mainstream method to characterize small molecules; however, it was not until the late 1970s that the same theoretical approach was implemented for imaging and used for clinical purposes [21]. Indeed, the repurposing of nuclear magnetic resonance with mass spectrometry (NMR-MS) as a modality to characterize "-omics"-based questions and opportunities in proteomics and metabolomics has been a vital step [19]. These instruments bridge our desire for inquiry and the technical capabilities in phenomics, which will help unravel the mysteries of the human condition in much the same way as we peel away the layers of an onion.

The pursuit of phenomics-based approaches also recognizes the practical needs and requirements of governments and policymakers to understand how humans interact with their environments so that they can make informed decisions regarding investments in health care. For example, as a globalized society, in which borders have become fluid, the relocation of populations from one country to another is highly disruptive at the level of population health [22]. Indeed, issues relating to forced migration become an issue when we think about the extent of unrest in some parts of the world and its impact on developed countries. If we accept the notion that individual diets, climates, and geographies impact human physiology, then, clearly, carrying forward this interaction to a new ecosystem further complicates how new immigrants respond to their new environments. These persistent changes in population dynamics and exposure to a variety of different foods and therapeutic innovations present unique challenges to scientific inquiry and to the management of health care systems because of the heterogeneity and the complex histories of native and foreign residents of a country. In this regard, what we capture and measure is impacted not only by time and technology but also by other dimensions of change in demographics. Phenomics is a call to action to governments and payers of health care to realize that there is a need to not only better understand population-level markers of disease but also share this information with other countries as a way to recognize the increasing diversity of our nations.

If we continue to formulate our health care policy and best practices on the basis of incomplete data, lack of standardization in how we collect and interpret them, and little capacity for global sharing of information, then we will be only adding to the risk we face with regard to the sustainability of our health care systems. The risk is that inaction will not adequately serve the needs of our complex society with unique health care needs. We need to begin capturing and measuring important trends in a cost-effective and time-sensitive manner to inform our health care practice and health care campaigns based on state-of-the-art basic and clinical science. Even though public health campaigns have increasingly helped promote smoking cessation because smoking is the main cause of lung cancer and is known to cause emphysema and other challenges to our pulmonary system [23]. But what about the long-term adverse effects of marijuana use, that have been demonstrated scientifically?[3] Why is it used to treat people with chronic disease [24]? Phenomics allows us to test this treatment option without the overarching prejudice of the consequences associated with tobacco consumption.

In an effort to curb the rising cost of delivering health care, our scientific approach must be more inclusive and sensitive to the inherent physiologic and molecular differences that reflect the individual ways in which we interact with our environments. This reflects the need for more personalized approaches to health care, which will impact the cost of health care and place a strain on budgets. In the backdrop of highly integrated and multicultural societies, we must develop new strategies that reflect the uniqueness of diseases rather than relying on mass messaging. In this vein, scientists, health care organizations, and governments must work together to adopt and integrate state-of-the-art platforms and modalities to further elucidate these new relationships that we share with our environments so as to reduce the burden of disease. We must not also forget the importance of sharing and integrating these data among different countries to ensure that our health care delivery compensates for increased globalization.

This is why the concept of establishing an international network of phenomics-based research is important. The sharing of phenomics data across different regions may help unearth country-specific metabolic biomarkers or signatures of disease. The sharing of this information among the centers in the network not only reduces the fixed cost of establishing multiple facilities but also helps different health care authorities to better understand how to deliver customized health care to immigrant communities. The setting up of a network of research in this manner will serve to create a worldwide resource that can provide opportunities for international experience for graduate students through internships across participating groups within the network.

Phenomics is a natural and progressive step in this regard because it is dedicated to the comprehensive assessment of human metabolism for understanding health risks and the optimization of disease treatments [14,17,25,26]. The focus on the metabolome, which is the equivalent of all genomic output by the cell, readily allows for the identification of metabolic processes that are selectively disrupted in diseased states [17]. A phenome center dedicated to the study of phenomics has tremendous capacity for characterizing and modeling the metabolic states of individuals and populations. Being transformational in terms of scale of population-level studies, it aims to understand the biochemical inner workings of human beings and the interplay between health and disease; and it seeks to achieve a balance among basic, translational, and clinical research, with a dual focus on population screening and personalized medicine covering all major disease areas. This will lead to new insights into the changing patterns of human disease, which will inform future health care policy and lead to the development of improved diagnostic tests and personalized therapies targeted to an individual's biochemistry. The long-term view of creating an international network of like-minded researchers with aspirations to create their own phenome centers where the analytical technologies are harmonized to be consistent with the need for global data to inform best practice in health care. The long-term goal is to create sharable tools, databases, and resources such as bio-banks with congruent analytical sets that are matched in terms of coverage, quality, and performance to provide an international metabolic bio-resource for the future. Standardization of all protocols is required to ensure that all data can be harmonized and linked in this fashion. Harmonization of data will enable governments, scientists, and policymakers to be informed about geographic and population-level differences to better formulate strategies to deliver personalized health care. The establishment of phenome centers will thus provide a new way to explain why no two hands are the same.

10.2 PHENOMICS: A FOCUS ON OUTPUTS

The study of *phenomes*, which are the measurable physical and chemical outcomes of the interactions between genes and the environments experienced by individuals [1], is very similar to what we see in effective organizations such as governments and businesses; that is, effective organizations focus on outputs as opposed to activities [27]. This is very relevant to the study of disease: disease is the result of the disequilibrium of cellular outputs caused by disrupted circuitry in our genome and proteome. Outputs, in effective organizations, for example, represent measureable outcomes that are of material value to the company and that support its overarching corporate strategy (eg, revenue, sales, profitability) [27]. The number of emails and the amount of related communications that constitute the activities required to generate these outcomes are not relevant. As another example, information technology workers are not paid for the amount of code they write but on the basis of how much value it creates. A focus on metabolic outputs such as modified phospholipids, which are shown to support the pathophysiologic state of the cell, is more helpful to interrogate which genomic and proteomic processes have been disrupted.

The relationship between phenomics and disease is very similar to the analogy used above to describe effective organizations. If we think about the cell as an organization and disease as an overarching condition, or strategy, that controls all of the communication such as the mechanisms and machinery that are internal to the cell (eg, activities), then as scientists we need to shift away from the study of these activities and focus on those molecules that actually cause the disease to progress (eg, output). In other words, phenomics helps shed light on the idea that events such as DNA replication, transcription, translation, and protein–protein interactions are all examples of internal activities that build themselves up in a coordinated fashion to generate specific patterns of metabolites or cellular outputs that, for all intended purposes, serve to reinforce the disease state of the cell (eg, strategy). Similarly, if we focus on key metabolites that are drivers of disease, such as taurine and glucose in breast cancer (eg, outputs) [25], then we should target these molecules and modulate their levels, rather than focusing on more upstream activities that produced these metabolites.

The relationship among phenomics, disease, and genomic output is very similar to the diabetes story above. Interestingly, when we visit our primary care physicians, we always focus on what the body does or does not do, which then lays the groundwork for follow-up tests and other procedures. In other words, there is always a natural tendency for humans to refer to outputs that cannot be readily explained by genomic and proteomic profiles. Metabolic outputs can also be used to identify patients born with phenylketonuria, a disorder in which the body cannot break down the amino acid phenylalanine [28]. This is an example of how a spectral look into the metabolic footprints in urine or blood would be a more robust and expeditious approach to identify this inborn error of metabolism than would complex genomic analyses [29]. Another classic example is hemophilia, wherein the diagnosis of the subtypes, A, B, and C can be readily deduced by analyzing small quantities of blood [30]. Phenome centers that focus on the scientific approach of phenomics can be used to screen infants with this disorder or other inborn errors of metabolism to mitigate any risks to their development [31]. Phenomics-based studies also provide a means for physicians to monitor patients and measure their response to tailored therapies.

In this spirit, phenomics is the next major step in the study of disease because it is more "reactive" in that it emphasizes the detection and measurement of cellular outcomes, as opposed to upstream cellular processes that generate them. For example, the use of metabolic profiling to detect and measure specific metabolic footprints such as modified lipids or sugars allow for indirect measurements of the intracellular activities that gave rise to these outputs at the level of which protein complexes were formed and which set of genes was activated. According to our earlier example, this means that if a corporate strategy is measured in terms of sales, then everyone in the company must be evaluated on this basis; it does not matter how many calls a sales representative makes, all that matters is that he or she is selling.

10.3 THE GENE–PROTEIN AXIS

Despite the accumulated wealth of knowledge about human genomes and proteomes, it is becoming clear that defining disease at the level of a singular disruption along the gene–protein axis is incomplete and insufficient to eradicate the onset and progression of chronic disease [9–13]. This is best illustrated by the case of the *p53* gene in 1979 [32–35], which created a great deal of attention when it was discovered; it was found that alterations of this gene predisposed individuals to cancer development [32,33,36]. Although basic and clinical research of this oncogene has offered new insights into its cellular function, the intense focus on this protein as a singular strategy to cure cancer has not come to fruition, as reflected by the continued rise in the incidence and prevalence rates of cancer and the lack of targeted *p53* therapies in the marketplace to reduce the onset and progression of the disease [37].

One of the reasons for this has to do with the number of *p53*-dependent protein interactions that have been characterized thus far; these interactions are also affected by the protein's propensity for extensive post-translational modifications (PTMs) by methylases, acetylases, and ubiquitylases [38]. In fact, these PTMs have also been shown to affect the nuclear activity of this oncoprotein [39]. The multifunctional nature of p53 through its own "interactome," as defined by a network comprising other effector proteins, showcases the complexity of this protein in cancer biology, as well as the overarching limitations of narrowly focusing on a single gene and its associated protein product to reduce the burden of human disease and improve population health.

Indeed, the gene–protein–disease paradigm is further strained when we expand our view beyond individual genotypes and begin to understand the dynamics of chronic disease at a population level. Although gene profiling through whole genome sequencing is a powerful approach to detect biomarkers of disease, especially with respect to a discrete set of genes such as *BRCA1* and *BRCA2* [40], it does not offer the opportunity to readily explore the consequences of how combinations of normal and mutated genes manifest themselves to affect the onset and progression of disease as we reflect on the history of the *p53* gene. Moreover, given the sheer number of the approximately 25,000 genes comprising the human genome [16], the families of factors that transcribe messages at different times for subsequent translation, and the increasing levels of genetic mutation as the extent of chronic disease progresses, all present significant technical challenges, which limit the opportunity to leverage these large genomic data sets to explore cellular response to external environment.

The same issues of data heterogeneity and complexity described earlier are similar for the use of proteomics as a strategy to curb the onset and progression of disease. The exquisite nature of how our 1,000,000 proteins come together in different cell types to form highly ordered complexes has fueled more than a decade worth of study to understand the structure–function relationships of proteins in isolation and in their complex forms. Although some of the findings from this research have been used to develop a correspondence with the onset and progression of disease, a proportion of this work has exclusively focused on the biophysics of protein folding and stability. The pursuit of this type of fundamental science is important, but there is a lag in capitalizing on this research. This is similar to the plethora of protein structures currently populating the Protein Data Bank (PDB), where in the future, the value of ab initio protein structure determination may be questioned given the inventory of 108,789 structures and the resources it consumes. These arguments are different in that they do not speak to data management issues per se but, rather, to the breadth and complexity of the proteome; this has slowed down efforts to capitalize on the use of the PDB database to inform downstream experimentation and assessment of its contribution to study cellular dysfunction in disease.

In addition to the vastness of our proteome, there are also technical limitations to how we capture structural and biochemical information of these entities. For example, protein expression (eg, membrane) and solubility issues often affect the opportunity to explore high-resolution structural studies of these proteins as well as existing hardware; this has been very well documented in publications describing NMR approaches to protein structure determination [41,42]. Solution NMR has limitations in terms of the molecular weight of the protein of interest that can be studied, thereby reducing the range of therapeutically interesting proteins [43]. Analogously, the use of protein X-ray crystallography is often limited by failure of the protein to fold properly, and although structural biologists are able to determine solution conditions that increase the chance for a protein solution to crystallize, this is only probabilistic and may never yield a diffraction quality protein crystal [44]. Cell biologists making use of similar protein expression and purification protocols may encounter similar challenges in identifying novel complexes that mediate specific cellular processes [42]. In this example, their pull-down experiments may not provide complete information owing to the solubility issues of recombinant proteins and the disruption of protein–protein complexes in the presence of low or high salts and detergents [45]. Not only do these technical limitations hamper our understanding of the proteome, manipulation of recombinant protein samples in this way also limits the opportunity to leverage a subset of proteins in drug discovery efforts. Taken together, this under-representation of our proteome limits our ability to fully make use of this tool in developing strategies to overcome disease threats.

The linear nature of how genes are transcribed and translated into proteins makes it highly desirable for understanding cellular function and to unearth viable targets in drug discovery efforts. Unfortunately, there are several limitations that preclude the use of this linearity at the intersection of how pathophysiology (eg, disease condition) affects cellular function. Aside from the technical limitations described earlier, there are issues associated with resolution that lessen the impact of the conclusions drawn from analyses. In this regard, a focus on genomic output (eg, metabolomics) not only reduces data complexity by honing in on the number of human metabolites [14] but also readily identifies novel targets to inform drug discovery efforts. This encourages efforts to capture and measure disease trends and provides scientists with unique markers to re-engage with genomic and proteomic data, and helps better understand how to modulate the underlying activities and reduce the prevalence of selected metabolic biomarkers. In this regard, a top-down approach (eg, metabolite to gene) works best to target those biochemical pathways that are differently affected in the disease state of the cell. This realization forces a new conversation about the need for new, more sophisticated modalities and customized approaches to this untapped area of research.

10.4 THE DIFFERENCE BETWEEN METABOLOMICS AND PHENOMICS

We have so far referred to metabolomics as the "output" of the cell, which, in principle, represents all the metabolites that our genome and proteome are committed to producing [19]. This is why the study of our metabolome is highly relevant to the study of disease in that from this information, we can trace back and identify the set of genes and proteins that were activated to produce a specific metabolite. The study of metabolites, or metabolomics, therefore, is very similar to forensic science, where a focus on the outcome can still provide scientists with a robust approach to reconstructing and investigating the causes, that is, the set of genes and proteins required to generate different metabolic signatures in different cell types and disease states.

The focus of metabolomics is very much on the study of metabolites, whereas phonemics introduces another layer of complexity and dimensionality by considering the impact of macro- and micro-environments on cellular processes [1,19]. Thus, phenomics is similar to molecular phenotyping. If we imagine the cell as a nanomachine that operates in a vacuum, then we can very easily appreciate the sequential relationship from gene to protein to metabolite. Unfortunately, we know this is not a realistic "-omics" view of the cell because the cell has nothing to respond to at the level of extracellular signals. For instance, proinsulin is not an intracellular molecule that is converted to insulin; rather, it is an extracellular signal that activates many intracellular pathways [46]. This is a classic example of how a changing micro-environment outside the cell produces dynamic shifts in the metabolic footprint in the cell to maintain the homeostatic state of the cell. Similarly, other stresses, such as inflammation, require a different metabolic complement that not only compensates for what the cell requires to handle this condition but must also manage other processes to maintain an appropriate level of cellular function to ensure homeostasis. As phenomics-based approaches become more commonplace through the proliferation of phenome centers, they will help establish how changes in our microbiome induced by either diet or infection contribute to inflammatory and disease processes that we can then attempt to modulate. Although the focus of phenomics is to explore the human condition in its environment, it has the potential to extend these analyses to include questions about how and why bacteria develop resistance to antibiotics. These answers will help inform the development of novel strategies to eradicate antibiotic resistance, which currently costs the United States an estimated 20 billion dollars per year.[4]

Phenomics is such a valuable branch of science because it captures the interplay between metabolomics and the environment. Hence, it is a powerful approach to monitor changing circumstances and fluctuations in metabolic outcomes, which is especially important in the context of early detection of disease and monitoring the events during its progression.

10.5   THE VALUE PROPOSITION OF PHENOMICS RESEARCH

It is worth noting at this time that a society's investment in a number of phenome centers is not duplicative or redundant in any way. It is especially important for governments and the private sector to appreciate this, as both would be important stakeholders in the phenome centers. Notwithstanding the economic climate that many regions are facing, there is a value proposition associated with phenome centers, as they not only will create value for health care and health research but will also serve to fuel new initiatives in drug discovery efforts for cancer, diabetes, and other chronic diseases, including inflammatory bowel disease and Crohn's disease. In this vein, phenome centers are quite different from clinical laboratories not only because of technologic and procedure-based structures but also in their attempts to redirect research from looking at cellular activities toward analyses on outputs. The focus on the human metabolome represents all of the genomic and proteomic circuitry required to generate these outputs, and this captures the essence of how our genes and environments intersect. Similar to other life science and health care initiatives, the phenome center will require highly qualified personnel who are well compensated for their efforts. Finally, sharing of information among members of the phenome center network will help accommodate local fluctuations in populations so that health care delivery and practice can be reoriented accordingly.

10.6   PHENOME CENTERS IN 21ST CENTURY HEALTH RESEARCH AND MEDICINE

The human phenome is a term that describes measurable physical and chemical outcomes of the interactions between genes and environments [1,19]. These interactions also determine both individual disease risks and those of whole populations; phenomic measurements provide insights into how these combinations produce specific biological consequences. Phenome centers, such as those already established and operationalized at Imperial College London, United Kingdom, as well as at Nanyang Technological University in Singapore and Chang Gung University in Taiwan, were built to develop large-scale broad and deep metabolic phenotyping capacity and are dedicated to the study of population-level metabolic profiles.

Briefly, their use of high-throughput and robust platforms such as NMR-MS allow for easy determination of metabolic signatures arising from human biofluids (eg, urine, blood) [47–49]. The enforcement of rigorous and standardized protocols to process and analyze data generated by these phenome centers allows for direct comparison of spectral patterns between "normal" and "diseased" specimens across different geographic regions. The evaluation of metabolic signatures in this way offers scientists and clinicians an unprecedented opportunity to identify salient differences that define the two states (normal and disease) and to design experiments that test hypotheses based on aggregate cellular function as opposed to those anchored around a single genetic mutation.

This simplifies the experimentation required to better understand normal and disease physiology because it focuses on specific metabolic signatures (or biomarkers) that may cause disease onset and progression, as opposed to a detailed understanding of more upstream processes or activities. The acceptance of the phenome center as a foundational tool dedicated to understanding health risks and to optimization of disease treatments not only will help enhance our understanding of the relationships among human genetics, diet, lifestyle, and environment but will also improve drug discovery efforts and strategies to reduce the incidence of chronic diseases at both the population level and the patient level.

10.6.1 Operational View of Phenome Centers

Metabolic phenotypes are the direct products of the interactions among genes, diet, and environment in individuals and are both statistically and biologically linked to disease risk and, ultimately, to response to therapies. To capture the vast range of metabolic products that define a phenotype, it is necessary to use a wide range of spectroscopic tools for exploratory and targeted purposes.

The MRC-NIHR Phenome Center (informally known as the National Phenome Center (NPC)) in the United Kingdom has developed two customizable offerings with underpinning informatics services for a full spectrum of research services to support broad and deep metabolic phenotyping and to deliver insight into complex biological processes. A comprehensive discovery service provides maximally broad and deep metabolome coverage of small molecule and complex lipid biofluid analytes, and guided monitoring services or targeted assays offer focused investigation of a subset of molecular species. Utilizing both predefined and customized approaches, the NPC is able to support the research needs of the scientific community in the United Kingdom. To maintain its dual focus on stratified medicine and epidemiology studies, the Centre is required to support comprehensive biofluid analysis, with an operating capacity of over 200,000 assays per year and the flexibility to accommodate both large (over 5000) and small (fewer than 300) sample sets.

The operational complexity and scale required to set up the NPC required a complete re-evaluation of the working constraints and previously applied methods in the academic sector, which had been primarily devised for the analysis of small- to moderate-sized studies. It was important to implement a highly efficient and robust workflow, from sample delivery to sample preparation and data acquisition, with emphasis on standardization and quality management. Development of methods, systems, and software were necessary to meet these anticipated needs in terms of throughput, efficiency, robustness, and long-term precision. In addition, the design specification of the NPC MS laboratory calls for high-throughput analysis of complex samples and analytical systems that were not designed specifically for this purpose.

Phenome centers capitalize on the strengths of two core analytical platforms: NMR spectroscopy, and ultra-performance liquid chromatography

(UPLC)–MS. Since no one technology platform or assay can provide the comprehensive coverage of the metabolome (because of chemical diversity and dynamic range considerations), from the different technologies available, the NPC utilizes NMR-MS in a robust workflow to provide high-quality exploratory broad and deep metabolic profiling and a focused investigation of a subset of molecular species in a number of high-value targeted assays. These are combined via an innovative and efficient workflow, underpinned by strong bioinformatics resources, to deliver a comprehensive end-to-end biomarker discovery and validation platform to researchers.

NMR spectroscopy generates inherently quantitative, highly robust profiles from each biofluid sample in a few minutes [47–49]. It is unequaled in its ability to provide detailed atom-centered information on molecular structures, facilitating the identification of unknown molecules. With detection limits in the low micromolar range, NMR spectroscopy permits characterization of organic molecules regardless of their physico-chemical properties. The platform is particularly advantageous for metabolic phenotyping, as it is rapid, nondestructive, inherently quantitative, and highly reproducible. It requires minimal sample preparation and no machine conditioning, so it allows the sequential measurement of different kinds of samples (urine, blood serum or plasma).

UPLC–MS offers an excellent combination of sensitivity and selectivity, making it an ideal analytical qualitative and quantitative platform for deep phenotyping [50–52]. UPLC–MS profiles are generated in a series of assays designed to minimize measurement redundancy while maximizing metabolite coverage within human biofluids, including urine and blood products. Using individually tailored chromatographic approaches for lipid profiling (in blood products), small molecule profiling (in urine), and high polarity metabolite profiling (in both sample types), the diverse range of metabolites present in complex biofluids is captured and separated for measurement [50–52]. A second dimension of separation is provided by high-resolution MS systems, conferring a high degree of chemical specificity to each assay. The superior chromatographic resolution and rapid separation of UPLC and the high sensitivity of MS allow or the detection and measurement of picomole to femtomole levels of thousands of metabolites in minutes by using multiple customized assays that capture lipidomic profiles as well as low-molecular-weight and polar molecules.

The analytical pipeline runs from sample registration through to biomarker identification and data modeling and represents a complete and distinctive offering to the UK and international research communities. Using this innovative suite of individually optimized technologies, the Centre is able to generate information-rich multiparameter data sets, providing maximally broad and deep coverage of the metabolic profile. The modular analysis framework integrating comprehensive exploratory metabolic profiling with a number of high value multiparameter targeted tests constitutes an unrivaled analytical service and permits a versatile and tailor-made solution for each collaborator to support biomarker discovery and validation on both large epidemiologic cohorts and smaller studies in stratified medicine.

10.6.2   Informatics—the Key Integrator of Metabolic Profiling and Disease

Notwithstanding the technical details associated with what constitutes a phenome center and how the various technologies work synergistically to assist in the metabolic profiling of patients, we must recognize that investment into these operations is not intended to create yet another clinical laboratory that has highly focused goals and serves very specific and acute needs of patients. Rather, the phenome center, as described earlier, and phenomics, as described in this chapter thus far, serve a much greater purpose because of the emphasis on biomedical informatics to integrate population-level studies with the personalized clinical pathways of patients.

As previously mentioned, the dedication of the phenome center in capturing the presence of thousands of different human metabolites helps strengthen our resolution to understand disease processes and, more importantly, enable us to trace outputs back to the causes to learn more about the delicate interplay between genes and proteins. In essence, phenomics provides scientists with the tools to "recreate the scene," not at the level of single genes and proteins but, rather to the sequential activities—how multiple genes are activated, in which sequence, and how protein–protein complexes are formed. Disease is like a puzzle in which the physical and chemical properties of cells are scrambled, and the study of phenomics in phenome centers would help solve the puzzle by focusing on genomic outputs.

In this regard, the inventory of technical capabilities of NMR-MS systems on their own is insufficient without strong informatics platforms and robust statistical methodologies. As with any "big data," it is key to ensure that the data are validated across multiple samples. For example, the processing of human tissues and related biopsy specimens may require validation by standard immunohistochemistry. With this in mind, being able to mine and interrogate the data with standardized approaches across multiple phenome centers (such as those in the United Kingdom, Singapore, and Taiwan) has the potential to provide unique population-level perspectives within their respective countries, but these harmonized strategies will also help, for the first time, to compare health risks across different geographic regions. In the backdrop of standardized and normalized approaches all the way from handling and processing to analysis of all sample types, the establishment of multiple phenome centers across the world will only enrich our understanding of the human condition. Where appropriate, this network of phenome centers may be well positioned to use data from all the sources to inform health care ministries on how to reduce health care costs.

10.7   THE UTILITY OF PHENOMICS IN HEALTH CARE

In view of the wealth of information and human data the phenome center can produce, expertly manipulate, and interrogate, it is clear that it is an invaluable resource that can be deployed to address key challenges in health care and several areas of health care research, which we will address later in this chapter.

The original concept behind the phenome center was that it would not be a clinical entity serving the same purpose and mission of traditional clinical laboratories. Rather, its focus would be on pursuing molecular epidemiologic studies at a population level to derive a set of statistically validated molecular signatures that correlate with the onset and progression of disease. Other phenome centers, for example, The Imperial College Clinical Phenome Centre, which are keen to use phenomics intraoperatively through the use of the "intelligent Knife" (i-Knife) [53,54], are much more clinically oriented. The execution of this task, in principle, would then provide a "database" that clinicians can use to interrogate the health of their patients as to whether these signatures are present in their urine or blood. This would help stratify patient populations and facilitate targeted interventions to accommodate for individual variations. This is most relevant for the registration of patients into clinical trials as a way to screen their eligibility, for example.

Context is everything when we think about phenomics because of the importance of environment in the modulation of our metabolome. A population-level analysis, therefore, does not mean that everyone needs to give blood and urine as inputs to the phenome center; this would be meaningless, given the lack of a specific environmental context to measure the impact on the metabolome. Rather, phenome centers would have to engage first in disease-oriented projects, where the focus would be on the metabolic profiling of diseases such as Alzheimer's disease and the deep profiling of various cancers and other inflammatory-induced disease states, and antibiotic resistance. The registering of such information, in relation to the patient journey through the health care system, facilitates longitudinal studies, which allow scientists to determine whether the patterns of molecular signatures change as a function of disease progression and to plan interventions accordingly. Indeed, the phenome center platform can also be used to investigate the difference between patients who respond to clinical trials and those who do not on the basis of their metabolic characteristics.

In Canada, as in the case of other countries, health care budgets consume a large portion of the national income, with predictions of future increases owing to an aging population and a rising incidence of chronic diseases [55]. For example, in the province of Ontario, planning has been oriented to reduce the extent of fragmentation by instituting different reimbursement schemes such as quality-based procedures (QBPs); these schemes are meant to improve efficiency and increase the volume of select surgical procedures (eg, knee, hip, and cataract surgeries).[5] These are long-awaited changes, as they begin to force hospitals and other health care organizations responsible for these procedures to conform to best practices and to identify other opportunities that will prevent costly re-admission for a repeat procedure. The utility of phenome centers, as described earlier, in knee replacements and hip replacements can help stratify patient risk and reduce the need for revision surgeries as a result of infection.

These phenomic data can also be used to suggest new treatment modalities to improve patient outcomes. Given the sensitivity of NMR-MS to detect early signals of disease and infection, there is value in the operationalization of phenome centers in that clinicians can intervene earlier to offset infection risks. This is one example wherein phenome centers align well with such change toward using best practices to improve care.

The conceptual framework of a phenome center also aligns itself with other payment schemes such as bundled payments. Briefly, bundled payments move away from paying on a per-procedure basis to an episode of care [56]. In this regard, bundled payments, in part, encourage the formation of interdisciplinary teams, whose collective functions minimize disruption to patient care over a select period. Bundled payments have led to increased integration and amalgamation of hospitals and rehabilitation institutes to structurally align with this method of payment for care. As with QBPs, these methods of payment are capturing more of the patient journey in the health care system. This is consistent with the intent of the phenome center, which seeks to explore human phenomes over the entire patient journey using the operational setup as previously described. Implicit in this is the intent of the center to explore the inter-relationship between genes and the environment not only over the time dimension as defined by the boundaries of payment, but also over the course of the entire patient experience. The phenome center artificially increases the window of episodic care, which allows governments to extract more value from every dollar spent to provide care.

As the boundaries of care between units within the health care system dissolve, as is the case between specialty care and primary care [57], there will be an increased requirement for standardization of patient-derived information. Indeed, this is one cost reduction strategy that governments and other bodies overseeing health care sustainability have implemented, as it is more cost effective to provide patient care in the primary care setting as opposed to more resource-intensive specialty-care venues. For example, innovative medicines have improved cancer survival, and many cancer patients are becoming increasingly reliant on primary care practitioners for support [57]. Traditionally, the primary care setting was not responsible for this episode of care. This poses several challenges, but one in particular is the need for ensuring that all physicians who are responsible for the clinical management of the patient are using the same data to inform their decision making and to determine when and how to intervene. This speaks to a broader need for phenome centers in the clinical management of patients with chronic diseases, considering that the use of NMR-MS provides a thorough and complete insight into the composition of cellular metabolites and fingerprints [17]. Moreover, all of the information that we can capture from our genomic output or from our metabolome not only increases our knowledge about the impact of the environment on cellular activities but it can also reduce the time of intervention and avoids unnecessary treatments; this would lead to more effective placement of patients in the health care system [17]. Phenome centers move away from the reliance on customized blood work that different clinicians require for their own purposes and areas of interest, and this is consistent with the need to align health care and health delivery to the patient journey in the health care system.

The establishment of phenome centers fulfills national and international needs for facilities with the capability of delivering large-scale, ultra-high-quality, broad, and deep metabolic phenotyping for both molecular epidemiology and experimental medicine studies, with the added capability of integrating data from other "-omics" platforms such as genomics and other clinical meta-data such as BMI. It is the most cost-effective way to provide the clinical community with access to world-class facilities and ensure access to innovative new metabolic screening technologies. In the future, researchers and clinicians will have unprecedented opportunity to augment their clinical decisions by integrating traditional clinical assays and scoring systems with metabolic profiles to gain a greater depth of information on which they can base their decisions, allowing for greater specificity and/or sensitivity of diagnostics. Such diagnostics range from novel ways of predicting anticancer drug treatment outcomes through to the prediction of cardiometabolic disease risks. Longitudinal metabolic profiling of a patient through time enables the patient journey to be mapped, and the response of each patient, good or bad, to both the disease process and to therapeutic intervention can be recorded for improved health outcomes. This database will provide greater opportunity for pharmaceutical and medical device companies to determine and declare the value of their innovations, thereby forcing governments to recognize these contributions and reduce barriers to reimbursement.

10.7.1   Phenome Centers and Chronic Disease Management

As it is explained elsewhere in this chapter, one of the key drivers of health care costs and utilization is the increasing prevalence of chronic diseases in both developed and developing countries [58]. The complexity and heterogeneity of the diseases impose unique needs on hospitals and health care organizations, as well as on patients and their families. It is, therefore, worth providing three examples of how the translation of metabolic profiling by phenome centers can be used to assist and reduce this burden on our society.

There are several opportunities for phenome centers to demonstrate their value to society by operating with context in mind. For example, phenome centers can provide reliable information on how innovative medicines and diet modulate cellular function. However, in of itself, the use of NMR-MS to collect as many metabolic signatures as possible has little scientific merit and little incremental benefit to society. Therefore, phenome centers must develop close affiliations, with cancer clinics and tumor bio-bank facilities, for example, which can be used as sources of investigative materials. If the activity of the phenome center is anchored around chronic diseases such as Alzheimer's disease, for example, then the profiling and subsequent analyses will facilitate the discovery of novel validated metabolic signatures that correlate with the onset and progression of disease to support new efforts in drug discovery. Another example is the use of phenomics to help stratify patients with cancer on the basis of their probability to respond to therapy [14,17,25]. Based on expected changes in the NMR-MS data, physicians can assess the effectiveness of new classes of therapeutics. The sensitivity and resolution with which metabolites can be measured offers unparalleled opportunities for physicians to change therapy regimens if a patient is not responding as expected. Clinicians and scientists can also use metabolic profiling in tumor prognostics, which would help predict patient survival. These analyses are necessary, as they may direct patient care or redirect the pattern of care, for example, to palliation.

In the backdrop of an integrated network of patient-level data that cross multiple points in the health care system, these data can be used to diagnose patients in different clinical settings. The detection of early stages of disease has been shown to improve patient outcomes [59]. Planned intervention through patient stratification is not only an effective strategy for the clinical management of patients but also provides health care organizations, and the broader system, an opportunity to better plan for the resources required to manage patients, thereby reducing the strain on the health care system and long-term care facilities.

The third context for phenome centers is surgery. Surgical intervention is a common episode of care in the clinical management of patients with cancer. These patients are generally subjected to an initial round of chemotherapy, which is followed by the surgical removal of the tumor. In spite of the preoperative planning, a risk for revision surgery as a result of positive margins [60] remains. Incomplete tumor resection leads to repeat surgery, which is usually much more aggressive and invasive than the first intervention, resulting in increased physical and psychological morbidity and, ultimately, greatly increased costs for all parties involved. Fewer repeat surgeries will reduce wait times and enable maintenance of operating room capacity.

Cancer surgery is a very delicate aspect of care given the risk for incomplete tumor removal [61]. Moreover, maintaining the integrity of the surrounding healthy tissue is one of the major challenges in removing brain tumors. Surgical oncologists aim to minimize removal of healthy tissue in breast cancer surgeries to maximize cosmesis [62]. Therefore, arming surgeons with technologies that assist them with their decision making in real time are invaluable for the success of operations, as defined by a reduction in repeat surgery rates.

The translation of metabolic information to improving patient outcomes will be accelerated through the commercialization of the "intelligent knife" (i-Knife). Briefly, the i-Knife is an electrocautery tool in rapid evaporative ionization mass spectrometry (REIMS) [53]. REIMS captures the metabolic information that is present in the plume of smoke that is generated by the i-Knife. Pioneering work led by Zoltan Takats at Imperial College, London, provided the i-Knife-REIMS instrument with unique capabilities for determining the tumor margins by using metabolic information collected from tumor biopsy specimens

[53]. Intraoperative diagnostics in the form of unique tumor-specific chemical signatures can also be used to limit the need for pathologist consults and lengthy histopathologic analyses. The use of real-time metabolomics in cancer surgery allows the surgeon to rely on validated unique genomic outputs and metabolic footprints, which can be used to guide intraoperative decision making by determining the boundaries between cancerous and healthy tissues.

10.8   THE VALUE OF PHENOME CENTERS IN HEALTH RESEARCH

We have emphasized throughout this chapter the need to reorient ourselves to the use of metabolic outputs as a strategy to reduce the burden of disease. The power of metabolomics has been described and illustrated in a variety of different contexts, but one area, in particular, may not have been so obvious—health research.

As we and others have pointed out, diseases are highly complex and occur because of a distortion that affects the delicate inter-relationship between biochemical pathways and the extent of cross-talk that exists both within and between them. This is why a single metabolite cannot cause the disequilibria of biochemical processes that force the disease state of the cell. Rather, it is a set of metabolites that either accumulate or are not present (eg, molecular signatures) that compromise the fidelity of cellular function [63]. As scientists, we can see that the exclusion of metabolic information in the context of understanding the onset and progression of disease limits opportunities to develop targeted therapies.

We can certainly appreciate use of phenomics in driving new and innovative research in the context of antibiotic resistance, as well as other threats to our microbiome. For example, stool transplants using artificial stools as well as fecal matter from healthy donors have been used to cure patients infected with *Clostridium difficile* [64]. These procedures work to restore the health of the large intestine [64]. In principle, medical microbiologists can use phenomics-based approaches to understand what is causing the disequilibrium in the microbiome of infected patients and create designer solutions to restore normal physiologic function. We can also use phenomics to explore the basis for bacteria developing antibiotic resistance and to look for other opportunities beyond antibiotics to manage this threat [65,66].

As health care becomes more patient-centric and begins to expand its view beyond procedural care to episodic care, so too will health research. In fact, a greater emphasis on knowledge translation by several granting agencies (eg, Canadian Institutes of Health Research, Medical Research Council, etc.) has forced many basic scientists to think about the applicability of their research to better align with patient-centric care. This has moved the value proposition of their research away from identifying novel chemical or physical properties of their molecules toward the need to exploit these characteristics to advance new innovations that can drive commercial initiatives or value to patients. The fact that phenomics takes a more holistic view of the patient, wherein metabolites can be used as probes to interrogate which biochemical and genetic pathways are disrupted in disease, raises the value of health research by focusing on the true underpinnings of chronic diseases and by crafting hypotheses that are better informed because they are based on human models.

To drive the development of new therapies, we must move away from our focus on one-to-one relationships between genes and proteins to the point where we are monitoring cellular endpoints, or metabolites, to capture the readout of all cellular output under a given set of conditions, which is the basis of metabolomics [19,25]. Hence metabolic profiling of chronic diseases by phenome centers is expected to provide a wealth of clinically relevant information, including the (1) identification of diagnostic, prognostic, and predictive biomarkers; (2) discovery of new therapeutic targets such as structural lipids in cancer cell membranes; and (3) identification of metabolites that characterize pathogenic activation or inactivation of critical cell signaling pathways. These discoveries will fuel additional research into the causes and modulators of chronic disease. This, in turn, will stimulate therapeutic discoveries that will generate commercialized outputs that will compete for the $29B market for predictive biomarkers and new therapeutics.[6]

The identification of metabolic biomarkers using human materials is certainly valuable to health research initiatives, but we would argue that there is another, perhaps more subtle benefit of phenomics, This has to do with the potential for using "big data" to better inform scientific investigation. This is analogous to corporations that collect market-derived data from consumers to learn about what is important to consumers and to leverage this insight to optimize the customer experience in their establishments. This competitive pressure means that price and product alone are not the only consideration that determines the basket size and conversion rate of those purchases into revenue for the company [67]. Companies, and also sports teams, have made incredible investments to strengthen their analytics capabilities because data are assets and can be used to increase profitability, if used correctly [67]. Those who do not subscribe to this philosophy will ultimately fail, unlike those organizations that inform their strategy based on the quality of how they collected and interpreted their data.

The competitive pressures faced by corporations to better customize the experience for its customers foreshadow the challenges that will be faced by health researchers in terms of how they use "big data" to make better use of resources to advance scientific initiatives. The benefit associated with the proliferation of phenomics-based research is that it recognizes the need for biomedical informatics, which is a training priority for the NPC in the United Kingdom, as stated on their website. This will help build the next generation of scientists with the necessary skill sets to populate such large datasets comprising patient-level genomic, proteomic, metabolomic, and image-level data. In the future, this caliber of scientists will be in high demand, as they will use their analyses to reduce the level of experimentation required to achieve the same result. This will increase the return on investment for research, which is currently marginalized owing to the indirect and direct costs associated with supporting it (eg, infrastructure and consumables). As is the case with successful Fortune-500 companies, funders will reward those scientists who can utilize data to focus their research efforts and make better assessments as to the success of their experiments.

10.9   THE LIMITATIONS AND CHALLENGES TO REALIZING THE POTENTIAL OF PHENOMICS AND PHENOME CENTERS

In spite of the many benefits of phenomics and of phenome centers, we must recognize, as with all scientific initiatives, that sustainability is key. In this regard, we must look for opportunities that will help create revenue for the center so that it is capable of continuing to support phenomics-based studies on human populations over the long term. This is to ensure that next-generation technologies are affordable and that the necessary talent is attracted and retained, and that the center is operationalized to manage their data requirements. We describe here several revenue-generating options that will help with the sustainability of phenome centers.

The richness and completion of regional databases allows pharmaceutical and medical device industries to demonstrate the value of their innovations. In this regard, companies that operate in this space should provide financial support to phenomic research. In a similar spirit, any opportunity for these data to be used to create new insights and value should be recognized accordingly. Phenomic research also creates value for governments and other payers for health care through reduced expenditures on the clinical management of chronic disease. These realized savings should be redirected to support phenomic research.

There is an intrinsic value associated with phenomics research in that metabolomics offers unique insight into the human condition. In this regard, public engagement efforts by phenome centers are not only important for philanthropic purposes but also to raise public support for phenomic research, which requires priority setting by politicians, governments, and other agencies.

Another challenge associated with phenomics and phenome centers is at the level of standardization and data linkage with patient-level electronic medical records. Scientists and physicians alike need to be trained on how to use these data to inform their practice. This is especially challenging for physicians who are generally unwilling to adopt new approaches and guidelines to help in the clinical management of their patients [68]. An interdisciplinary approach involving a team of scientists, data scientists, and physicians will be required to reduce the resistance and anxiety about this shift to rely on "big data."

CONCLUSIONS

The study of phenomics is a major step forward to help us unravel the mystery of the human condition on earth. It aligns well with our inherent nature as humans to focus on outputs. This is evidenced in how we report macro-changes to our primary care physicians and other health care practitioners. As with any effective organization, a focus on outputs (eg, our metabolic response) can help better understand the underlying conditions that cause our bodies to behave the way they do. Phenome centers are necessary to follow our observational curiosity and acuity because they are designed to extract as much information as possible from scant human samples. They are able to expand our knowledge of the host response and, in some cases, can even detect underlying differences when the response is the same (eg, "sweet" urine in patients with diabetes). Deep metabolic profiling can unearth such salient differences and more appropriately design therapeutic strategies customized to patient-level needs. The fewer number of human metabolites should increase our resolution to read and interpret information to draw conclusions. Finally, the accuracy with which we can derive our conclusions is risk mitigated because we are looking for patterns of metabolites instead of an absolute marker of disease onset and progression. In summary, there are strong grounds for phenomics research to flourish and for obvious synergies between consortia to be realized so that better care and treatment options can be delivered to patients living with chronic diseases.

ENDNOTES

1. Cannon WB. Physiological regulation of normal states: some tentative postulates concerning biological homeostatics. In: Pettit A, editor. A Charles Richet: ses amis, ses collègues, ses élèves (in French). Paris: Les Éditions Médicales; 1926. p. 91.
2. Centers for Disease Control and Prevention.
3. Medical Marijuana Policy in the United States. Stanford.edu; May 15, 2012 [retrieved 15.1.13].
4. Centers for Disease Control and Prevention (April 2011).
5. Health System Funding Reform, Quality-Based Procedures.
6. Biomarkers: Technologies and Global Markets, Marketwatch; 2014.

REFERENCES

[1] Davis BD. The isolation of biochemically deficient mutants of bacteria by means of Penicillin. Proc Natl Acad Sci USA 1949;35(1):1–10.
[2] Freimer N, Sabatti C. The human phenome project. Nat Genet 2003;34(1):15–21.
[3] Gerlai R. Phenomics: fiction or the future? Trends Neurosci 2002;25(10):506–9.
[4] Schilling CH, Edwards JS, Palsson BO. Toward metabolic phenomics: analysis of genomic data using flux balances. Biotechnol Prog 1999;15(3):288–95.
[5] Warringer J, Ericson E, Fernandez L, Nerman O, Blomberg A. High-resolution yeast phenomics resolves different physiological features in the saline response. Proc Natl Acad Sci USA 2003;100(26):15724–9.
[6] Zbuk KM, Eng C. Cancer phenomics: RET and PTEN as illustrative models. Nat Rev Cancer 2007;7(1):35–45.

[7] Newton I. Axioms or Laws of Motion. Principia 1729:19–20.

[8] Keys A, Fidanza F, Karvonen MJ, Kimura N, Taylor HL. Indices of relative weight and obesity. J Chronic Dis 1972;25(6):329–43.

[9] Delneri D, Brancia FL, Oliver SG. Towards a truly integrative biology through the functional genomics of yeast. Curr Opin Biotechnol 2001;12(1):87–91.

[10] Lindon JC, Nicholson JK, Holmes E, Everett JR. Metabonomics: metabolic processes studied by NMR spectroscopy of biofluids. Concepts Magnetic Res 2000;12:289–320.

[11] Nicholson JK, Lindon JC, Holmes E. Metabonomics: understanding the metabolic responses of living systems to pathophysiological stimuli via multivariate statistical analysis of biological NMR spectroscopic data. Xenobiotica 1999;29(11):1181–9.

[12] Raamsdonk LM, Teusink B, Broadhurst D, Zhang N, Hayes A, Walsh MC, et al. A functional genomics strategy that uses metabolome data to reveal the phenotype of silent mutations. Nat Biotechnol 2001;19(1):45–50.

[13] Smith LL. Key challenges for toxicologists in the 21st century. Trends Pharmacol Sci 2001;22(6):281–5.

[14] Davis VW, Bathe OF, Schiller DE, Slupsky CM, Sawyer MB. Metabolomics and surgical oncology: potential role for small molecule biomarkers. J Surg Oncol 2011;103(5):451–9.

[15] Di Leo A, Claudino W, Colangiuli D, Bessi S, Pestrin M, Biganzoli L. New strategies to identify molecular markers predicting chemotherapy activity and toxicity in breast cancer. Ann Oncol 2007;18(Suppl. 12):xii8–14.

[16] Malandrino N, Smith RJ. Personalized medicine in diabetes. Clin Chem 2011;57(2):231–40.

[17] Mirnezami R, Kinross JM, Vorkas PA, Goldin R, Holmes E, Nicholson J, et al. Implementation of molecular phenotyping approaches in the personalized surgical patient journey. Ann Surg 2012;255(5):881–9.

[18] Joy JE, Penhoet EE, Petitti DB, editors. Saving women's lives: strategies for improving breast cancer detection and diagnosis. Washington, DC: National Academies Press; 2005.

[19] Nicholson JK, Connelly J, Lindon JC, Holmes E. Metabonomics: a platform for studying drug toxicity and gene function. Nat Rev Drug Discov 2002;1(2):153–61.

[20] Poretsky L, editor. Principles of diabetes mellitus (2nd ed). New York, NY: Springer; 2009.

[21] Filler AG. Neurography and diffusion tensor imaging: origins, history & clinical impact. Nat Proceedings 2009.

[22] MacPherson DW, Gushulak BD, Macdonald L. Health and foreign policy: influences of migration and population mobility. Bull World Health Org 2007;85(3):200–6.

[23] Warner KE. Effects of the antismoking campaign: an update. Am J Public Health 1989;79(2):141–51.

[24] Naftali T, Mechulam R, Lev LB, Konikoff FM. Cannabis for inflammatory bowel disease. Dig Dis 2014;32(4):468–74.

[25] Goldsmith P, Fenton H, Morris-Stiff G, Ahmad N, Fisher J, Prasad KR. Metabonomics: a useful tool for the future surgeon. J Surg Res 2010;160(1):122–32.

[26] Mirnezami R, Nicholson J, Darzi A. Preparing for precision medicine. N Engl J Med 2012;366(6):489–91.

[27] Reddin B. The output oriented organization. Aldershot, UK: Gower Publishing; 1988.

[28] Strisciuglio P, Concolino D. New strategies for the treatment of phenylketonuria (PKU). Metabolites 2014;4(4):1007–17.

[29] Blau N, Shen N, Carducci C. Molecular genetics and diagnosis of phenylketonuria: state of the art. Expert Rev Mol Diagn 2014;14(6):655–71.

[30] Palla R, Peyvandi F, Shapiro AD. Rare bleeding disorders: diagnosis and treatment. Blood 2015;125(13):2052–61.

[31] Touw CM, Derks TG, Bakker BM, Groen AK, Smit GP, Reijngoud DJ. From genome to phenome—Simple inborn errors of metabolism as complex traits. Biochim Biophys Acta 2014;1842(10):2021–9.

[32] DeLeo AB, Jay G, Appella E, Dubois GC, Law LW, Old LJ. Detection of a transformation-related antigen in chemically induced sarcomas and other transformed cells of the mouse. Proc Natl Acad Sci USA 1979;76(5):2420–4.

[33] Kress M, May E, Cassingena R, May P. Simian virus 40-transformed cells express new species of proteins precipitable by anti-Simian virus 40 tumor serum. J Virol 1979;31(2):472–83.

[34] Linzer DI, Levine AJ. Characterization of a 54K dalton cellular SV40 tumor antigen present in SV40-transformed cells and uninfected embryonal carcinoma cells. Cell 1979;17(1):43–52.

[35] Linzer DI, Maltzman W, Levine AJ. The SV40 A gene product is required for the production of a 54,000 MW cellular tumor antigen. Virology 1979;98(2):308–18.

[36] Rotter V, Witte ON, Coffman R, Baltimore D. Abelson murine leukemia virus-induced tumors elicit antibodies against a host cell protein, P50. J Virol 1980;36(2):547–55.

[37] Khoo KH, Verma CS, Lane DP. Drugging the p53 pathway: understanding the route to clinical efficacy. Nat Rev Drug Discov 2014;13(3):217–36.

[38] Kruse JP, Gu W. Modes of p53 regulation. Cell 2009;137(4):609–22.

[39] Meek DW, Anderson CW. Posttranslational modification of p53: cooperative integrators of function. Cold Spring Harb Perspect Biol 2009;1(6):a000950.

[40] Rubinstein WS. Hereditary breast cancer in Jews. Fam Cancer 2004;3(3-4):249–57.

[41] Frueh DP, Goodrich AC, Mishra SH, Nichols SR. NMR methods for structural studies of large monomeric and multimeric proteins. Curr Opin Struct Biol 2013;23(5):734–9.

[42] Rosano GL, Ceccarelli EA. Recombinant protein expression in *Escherichia coli*: advances and challenges. Front Microbiol 2014;5:172.

[43] Gardner KH, Kay LE. The use of ^{2}H, ^{13}C, ^{15}N multidimensional NMR to study the structure and dynamics of proteins. Annu Rev Biophys Biomol Struct 1998;27:357–406.

[44] Saridakis E, Chayen NE. Systematic improvement of protein crystals by determining the supersolubility curves of phase diagrams. Biophys J 2003;84(2 Pt 1):1218–22.

[45] Tsumoto K, Ejima D, Senczuk AM, Kita Y, Arakawa T. Effects of salts on protein–surface interactions: applications for column chromatography. J Pharm Sci 2007;96(7):1677–90.

[46] Cheatham B, Kahn CR. Insulin action and the insulin signaling network. Endocr Rev 1995;16(2):117–42.

[47] Crockford DJ, Maher AD, Ahmadi KR, Barrett A, Plumb RS, Wilson ID, et al. ^{1}H NMR and UPLC–MS(E) statistical heterospectroscopy: characterization of drug metabolites (xenometabolome) in epidemiological studies. Anal Chem 2008;80(18):6835–44.

[48] Maher AD, Fonville JM, Coen M, Lindon JC, Rae CD, Nicholson JK. Statistical total correlation spectroscopy scaling for enhancement of metabolic information recovery in biological NMR spectra. Anal Chem 2012;84(2):1083–91.

[49] Sands CJ, Coen M, Maher AD, Ebbels TM, Holmes E, Lindon JC, et al. Statistical total correlation spectroscopy editing of ^{1}H NMR spectra of biofluids: application to drug metabolite profile identification and enhanced information recovery. Anal Chem 2009;81(15):6458–66.

[50] Gray N, Lewis MR, Plumb RS, Wilson ID, Nicholson JK. High-throughput microbore UPLC–MS metabolic phenotyping of urine for large-scale epidemiology studies. J Proteome Res 2015;14:2714–2721.

[51] Want EJ, Masson P, Michopoulos F, Wilson ID, Theodoridis G, Plumb RS, et al. Global metabolic profiling of animal and human tissues via UPLC–MS. Nat Protoc 2013;8(1):17–32.

[52] Want EJ, Wilson ID, Gika H, Theodoridis G, Plumb RS, Shockcor J, et al. Global metabolic profiling procedures for urine using UPLC–MS. Nat Protoc 2010;5(6):1005–18.

[53] Balog J, Sasi-Szabo L, Kinross J, Lewis MR, Muirhead LJ, Veselkov K, et al. Intraoperative tissue identification using rapid evaporative ionization mass spectrometry. Sci Transl Med 2013;5(194) 194ra93.

[54] Nicholson JK, Holmes E, Kinross JM, Darzi AW, Takats Z, Lindon JC. Metabolic phenotyping in clinical and surgical environments. Nature 2012;491(7424):384–92.

[55] Morgan MW, Zamora NE, Hindmarsh MF. An inconvenient truth: a sustainable healthcare system requires chronic disease prevention and management transformation. Healthc Pap 2007;7(4):6–23.

[56] Hirsch JA, Leslie-Mazwi TM, Barr RM, McGinty G, Nicola GN, Silva 3rd E, et al. The bundled payments for care improvement initiative. J Neurointerv Surg 2015.

[57] Grunfeld EaE CC. The interface between primary care and oncology specialty care: treatment through survivorship. J Natl Cancer Inst Monogr 2010;40:25–30.

[58] Rich PB, Adams SD. Health care: economic impact of caring for geriatric patients. Surg Clin North Am 2015;95(1):11–21.

[59] Heidari B. Rheumatoid arthritis: early diagnoses and treatment outcomes. Caspian J Int Med 2011;2(1):161–70.

[60] Raziee HR, Cardoso R, Seevaratnam R, Mahar A, Helyer L, Law C, et al. Systematic review of the predictors of positive margins in gastric cancer surgery and the effect on survival. Gastric Cancer 2012;15(Suppl. 1):S116–24.

[61] Rosenthal EL, Warram JM, Bland KI, Zinn KR. The status of contemporary image-guided modalities in oncologic surgery. Ann Surg 2015;261(1):46–55.

[62] Law TT, Kwong A. Surgical margins in breast conservation therapy: how much should we excise? South Med J 2009;102(12):1234–7.

[63] Coen M, Holmes E, Lindon JC, Nicholson JK. NMR-based metabolic profiling and metabonomic approaches to problems in molecular toxicology. Chem Res Toxicol 2008;21(1):9–27.

[64] Petrof EO, Khoruts A. From stool transplants to next-generation microbiota therapeutics. Gastroenterology 2014;146(6):1573–82.

[65] Jump RL, Polinkovsky A, Hurless K, Sitzlar B, Eckart K, Tomas M, et al. Metabolomics analysis identifies intestinal microbiota-derived biomarkers of colonization resistance in clindamycin-treated mice. PLoS One 2014;9(7):e101267.

[66] Theriot CM, Koenigsknecht MJ, Carlson Jr. PE, Hatton GE, Nelson AM, Li B, et al. Antibiotic-induced shifts in the mouse gut microbiome and metabolome increase susceptibility to *Clostridium difficile* infection. Nat Commun 2014;5:3114.

[67] Soman D, N-Marandi S. Managing Customer Value. Singapore: Wold Scientific Publishing Company; 2010.

[68] Cabana MD, Rand CS, Powe NR, Wu AW, Wilson MH, Abboud PA, et al. Why don't physicians follow clinical practice guidelines? A framework for improvement. JAMA 1999;282(15):1458–65.

Chapter 11

From Databases to Big Data

Nadine Levin[1], Reza M. Salek[2] and Christoph Steinbeck[2]

[1]*Institute for Society and Genetics, University of California, Los Angeles, CA, USA* [2]*The European Bioinformatics Institute, Wellcome Genome Campus, Hinxton, Cambridge, UK*

Chapter Outline

11.1 INTRODUCTION

Biomedical sciences have arrived in the age of big data. The rate at which data are being produced is increasing exponentially, and there is mounting enthusiasm over the generation, collection, and use of data to address longstanding questions about human health and disease. In 2016, mankind will approach the milestone of having sequenced half a million genomes [1]. With innovations in data generation and analysis, the biomedical community is moving toward a better understanding of how the fate of an individual human organism is not only determined by its own genome but also by its many other "-omes": the DNA and metabolic activity of the microbiome, and the myriad interactions from its exposome (see Chapter 7 for a description of the work being undertaken in determining metabolic consequences of exposure to various agents), all of which can be measured and quantified with metabolic phenotyping (see Chapter 2 for a review of the development of the field).

E. Holmes, J.K. Nicholson, A.W. Darzi & J.C. Lindon (Eds): Metabolic Phenotyping in Personalized and Public Healthcare. DOI: http://dx.doi.org/10.1016/B978-0-12-800344-2.00011-2

Recently, there has been an explosion of interest in metabolomics research, facilitated by instruments that can provide high-throughput and high-resolution data. The growth of metabolomics as a field, particularly its use for metabolic phenotyping, has generated increasingly large data sets, and has also increased the need to develop technologies and methods for handling and dealing with metabolomics data. Because such data can be analyzed and interpreted in multiple ways, there is also a need to ensure that the data which are already being produced can circulate freely and openly, maximizing the value of existing resources, data, and expertise.

Big data in biology can be defined as any volume of data for which the management, curation, processing, and analysis pose difficulties when conventional technologies of a given point in history are used. In other words, big data are data too big to fit on a spreadsheet. This concept is typically defined in relation to the "3Vs"—volume, variety, and velocity—which was developed to explain the rise of big data corporations such as Google and Facebook [2]. These three defining properties, or dimensions, of big data refer to the amount of data, number of types of data, and speed of data analysis and processing. More recently, additional Vs—including *variability* in the range of values typical of a large data set, and *value* in finding ways to assess data—have been proposed to account for the challenge of not only generating data but also managing and interpreting it [3].

In biomedical "big data" experiments, biologists often need to perform an analysis of data arising from various different "-omics" sources, resulting from different analytical instruments, annotated according to standards covering different time scales and molecular domains. In population and medical phenotyping, this is complemented by information on exposome information, demographics, and clinical symptoms, yielding complex mixtures of data to analyze and interpret information. Furthermore, with the increasing scale and scope of research, many groups and institutions are often working on similar problems but remain distributed and unconnected throughout the community.

Ultimately, working with big data requires significant resources and investments in time, funding, and labor. There is not only a need to develop computational and analytical technologies for dealing with big data but also a need to ensure that data can circulate freely and openly and can be used to the maximum effect by the wider research community. Contemporary biomedicine is at a historic moment in which researchers are being encouraged to circulate data throughout society so as to increase knowledge via reuse and to increase transparency via peer review. Despite the rise of government policies and guidelines to facilitate Open Access and Open Data [4], as well as community norms and ideals for achieving Open Science, only a fraction of these data are made available to the general scientific community without barriers to access, either in the form of paywalls or restrictive intellectual property regimes. Of the many metabolic phenotype data sets collected to date, only a few have been made generally available.

Because of the challenges of accessing and reusing data, the development of tools and resources for making data available and useful to the research community is becoming increasingly important and relevant. To facilitate working with increasingly large, heterogeneous, and complex data sets, researchers have begun to develop tools, resources, and services for storing, exchanging, and making sense of data. Through the combined efforts of government initiatives, institutions, and local communities, databases have been developed to combine data collected with similar "-omics" platforms, or under similar topical areas. For the large "-omics" areas such as genomics, proteomics, and metabolomics, primary research data have been collected in centralized repositories maintained by specialized institutions such as the European Bioinformatics Institute (EMBL-EBI) [5] and the National Center for Biotechnology Information (NCBI) [6]. Although genomics researchers established guidelines for the deposition of sequence data in 1996 with the creation of the Bermuda Principles [7], in metabolomics, there remains no primary database or format for data, and no obligation for authors to make data available, although the situation is beginning to change.

Beyond "-omics" and other forms of digital data, biomedical researchers have also begun to focus on the storage of biological and patient samples to facilitate the longitudinal analysis of multiple types of data about health and disease. For example, in 2007 the UK Government established the UK Biobank [8] as a national health resource, with the long-term aim of improving the prevention, diagnosis, and treatment of a wide range of serious and life-threatening illnesses, including cancer, heart diseases, stroke, diabetes, arthritis, osteoporosis, eye disorders, depression, and forms of dementia. The UK Biobank recruited 500,000 people aged between 40 and 69 years in 2006–2010 to take part, with the subjects providing blood, urine, and saliva samples for future analysis and agreeing to have their health monitored. The UK Biobank has also invited participants to wear a wrist-worn activity monitor, has undertaken repeat measures on 20,000 participants, and is currently undertaking magnetic resonance imaging (MRI) body and dual energy X-ray absorptiometry (DEXA) bone scanning [9]. Many participants have also completed detailed web-based questionnaires on their diet, cognitive function, and work history. All data, including genetic, biochemical, and imaging data, are offered to researchers as they become available. Over many years, the UK Biobank should provide a powerful resource for tracking peoples' phenotypes and for helping scientists discover why some people develop particular diseases and others do not.

Given the opportunities and challenges associated with big data, in this chapter, we provide an overview of the resources and tools that exist for facilitating big data in metabolomics. The databases for metabolomics can often be categorized into those containing reference chemical data and a separate type containing metabolomics-based experimental data. We first discuss the pathway-centric and compound-centric databases that make up metabolomics reference chemical databases, and then we discuss the experimental databases and data warehouses that make up metabolomics experimental databases.

11.2 REFERENCE CHEMICAL DATABASES

Reference databases mainly consist of publically or commercially available compound databases with information on chemical structures, physicochemical properties, biological functions, pathway network, and, most importantly, reference spectral data. Fiehn et al. [10] have suggested two different categories for such a resource: pathway-centric and compound-centric databases. Examples for a pathway-centric most commonly used in metabolomics are: Biocyc [11], Kyoto Encyclopedia of Genes and Genomes (KEGG) [12], Reactome [13], and Wikipathways [14]. Examples for compound-centric databases are BioMagResBank (BMRB) [15], Chemical Entities of Biological Interest (ChEBI) [16], ChemSpider [17], Golm Metabolome Database (GMD) [18], Human Metabolome Database (HMDB) [19], MassBank [20], METLIN [21], National Institute of Standards and Technology (NIST) [22], and PubChem [23]. The compound-centric resources often contain nuclear magnetic resonance (NMR) data, as well as gas chromatographymass spectrometry and liquid chromatography–mass spectrometry (GC-MS/LC-MS) spectral data. In metabolomics, references compounds are often used for metabolite identification by matching NMR resonance or MS features to those of an unknown compound.

11.2.1 Pathway-Centric Databases

11.2.1.1 BioCyc (www.biocyc.org)

BioCyc is a collection of 5711 Pathway/Genome Databases (PGDBs), which is accompanied by software tools for searching, visualizing, comparing, and analyzing pathway data. BioCyc comprises a genome browser, tools for display of individual metabolic pathways and of full metabolic maps, and multiple analysis methods for user-supplied "-omics" and multi-omics data sets. The BioCyc databases are divided into three tiers based on their quality. Tier 1 databases have received at least one person-year of literature-based curation and are the most accurate. Tier 2 and Tier 3 databases contain computationally predicted metabolic pathways, predictions as to which genes code for missing enzymes in metabolic pathways, and predicted operons. BioCyc also contains separated home pages for 12 different organisms, including HumanCyc for humans and YeastCyc for yeast [24, 25].

11.2.1.2 Kyoto Encyclopedia of Genes and Genomes (www.genome.jp/kegg/pathway.html)

The KEGG is a free and open-source collection of 17 main databases (Table 11.1) for understanding the high-level functioning of biological systems—such as the cell, the organism, and the ecosystem—using genomic and other molecular-level information. It was developed in 1995 by Minoru Kanehisa at Kyoto University, under the then-ongoing Japanese Human Genome Project. KEGG provides a computer representation of the biological system, consisting of

TABLE 11.1 KEGG Databases

Category	Database	Content
Systems information	KEGG PATHWAY	KEGG pathway maps
	KEGG BRITE	BRITE functional hierarchies
	KEGG MODULE	KEGG modules of functional units
Genomic information	KEGG ORTHOLOGY	KEGG Orthology (KO) groups
	KEGG GENOME	KEGG organisms with complete genomes
	KEGG GENES	Gene catalogs of complete genomes
	KEGG SSDB	Sequence similarity database for GENES
Chemical information	KEGG COMPOUND	Metabolites and other small molecules
	KEGG GLYCAN	Glycans
	KEGG REACTION	Biochemical reactions
	KEGG RPAIR	Reactant pair chemical transformations
	KEGG RCLASS	Reaction class defined by RPAIR
	KEGG ENZYME	Enzyme nomenclature
Health information	KEGG DISEASE	Human diseases
	KEGG DRUG	Drugs
	KEGG DGROUP	Drug groups
	KEGG ENVIRON	Crude drugs and health-related substances

molecular building blocks and diagrams of genes and proteins (genomic information) and chemical substances (chemical information), which are integrated with the knowledge of molecular connections, reactions, and relational networks (systems information). It also contains disease and drug information as perturbations of the biological system. The KEGG database is categorized into systems, genomic, chemical, and health information [see also 26].

11.2.1.3 Reactome (www.reactome.org)

Reactome is a free and open source database of biological pathways, which also includes bioinformatics tools for analyzing, interpreting, and visualizing the mechanisms and small molecules involved in biological pathways. Several "reactomes" are represented within the database, the largest of which is focused on human biology and the human reactome. The core unit of Reactome is the pathway: Reactome visually represents how various entities such as nucleic acids, proteins, complexes, and other small molecules are involved in biological pathways, including those in classic intermediary metabolism, signaling, innate and adapted immunity, transcriptional regulation, apoptosis, and various diseases. Reactome visualizes these pathways similar to classic chemistry, specifying "substrates," which interact with "catalysts" to form "products." The pathways represented in Reactome are species-specific. Reactome can also be used for data mining and large-scale analysis of gene, protein, and metabolite lists, by providing several tools for pathway browsing and data analysis.

11.2.1.4 Wikipathways (www.wikipathways.org)

Wikipathways is a community resource for contributing, annotating, and maintaining information about biological pathways. Building on the same MediaWiki software that powers Wikipedia, it enhances and complements other pathway-oriented databases such as KEGG and Reactome by enabling users to graphically edit integrated pathway information about genes, proteins, and small molecules. Within Wikipathways, each pathway has a dedicated Wiki page, displaying a diagram, description, references, version history, component lists, and download options [see also 27].

11.2.2 Compound-Centric Databases

11.2.2.1 Databases with Reference Spectra

BioMagResBank (www.bmrb.wisc.edu)

BMRB, a public repository for experimental and derived data collected using NMR spectroscopy on biological molecules, was developed at the University of Wisconsin-Madison and by the National Library of Medicine in September 1996. BRMB consists of four different NMR-based data depositories, one of which is a metabolomics or small molecule repository containing one- and two-dimensional ^{1}H and ^{13}C NMR spectra for over metabolites supplied by the Madison Metabolomic Consortium (MMC) and the NMR community. Where possible, one-dimensional ^{1}H and ^{13}C NMR spectra as well as ^{1}H-^{1}H TOCSY and ^{1}H-^{13}C HSQC two-dimensional NMR spectra are collected for each metabolite, along with NMR peak lists, spectral transition lists, images of the spectra, and assigned peak lists. BMRB provides tools for querying NMR chemical shifts in order to retrieve the closest possible matching metabolite.

Golm Metabolome Database (gmd.mpimp-golm.mpg.de)

The GMD is a repository of GC-MS data, developed for the search and dissemination of reference mass spectra from biologically active metabolites. It provides information about mass spectra and retention times, to aid in the identification of metabolites from complex biological mixtures. It was developed in 2005 by the Max Planck Institute of Molecular Plant Physiology in Golm district of Potsdam, Germany. GMD supports text-based functionality to search for particular metabolites, reference compounds, respective chemical derivatives, sum formula, molecular weight, functional chemical groups, KEGG identifiers, and other properties.

Human Metabolome Database (www.hmdb.ca)

The HMDB is one of the most comprehensive web-accessible, organism-specific metabolomics databases and is an outgrowth of the Human Metabolome Project funded by Genome Canada in 2004. It contains detailed information on the metabolites found in the human body and is organized in the "MetaboCard" layout format, which includes different fields that contain clinical, chemical, spectral, biochemical, and enzymatic data. It was first introduced in 2007, and, as of May 2015, is at version 3.6, accounting for more than 40,000 metabolite structures accompanied by nearly 10,000 NMR, GC-MS, and LC-MS spectra. In HMDB, the term *metabolites* refers to those molecular entities that can be detected in human samples such as biofluids and whose biochemical pathways are known to be involved in human intake and exposure, such as dipeptides, drug metabolites, and food-derived compounds.

LIPID MAPS (www.lipidmaps.org)

The LIPID MAPS Structure Database encompasses structures and annotations of biologically relevant lipids for more than 40,000 unique lipid structures, making it the largest publically available lipid-only database. The LIPID MAPS classification system comprises eight lipid categories, each with its own subclassification hierarchy and unique LIPID MAPS ID. Users can search the database by lipid class, common name, systematic name or synonym, mass, InChIKey, or LIPID MAPS ID using the "Quick Search" tool. In addition, LIPID MAPS contains several MS analysis tools for *in silico* predict of possible lipid structures from precursor or product-ion MS experimental data.

MassBank (www.massbank.jp)

MassBank is the first public, open-community repository of MS data. MassBank data are contributed by consortium members mainly from Japan but also from the European Union (EU), Switzerland, the United States, Brazil, and China. It contains detailed MS data—in various ionization modes, as well as high and low resolutions—for more than 13,000 standard compounds with more than 43,000 spectra (in both positive and negative modes), which were mainly obtained by Keio University, RIKEN PSC, and other Japanese and some international research institutions. It is officially sanctioned by the Mass Spectrometry Society of Japan and includes spectral and structural searching utilities.

METLIN (metlin.scripps.edu)

The METLIN Metabolomics Database is a repository for mass spectrometry metabolomics data, which is designed to aid in metabolite identification. It was developed in 2005 by The Scripps Research Institute and contains MS-MS, LC-MS, and Fourier transform mass spectrometry (FTMS) data that can be searched by peak lists, mass range, biological source, or disease. The database contains over 240,000 compounds, which include endogenous metabolites from different organisms and exogenous compounds such as pharmaceutical drugs and other synthetic organic compounds. Currently, METLIN has high-resolution MS-MS data for more than 13,000 authentic chemical standards, resulting in over 68,000 high-resolution MS-MS spectra, including structures with accurate masses and chemical formulas, and over 160,000 *in silico* predicted unique fragment structures.

National Institute of Standards and Technology (www.nist.gov/srd)

The NIST is a nonregulatory federal agency within the US Department of Commerce, whose Mass Spectrometry Data Center compiles, evaluates, correlates, and measures standard reference data. Founded in 1901, the NIST develops and disseminates associated electronic databases and analysis software for the chemical identification and quality control for (1) electron ionization mass spectra and associated retention properties by the technique of GC-MS, (2) reference tandem mass spectra of "small molecules" generated by electrospray and matrix-assisted laser desorption ionization (MALDI)-MS widely used for biological analysis, and (3) peptide fragmentation libraries that enable the identification of products of digestion of complex protein mixtures in the field of proteomics. This also involves the development of structure-based estimation methods and reference data quality control methods for the more effective use of this reference data. The NIST's seven laboratories include the Physical Measurement Laboratory, the Material Measurement Laboratory, the Engineering Laboratory, the Information Technology Laboratory, and the Communications Technology Laboratory. The NIST consists of approximately 280,000 mass spectra from 243,000 unique compounds. Besides mass spectra, other data include nomenclature, formula, molecular structure, molecular weight, Chemical Abstracts Service (CAS) (number, contributor name, list of peaks, synonyms, and measured retention index.

11.2.2.2 Databases Without Reference Spectra

Chemical Entities of Biological Interest (www.ebi.ac.uk/chebi)

ChEBI is a freely available dictionary of the molecular entities—atoms, molecules, ions, and complexes—derived from "small" chemical compounds. ChEBI includes chemical entities that are products of nature (ie, metabolites) and those that are products of synthetic compounds (ie, drugs or toxins) that interact with organisms. ChEBI includes structure and nomenclature information, as well as an ontologic classification, whereby the relationships between molecular entities or classes of entities are precisely specified. ChEBI has more than 15,500 chemical entities in its database and conforms to the International Union of Pure and Applied Chemistry (IUPAC) nomenclature.

ChemSpider (www.chemspider.com)

ChemSpider is a chemical structure database that provides access to the structures, properties, and molecular information for more than 35 million organic molecules. It is owned by the Royal Society of Chemistry, following its acquisition from ChemZoo in 2009, and contains information from diverse sources such as a marine natural products database, ACD-Labs chemical databases, the DSSTox databases of the Environment Protection Agency, and a series of chemical vendors.

PubChem (pubchem.ncbi.nlm.nih.gov)

PubChem is a freely available database that contains the chemical structures and biological activities for over 19 million small organic molecules. Developed in 2004 by the NCBI, it contains structure, nomenclature, and physico-chemical data. PubChem is organized into three linked databases: PubChem Substance, PubChem Compound, and PubChem BioAssay.

11.3   DATABASES CONTAINING METABOLOMICS-BASED EXPERIMENTS

Over the past few years, there has been steady growth in the number of databases and resources dedicated to storing and organizing metabolomics, proteomics and other related "-omics" datasets. Because of the increasing need to combine and compare data sets, the biological community has also begun to develop data warehouses made up of integrated data from one or more sources [28]. Developments in data capture, processing power, data transfer, and storage capabilities have enabled organizations to provide data warehouses for centralized data management and retrieval [29].

11.3.1   Experimental Databases

11.3.1.1   MetaboLights

The MetaboLights database and repository is the first cross-species, general-purpose repository for metabolomics data. Launched in 2012 by the EMBL-EBI [30], it has seen steady growth in number of submissions, with each submission currently averaging about 20 GB in size. The database is cross-species, and cross-technique. It covers metabolite structures and their reference spectra, as well as the biological roles, locations, concentrations, and experimental data from metabolic experiments [31]. MetaboLights includes user submission tools, and incorporates de facto standard formats for encoded spectral and chromatographic data, associated information about chemical structures, and meta-data for describing assays and studies as a whole.

The experimental data that scientists submit to MetaboLights have been used to justify findings in scientific studies and to verify experimental methods in peer-reviewed publications. Many funders and journals now require data

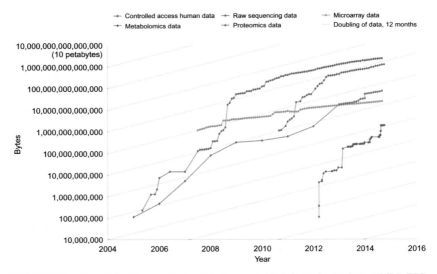

FIGURE 11.1 Growth in data repositories at the European Bioinformatics Institute (EMBL-EBI). Shown are repositories for controlled access human data, raw sequencing data, microarray, proteomics, and metabolomics data. Archives were started at different point in history. Metabolomics shows the steepest growth of all repositories at EMBL-EBI.

arising from publicly funded organizations to be made freely accessible, such MetaboLights and its sister databases play an important role in enabling the transparent reproduction and reuse of metabolomics results. MetaboLights is now the fastest growing in repository at the EMBL-EBI, with a 3-month doubling time in 2014 (Fig. 11.1).

11.3.1.2 The Metabolomics Workbench

The Metabolomics Workbench serves as a national and international repository for metabolomics data and meta-data and also includes data analysis tools and access to metabolite standards, protocols, tutorials, and training. The database was funded by the National Institutes of Health (NIH) Metabolomics Common Fund, with the aim to increase US national capacity in metabolomics by supporting the development of next generation technologies, providing training, enhancing the availability of high quality reference standards, and promoting data sharing and collaboration [32].

The Metabolomics Workbench is hosted at the Metabolomics Program's Data Repository and Coordinating Center (DRCC), which is housed at the San Diego Supercomputer Center (SDSC) at the University of California, San Diego. The DRCC acts as a North American hub for metabolomics-related research carried out at each of the six Regional Comprehensive Metabolomics Research Cores (RCMRCs) [33]. All metabolomics research carried out at these centers and funded by the NIH Metabolomics Common Fund must be made publically available via the Metabolomics Workbench.

11.3.2   Data Warehouses

11.3.2.1   MetabolomeXchange

MetabolomeXchange aggregates data from four different data providers—MetaboLights, Metabolomics Workbench, GMD, and Metabolomic Repository Bordeaux—which together make up the MetabolomeXchange Consortium [34]. The goal of MetabolomeXchange is increase the accessibility of and awareness about newly released, publicly available metabolomics data sets from verified members of the Consortium. MetabolomeXchange aims to provide a network of stable and coordinated metabolomics data, while also ensuring that both the scientific community and the commercial user community have access to high-quality reference data. The data "exchanged" through MetabolomeXchange consists of both experimental data and metadata for individual metabolites and metabolomic profiles.

MetabolomeXchange enables researchers to submit data either by submitting to the existing data repositories within the MetabolomeXchange Consortium or by becoming a data provider and member of the consortium. MetabolomeXchange was launched in 2014 and is coordinated by the EMBL-EBI. It is the outcome of the European Commission–funded Coordination of Standards in Metabolomics (COSMOS) project [34], which ran from 2012 to 2015 and gathered the European metabolomics data providers to establish and promote community standards for metabolomics data and experiments [35]. MetabolomeXchange is modeled on the ProteomeXchange [36], a consortium established in 2012 to provide a coordinated submission of MS proteomics data to the main existing proteomics repositories and to encourage optimal data dissemination.

11.4   CONCLUSIONS: THE CHALLENGES OF METABOLOMICS BIG DATA

In this chapter, we have provided an overview of the available databases for the metabolomics community, which can aid in the interpretation of the metabolomics data generated during metabolic phenotyping experiments. However, many challenges, both technical and social, still remain for harnessing the value of metabolomics big data. Although databases aid in the identification of compounds and structures, and enable access to existing metabolomics data sets, questions remain around the quality and reusability of data. Given that success and credit are still evaluated in relation to published papers, how can researchers be encouraged to devote valuable time and resources toward donation and annotation of data? Given that instrument vendors continue to produce many different proprietary formats for data, how can researchers be encouraged to adopt and adhere to existing standards? As the community gets bigger and bigger, and as databases and repositories proliferate, how can researchers get the most use out of experimental data, as they are being generated and published?

To address such questions, the scientific community at large has begun to recognize the need for credit attribution for scientific outputs beyond journal articles, which include the creation and maintenance of databases and the submission and curation of data. Credit attribution mechanisms influence hiring, promotion, tenure, and competitive grant processes and shape how the research community engages in various activities [37]. Researchers have begun to adopt various "altmetrics" for judging and crediting data downloads, use, reuse, citation, and other metrics [38]. Such altmetrics encourage the research community to devote valuable time and resources to database-related activities.

Metabolomics researchers have also begun to develop standard file formats and minimum reporting standards to guide annotation and meta-data, both of which aim to make the data not only accessible but also useful and reusable by third parties. In metabolomics, researchers have developed open-source Extensible Markup Language (XML)–based file formats for instrument-derived raw data, through consortia such as the COSMOS [39], which is based on previous works by the Metabolomics Standards Initiative (MSI) [40, 41], and the Proteomics Standards Initiative (PSI) communities [42]. These open-source formats, which include mzML and nmrML [43], make data generated with various commercial and proprietary platforms interoperable. Moreover, databases such as MetaboLights encourage the submission of standardized metadata alongside experimental data to enhance the reusability and ensure the quality of metabolomics data.

In addition, metabolomics researchers have begun to integrate big data analytics into existing database resources, including tools for the processing and analysis of data alongside spectral data and metabolomics data sets. In genomics, for example, the algorithm Basic Local Alignment Search Tool (BLAST) has been made available via the NCBI since 1990 to aid in the comparison of primary biological sequence data [44]. For metabolic profiling, tools such as MetaboAnalyst [45], MZmine [46], and XCMS Online [47] have been developed to aid in the processing and analysis of MS data. XCMS Online, for example, draws upon the MS database METLIN to help with compound identification and is capable of accepting data in 11 different formats.

In the end, to make metabolomics truly ready for "big data" and metabolic phenotyping, the community will have to address the challenge not only of dealing with complex metabolomics data but also of integrating metabolomics data with many other types of data: data generated from genomics, transcriptomics, and proteomics, data generated from patient samples such as those contained within the UK Biobank, and data generated from patient health care records. Researchers will not only face increasing computational complexity but will also face increasing ethical challenges surrounding the use and access to confidential patient information [48]. However, by combining many different types of data, researchers will have the ability to truly analyze, understand, and visualize the human phenome as it changes in space and time.

REFERENCES

[1] Technology Review. Illumina says 228,000 human genomes will be sequenced this year. Available from: <http://www.technologyreview.com/news/531091/emtech-illumina-says-228000-human-genomes-will-be-sequenced-this-year/>; 2014 [September 4, 2015].

[2] Laney D. 3D data management: controlling data volume, velocity and variety. META Group Research Note 2001:6.

[3] Mark van Rijmenam. Why the 3V's are not sufficient to describe big data. Available from: <https://datafloq.com/read/3vs-sufficient-describe-big-data/166>; 2015 [September 4, 2015].

[4] Levin N. Open access, open data, open science...what does "openness" mean in the first place? Available from: <http://somatosphere.net/2015/02/open-science.html>; 2015.

[5] European Molecular Biology Laboratory. The European Bioinformatics Institute. Available from: <http://www.ebi.ac.uk/>; 2015 [September 4, 2015].

[6] National Institutes of Health. National Center for Biotechnology Information. Available from: <http://www.ncbi.nlm.nih.gov/home/about/>; 2015 [September 4, 2015].

[7] HUGO. Summary of principles agreed at the first international strategy meeting on human genome sequencing, Bermuda. Available from: <http://www.ornl.gov/sci/techresources/Human_Genome/research/bermuda.shtml>;1996 [August 25, 2015].

[8] Elliott P, Peakman TC. The UK Biobank sample handling and storage protocol for the collection, processing and archiving of human blood and urine. Int. J Epidemiol. 2008;37(2): 234–44.

[9] Petersen SE, Matthews PM, Bamberg F, Bluemke DA, Francis JM, Friedrich MG, et al. Imaging in population science: cardiovascular magnetic resonance in 100,000 participants of UK Biobank-rationale, challenges and approaches. J. Cardiovasc. Magn. Reson. 2013;15(1):46.

[10] Fiehn O, Barupal DK, Kind T. Extending biochemical databases by metabolomic surveys. J. Biol. Chem. 2011;286(27):23637–43.

[11] Karp PD, Ouzounis CA, Moore-Kochlacs C, Goldovsky L, Kaipa P, Ahren D, et al. Expansion of the BioCyc collection of pathway/genome databases to 160 genomes. Nucleic Acids Res. 2005;33(19):6083–9.

[12] Kanehisa M, Goto S. KEGG: kyoto encyclopedia of genes and genomes. Nucleic Acids Res. 2000;28(1):27–30.

[13] Joshi-Tope G, Gillespie M, Vastrik I, D'Eustachio P, Schmidt E, de Bono B, et al. Reactome: a knowledgebase of biological pathways. Nucleic Acids Res. 2005;33(Suppl. 1):D428–32.

[14] Pico AR, Kelder T, van Iersel MP, Hanspers K, Conklin BR, Evelo C. WikiPathways: pathway editing for the people. PLoS Biol. 2008;6(7):e184.

[15] Markley JL, Ulrich EL, Berman HM, Henrick K, Nakamura H, Akutsu H. BioMagResBank (BMRB) as a partner in the Worldwide Protein Data Bank (wwPDB): new policies affecting biomolecular NMR depositions. J. Biomol. NMR 2008;40(3):153–5.

[16] Degtyarenko K, de Matos P, Ennis M, Hastings J, Zbinden M, McNaught A, et al. ChEBI: a database and ontology for chemical entities of biological interest. Nucleic Acids Res. 2008;36(Suppl. 1):D344–50.

[17] Pence HE, Williams A. ChemSpider: an online chemical information resource. J. Chem. Educ. 2010;87(11):1123–4.

[18] Smith CA, O'Maille G, Wang EJ, Qin C, Trauger SA, Brandon TR, et al. METLIN: a metabolite mass spectral database. Ther. Drug Monit. 2005;27(6):747–51.

[19] Horai H, Arita M, Kanaya S, Nihei Y, Ikeda T, Suwa K, et al. MassBank: a public repository for sharing mass spectral data for life sciences. J. Mass Spectrom. 2010;45(7):703–14.

[20] Wishart DS, Tzur D, Knox C, Eisner R, Guo AC, Young N, et al. HMDB: the human metabolome database. Nucleic Acids Res. 2007;35(Suppl. 1):D521–6.

[21] Kopka J, Schauer N, Krueger S, Birkemeyer C, Usadel B, Bergmüller E, et al. GMD@ CSB. DB: the Golm metabolome database. Bioinformatics 2005;21(8):1635–8.

[22] U.S. Food and Drug Administration. Table of pharmacogenomic biomarkers in drug labels. Available from: <http://www.fda.gov/drugs/scienceresearch/researchareas/pharmacogenetics/ucm083378.htm>; 2013.

[23] Wang Y, Xiao J, Suzek TO, Zhang J, Wang J, Bryant SH. PubChem: a public information system for analyzing bioactivities of small molecules. Nucleic Acids Res. 2009;37(Suppl. 2): W623–33.

[24] Caspi R, Altman T, Billington R, Dreher K, Foerster H, Fulcher CA, et al. The MetaCyc database of metabolic pathways and enzymes and the BioCyc collection of Pathway/Genome Databases. Nucleic Acids Res 2014;42(D1):D459–71.

[25] Karp PD, Paley SM, Krummenacker M, Latendresse M, Dale JM, Lee TJ, et al. Pathway Tools version 13.0: integrated software for pathway/genome informatics and systems biology. Brief. Bioinform. 2009:bbp043.

[26] Kanehisa M, Goto S, Sato Y, Kawashima M, Furumichi M, Tanabe M. Data, information, knowledge and principle: back to metabolism in KEGG. Nucleic Acids Res. 2014;42(D1):D199–205.

[27] Kelder T, van Iersel MP, Hanspers K, Kutmon M, Conklin BR, Evelo CT, et al. WikiPathways: building research communities on biological pathways. Nucleic Acids Res. 2012;40(D1):D1301–7.

[28] Wikipedia. Data warehouse. Available from: <http://en.wikipedia.org/wiki/Data_warehouse>; 2003 [June 20, 2015].

[29] Wikipedia. Data mining. Available from: <https://en.wikipedia.org/wiki/Data_mining>; 2015 [September 4, 2015].

[30] Haug K, Salek RM, Conesa P, Hastings J, de Matos P, Rijnbeek M, et al. MetaboLights—an open-access general-purpose repository for metabolomics studies and associated meta-data. Nucleic Acids Res. 2012:gks1004.

[31] Salek RM, Haug K, Conesa P, Hastings J, Williams M, Mahendraker T, et al. The MetaboLights repository: curation challenges in metabolomics. Database 2013;2013.

[32] NIH Common Fund's Data Repository and Coordinating Center. Metabolomics workbench. Available from: <http://www.metabolomicsworkbench.org/>; 2015 [cited September 4, 2015].

[33] National Institute of Health. NIH announces new program in metabolomics. Available from: <http://www.nih.gov/news/health/sep2012/od-19.htm>; 2012 [August 20, 2013].

[34] COSMOS. MetabolomeXchange. Available from: <http://metabolomexchange.org/site/>; 2015 [September 4, 2015].

[35] Salek RM, Haug K, Steinbeck C. Dissemination of metabolomics results: role of MetaboLights and COSMOS. GigaScience 2013;2(8.10):1186.

[36] EMBL-EBI. ProteomeXchange. Available from: <http://metabolomexchange.org/>; 2015 [September 4, 2015].

[37] Ankeny RA, Leonelli S. Valuing data in postgenomic biology: how data donation and curation practices challenge the scientific publication system Stevens H, Richardson S, editors. Postgenomics. Durham: Duke University Press; 2015. p. 126–49.

[38] Piwowar H. Altmetrics: value all research products. Nature 2013;493(7431) 159–9.

[39] Salek R, Neumann S, Schober D, Hummel J, Billiau K, Kopka J, et al. Coordination of Standards in MetabOlomicS (COSMOS): facilitating integrated metabolomics data access. Metabolomics 2015:1–11.

[40] Fiehn O, Robertson D, Griffin J, van der Werf M, Nikolau B, Morrison N, et al. The metabolomics standards initiative (MSI). Metabolomics 2007;3(3):175–8.

[41] Castle AL, Fiehn O, Kaddurah-Daouk R, Lindon JC. Metabolomics standards workshop and the development of international standards for reporting metabolomics experimental results. Brief. Bioinform. 2006;7(2):159–65.

[42] Orchard S, Kersey P, Hermjakob H, Apweiler R. The HUPO proteomics standards initiative meeting: towards common standards for exchanging proteomics data. Comp. Funct. Genomics. 2003;4(1):16–19.

[43] Salek RM, Arita M, Dayalan S, Ebbels T, Jones AR, Neumann S, et al. Embedding standards in metabolomics: the Metabolomics Society data standards task group. Metabolomics 2015;11(4):782–3.

[44] Altschul SF, Gish W, Miller W, Myers EW, Lipman DJ. Basic local alignment search tool. J. Mol. Biol. 1990;215(3):403–10.

[45] Xia J, Psychogios N, Young N, Wishart DS. MetaboAnalyst: a web server for metabolomic data analysis and interpretation. Nucleic Acids Res. 2009;37(Suppl. 2):W652–60.

[46] Katajamaa M, Miettinen J, Orešič M. MZmine: toolbox for processing and visualization of mass spectrometry based molecular profile data. Bioinformatics 2006;22(5):634–6.

[47] Tautenhahn R, Patti GJ, Rinehart D, Siuzdak G. XCMS Online: a web-based platform to process untargeted metabolomic data. Anal. Chem. 2012;84(11):5035–9.

[48] Cambon-Thomsen A. The social and ethical issues of post-genomic human biobanks. Nat. Rev. Genet. 2004;5(11):866–73.

Chapter 12

Modeling People and Populations: Exploring Medical Visualization Through Immersive Interactive Virtual Environments

Sarah Kenderine[1], Jeremy K. Nicholson[2] and Ingrid Mason[3]

[1]*Expanded Perception and Interaction Centre, University of New South Wales, Sydney, New South Wales, Australia* [2]*Department of Surgery and Cancer, Imperial College London, London, UK* [3]*Intersect Australia Pty Ltd, Sydney, New South Wales, Australia*

Chapter Outline

E. Holmes, J.K. Nicholson, A.W. Darzi & J.C. Lindon (Eds): Metabolic Phenotyping in Personalized and Public Healthcare. DOI: http://dx.doi.org/10.1016/B978-0-12-800344-2.00012-4

12.1 INTRODUCTION

The increasingly complex, diverse, high-dimensional, and large-scale of data accrued in biomedicine, health care, and the life sciences are opening up opportunities for improved care and better and faster clinical research. However, the pace, scale, density, and complexity of these data are quickly rendering current information visualization benchmarks and paradigms inadequate. Within this landscape, immersive interactive virtual environments are changing the visual language of medicine by providing high-resolution display systems with interactive, immersive, stereoscopic, and omnidirectional properties—to establish new paradigms in medical data visualization and interpretation. Issues necessitating improved resolution, dynamic modeling, data immersion, collaboration, and co-location are thus steering the future of medical data visualization.

Immersive interactive virtual environments are becoming established across diverse fields in medicine not only as training environments for simulated surgeries but also as potential therapeutic spaces. The full immersive and interactive visual potential of these systems will be pushed by the challenges of biomedical data itself. For example, complex "-omics" data require an equally complex and visually rich multirelational model. The density and volume of data accrued in human genome mapping far exceeds the limitations in conventional computer monitor–based research. As visualization researcher Khairi Reda states "HR [high resolution] environments allow for a new type of high-density genome visualization, making it possible to view genomic data from hundreds to thousands of genomes in a single view" [1]. The use of wall-sized high-resolution displays in the comparative study of genomics data was shown to "help scientists to find patterns in their data, by using the display real estate to show a large number of samples at once and avoid screen-thrashing" [2].

This chapter undertakes an overview of medical imaging and selected visualization paradigms as the necessary background to understand trends in medical big data and the relevance of immersive interactive virtual environments for solving the challenges presented by large, complex, high-dimensional (often weakly structured) data sets and massive amounts of unstructured information. The challenge of "big data" has been described as encompassing the issues of volume, variety, velocity and veracity. With increasingly integrative approaches that combine data sets using rich networks of specific relationships, the medical doctor, the surgeon, or the biomedical researcher of today are incapable of interpreting and interacting with all these data in a coherent way [3].

Modern biomedicine will not develop any further without new computational and visualization solutions for interacting and analyzing these data. A synergistic combination of methodologies and approaches offer ideal conditions for solving these aforementioned problems: human–computer interaction, information design, knowledge discovery, and data mining. Human–computer interaction (HCI) and information design emphasize human issues, including perception, cognition, interaction, reasoning, decision making, human learning, and human intelligence.

Knowledge creation is similarly supported through data integration, fusion, pre-processing, data mining, visualization, and simulation, including computational statistics, machine learning, and artificial intelligence [3]. By enabling scientists to see data through "fresh eyes" and to "see more," the approach taken in immersive interactive virtual environments attempts not only to provide scientific breakthroughs but also to reframe how research is perceived throughout related communities.

Essential to the creation of new knowledge, the immersive interactive virtual environments described in this chapter facilitate human-to-human as well as human–computer interaction, information design, knowledge discovery, and data mining in the processes of analytical visualization. Importantly, the chapter also addresses the building blocks of visualization that is collaborative, interoperable, and distributed, establishing best practices in data management with widespread implications for all forms of advanced visualization in the medical sector.

12.1.1 An Overview of Medical Imaging

Since the 1970s, the field of medical "visualization" has undergone significant improvements, and medical image acquisition technology has undergone continuous and rapid development. In 2012, Botha et al. summarized the major developments and trends in visualization, showing the progression from scalar volume data sets through time-dependent data to multifield and finally multi-subject data sets, as representative of various trends in the development of the field [4]. Originating in 1978 with unimodal visualization of computed tomography (CT) data using three-dimensional (3D) surfaces for diagnosis and visual radiotherapy planning, the use of multimodal volume visualization (combining CT with magnetic resonance imaging [MRI], for example) was followed in 1986 by monochromatic shaded images of 3D CT data by Hohne and Bernstein. Multimodal volume rendering for the planning of neurosurgical interventions, where MRI, CT, functional MRI (fMRI), positron emission tomography (PET), and dynamic systems angiography (DSA) data are all combined in an interactive but high-quality visualization for the planning of brain tumor resection. This rapidly advanced to the extraction of 3D isosurfaces, with the Marching Cubes isosurface extraction algorithm (Lorensen and Cline, 1987), and shortly thereafter to volume raycasting [4]. In 1993, Altobelli et al. published a report on the use of CT data to visualize the possible outcome of complex craniofacial surgery and in 1996, Hong et al. on 3D virtual colonoscopy (VC). VC combined CT and volume visualization technologies for diagnosis laying the basis for computer-aided design diagnostics. Botha et al. noted the importance of VC as:

> … *rapidly gaining popularity and … poised to become the procedure of choice in lieu of the conventional optical colonoscopy for mass screening for colon polyps—the precursor of colorectal cancer. Unlike optical colonoscopy, VC is patient friendly, fast, non-invasive, more accurate, and cost-effective procedure for mass screening for colorectal cancer [4].*

Multifield MRI data include diffusion tensor imaging (DTI), which was introduced by Basser et al. in 1994, and they were also some of the first to extract and visualize fibertract trajectories from DTI data of the brain, thus linking together the point diffusion measurements to get an impression of the global connective structures in the brain. DTI serves as one of the first examples of natively multifield medical data, that is, medical data with multiple parameters defined over the same spatiotemporal domain. The advent of DTI initiated a whole body of medical visualization research dedicated to the question of how best to visually represent and interact with diffusion tensor data in particular and multifield medical data in general. Blaas (2010) used a visual analysis—inspired solution based on linked physical and feature space views and exemplifies this area of visualization research [5].

Time-varying medical volume data visualization began in 1996 with work by Behrens et al., who used dynamic contrast-enhanced MRI mammography data in combination with the display of parameter maps, the selection of regions of interest, the calculation of time-intensity curves, and the quantitative analysis of these curves. In 2001, Tory et al. presented methods for visualizing multitime point (1-month interval) MRI data of a patient with multiple sclerosis, where the goal was to study the evolution of brain white matter lesions over time. Methods used included glyphs, multiple isosurfaces, direct volume rendering, and animation [4].

12.1.2 Accelerating Data Acquisition

Toshiba's Aquilion One CT scanner that can slice volumes at 320 per second (fast enough to image a beating heart), high angular resolution diffusion imaging, and diffusion spectrum imaging in MRI present various challenges for visualization. In addition, molecular imaging enables the imaging of biochemical processes at the macroscopic level in vivo, yielding data sets that vary greatly in scale, sensitivity, and spatial temporality. Each permutation brings with it new domain-specific questions and visualization challenges. Microscopy is accelerating rapidly, and recent examples include techniques for interactively visualizing large-scale biomedical image stacks demonstrated on data sets of up to 160 gigapixels and tools for the interactive segmentation and visualization of large-scale 3D neuroscience data sets, demonstrated on a 43-gigabyte electron microscopy volume data set of the hippocampus [4].

These rapid, continuous advances in imaging yield data sets that are both dynamic in nature and large in magnitude and demand improvements to computational and perceptual scalability. Indeed, as Botha et al. summarized, "Completely new visual metaphors are required to cope with the highly multivariate and 3D data of diffusion weighted imaging in particular and many other new imaging modalities in general" [4]. Advancing image acquisition, data diversification, and greater and greater magnification demand correlated advances in display and their interaction. Against this background, the remainder of this chapter focuses on the typical ways in which these data are visualized and in particular the ways in which large-scale immersive interactive visualization systems are being used to provide researchers with insights that they may not be able to achieve otherwise.

12.2   MODALITIES AND METHODS IN MEDICAL VISUALIZATION

The modalities and methods for medical imaging and data analysis take a myriad of forms, from topologic through to population modeling and "-omic" visualization. Several visualization types are introduced here before discussing more extensive "big data" challenges.

12.2.1   Topologic Methods

Topologic data representation has played an important role in medical visualization, since it can allow us to segment specific features such as human organs and bones from the input medical data sets systematically and identify their spatial relationships consistently. The Visible Human Project 1994 [6], for example, used advanced cryogenic serial sectioning methods, by which a whole cadaver was sliced at 1 mm thickness, and high-resolution digital images were taken for each slice. This very large digital data collection was then reconstructed into a 3D stack, which provided the first high-resolution mapping of the human anatomy. Serially sectioned images of whole cadavers have become available through the work of four projects: the Visible Human Project, the Visible Korean Project, the Chinese Visible Human Project, and the Virtual Chinese Human Project [7].

Medical visualization approaches become significantly more valuable when enhanced with simulated models, such as blood flow simulations, interactive skeletal range of motion and biomechanical stress simulation models for implant planning in orthopedics, which help predict the outcome of a disease process or therapeutic procedure or that enrich measured data with expected physiologic phenomena.

While the aforementioned Visible Human Project was considered a turning point when data were released and was doing pioneering work in the field of digital anatomy, it remained constrained to a limited number of single humans. "We do not know how he breathed, walked, swallowed, digested—the virtual human (VH) data totally lacks multiplicity and functionality" [8]. Over time, topologic approaches have also been extended to analyze 3D medical volume data and are now being developed for visualizing multivariate and high-dimensional data sets and thus potentially for analyzing tensor fields obtained through DTI, multisubject data in group fMRI studies, and time-varying data measured by high-speed CT. Potentially the best way to provide VH with these features involves the development of the so-called Virtual Physiologic Human (VPH), intended as a framework of methods and technologies that will make it possible to describe human physiology and pathology in a complete and integrated way.

12.2.2 Virtual Physiologic Human

In 2005, VPH was introduced to indicate "a framework of methods and technologies that, once established, will make possible the collaborative investigation of the human body as a single complex system." The project has a twofold strategy:

1. *Decompose*: reducing the complexity of living organisms, decomposing them into parts (cells, tissues, organs, organ systems) and investigating one part in isolation from the others. This approach has produced, for example, the medical specialties, where the nephrologist looks only at the kidneys and the dermatologist only at the skin; this makes it very difficult to cope with multi-organ or systemic diseases, to treat multiple diseases (so common in the aging population), and in general to unravel systemic emergence due to genotype–phenotype interactions.

2. *Recompose*: with computer models using all the data and all the knowledge obtained about each part, simulations can investigate how these parts interact with one another, across space and time and across organ systems. Human anatomy is central to the study and practice of medicine. A significant proportion of clinical symptoms, signs, phenotypes, diagnostic tests, and medical procedures are recorded and described with reference to anatomical location. The VPH domain thus focuses on the mechanistic modeling of a significant number of functional relationships, as well as the statistical correlation of biomedical data sets that pertain to distinct anatomical locations.

The VPH vision contains a tremendous challenge, namely, the development of mathematical models capable of accurately predicting what will happen to a biological system. To tackle this huge challenge, multifaceted research is necessary: around medical imaging and sensing technologies (to produce quantitative data about the patient's anatomy and physiology), data processing to extract from such data information that in some cases is not immediately available, biomedical modeling to capture the available knowledge into predictive simulations, and computational science and engineering to run hypermodels (orchestrations of multiple models) under the operational conditions imposed by clinical usage [9].

12.2.3 Population Models

Population imaging involves the collection of medical image data and other measurements of a sample size typically over a thousand people, over a period of years, collated to study the onset and progression of disease, integrating large quantities of heterogeneous, multi modal and multi-time point data acquired of a large group of subjects. Examples of population modeling include the Rotterdam Scan Study and the Study of Health in Pomerania (SHIP) focusing on general health." The SHIP investigates common risk factors, subclinical disorders, and manifest diseases with highly innovative noninvasive methods in the high-risk population of northeast Germany, while the Rotterdam study investigated factors that determined the occurrence of cardiovascular, neurologic, ophthalmologic, endocrinologic, and psychiatric diseases in older people. This domain is characterized by the extreme heterogeneity and magnitude of the data, coupled with the explorative nature of the research, renders this a promising long-term application domain for visual analysis and medical visualization [4].

12.2.4 Predictive Models and Visualization in Systems Biology

In recent years, the study of such systems has been profoundly influenced by the development of a wide range of experimental methods resulting in a greatly increased volume of complex, interconnected data. There has been a corresponding increase in the development of visualization tools for systems biology data. These tools are very diverse, but they can be broadly divided into two partly overlapping categories, the first consisting of tools focused on automated methods for interpreting and exploring large biological networks and the second consisting of tools focused on assembly and curation of pathways. Many of these tools are tightly integrated with public databases, thus allowing users to visualize and interpret their own data in the context of previous knowledge. Gehlenborg et al. in 2010 collated over 60 visualization tools focused on *interaction networks*, *pathways*, and *multivariate "-omics"* data [10]. These are available, often free, in cross-platform formats as stand-alone software, through web browsers as downloadable plug-in or as fully integrated online services.

12.2.5 Metabolic Profiling

> **Box 12.1.** *Technologies* **extract from "Metabolic Phenotyping in Clinical and Surgical Environments"** [11]
>
> Nuclear magnetic resonance (NMR) spectroscopy and mass spectrometry (MS) are the main techniques that are used for the metabolic profiling of biofluids (eg, urine, blood plasma, amniotic fluid, and cerebrospinal fluid) and tissues in the form of extracts or intact biopsy specimens [12,13]. New solid-state NMR methods allow analysis of samples of less than 1 mg, thereby enabling detailed studies of tissue heterogeneity, such as between tumor tissue and tumor margins [14].
>
> Both NMR and MS can simultaneously identify and quantify information on a wide range of small molecules with good analytical precision and accuracy and require only a small amount of sample (typically 10–400 μL). NMR spectroscopy is highly reproducible, with a detection limit in the submicromolar range. All hydrogen-containing metabolites in a biofluid are detected simultaneously and nondestructively, with little sample preparation. MS has much lower detection limits, but it is destructive, and a more targeted approach is often needed with prior separation of metabolites, using either chromatography or capillary electrophoresis.

Thus, MS approaches tend to be less reproducible, more platform dependent, and susceptible to variability. Some of the earliest clinical studies used gas chromatography–mass spectrometry (GC-MS), especially for detection of inborn errors of metabolism [15], and although still widely used, the requirement for chemical derivation of the sample to allow metabolite volatilization imposes limitations on its widespread use in clinical diagnostics. Liquid chromatography (LC), particularly ultra-high-performance liquid chromatography (UPLC), is being used increasingly for metabolic profiling.

Chemometrics—multivariate statistics applied to chemical data—are used in clinical metabonomics to reduce the dimensionality of complex spectroscopic data sets, and to identify biochemical patterns that relate to a disease or an intervention. Linear projection methods such as principal components analysis and partial-least-squares discriminant analysis are commonly used to map samples on the basis of their biochemical similarity and to extract patterns of metabolites that relate to a particular disease [16]. Principal-components analysis is used extensively in metabonomics. This technique transforms the data descriptors into a set of linear combinations of the original features based on decreasing levels of variance. Any clustering seen is based on the data alone, and there is no preassignment of sample classes. Alternatively, in "supervised" methods, multiparametric data sets can be modeled so that the class of separate samples (a "validation" set) can be predicted on the basis of a series of mathematical models derived from the original data or "training" set. Partial-least-squares analysis is a widely used supervised method (using a training set of data with known endpoints). This method relates a data matrix containing independent variables from samples, such as spectral intensity values (an X matrix) to a matrix containing dependent variables (eg, measurements of response) for those samples (a Y matrix). Partial-least-squares analysis can also be combined with discriminant analysis to establish the optimal position to place a surface that best separates classes. Other popular chemometric methods include hierarchical clustering, self-organizing maps, and neural networks [16, 17].

Statistical spectroscopy is a form of computational modeling that is used to enhance biomarker recovery, allowing improved information extraction from a set of spectra. This method generally operates on defining correlation structures between variables (signals) that are found to be discriminatory between sample classes. Highly correlated signals are likely to come from the same molecule or from molecules regulated by the same metabolic pathway. Statistical total correlation spectroscopy is used to identify correlated signals within a data set for biomarker identification. NMR and MS possess high complementarity in molecular-structure elucidation studies. In many metabonomic studies, multiple samples with a wide range of biochemical variation are available for both NMR and MS analysis, creating an opportunity for statistical analysis of signal amplitude co-variation between the two sets of data. Statistical heterospectroscopy is an extension of statistical total correlation spectroscopy for the co-analysis of multispectroscopic data sets, which have been acquired from multiple samples. The statistical heterospectroscopy approach, originally developed for NMR and MS correlation, can be used if any two or more independent spectroscopic data sets from any source are available for any sample cohort [18].

A new and exciting approach that uses MS is the analysis of smoke from a cauterization device used in surgery to identify the exact type of tissue being investigated [19,20]. By using a combination of new sampling methods, high-speed MS and chemometrics for classification purposes, it is possible to identify different types of tissue in real time during a surgical procedure a development known as the intelligent knife (i-knife). A parallel application of MS is to locate molecules within a sample as an imaging technique by using ionization methods based on laser ablation from the tissue surface [21]. In combination with chemometric-enhanced information recovery, this has led to the possibility of an augmented histological assessment [22]

12.3   BIG DATA AND THEIR MANAGEMENT

Due to the increasing trend towards personalized and precision medicine (P4 medicine: Predictive, Preventive, Participatory, Personalized), biomedical data today results from various sources in different structural dimensions, ranging from the microscopic world, and in particular from the omics world (e.g., from genomics, proteomics, metabolomics, lipidomics, transcriptomics, epigenetics, microbiomics, fluxomics, phenomics, etc.) to the macroscopic world (e.g., disease spreading data of populations in public health informatics) [3].

Data are growing and moving faster than health care organizations can consume them; getting access to these valuable data and factoring them into clinical and advanced analytics is critical to improving care and outcomes, incentivizing behavior, and driving efficiencies. Health care organizations are leveraging big data technology to capture all of the information about a patient to get a more complete view for insight into care coordination and outcomes-based reimbursement models, population health management, and patient engagement and outreach.

The nature of these health data has evolved to the point where 80% is unstructured but clinically relevant. These data reside in multiple places such as individual electronic medical records (EMRs), laboratory and imaging systems, physician notes, medical correspondence, claims, content management systems, and financial systems. They include clinical data from computerized provider order entries and clinical decision support systems (physician's written notes and prescriptions, medical imaging, laboratory, pharmacy, insurance, and other administrative data); patient data in electronic patient records; machine generated/sensor data such as from monitoring vital signs; social media posts, including Twitter feeds (so-called tweets), blogs, status updates on Facebook and other platforms, and web pages; and less patient-specific information, including emergency care data, news feeds, and articles in medical journals. Analytical techniques scaled up to the complex and sophisticated analytics are necessary to accommodate volume, velocity, and variety. Co-related to the visualization challenges described in this chapter, the enormous variety of data—structured, unstructured, and semistructured—is a dimension that makes health care data both interesting and challenging. By definition, "big data" in health care refer to electronic health data sets so large and complex that they are difficult (or impossible) to manage with traditional software and/or hardware; nor can they be easily managed with traditional or common data management tools and methods [23].

Holzinger et al. described the grand challenge of "big data" as their usefulness to the end user, suggesting that the solution may lay with the intelligent human user interaction, rather than dependence on computer analysis [3]. The current problem in this process is the complexity of modeling in biomedical data, which renders manual analysis extremely difficult. To draw out the unique insights of human problem solving and analysis to this highly complex data climate, a new advanced visual analytic paradigm is necessary. The answer to the optimized human—data interaction is in a synergy of HCI and "knowledge discovery from data," which would allow an untrained domain expert to interactively handle data sets in order to ask questions [3]. To engage with this more complex scale, data reduction is necessary: first, in the reduction the number of dimensions of samples (reducing high-dimensional samples to scalar values); second, in the reduction of the number of data points (identifying clusters as single entities to engage on a broader scale) [5].

12.3.1 Data Fusion

Fusing data from multiple sources has proved useful in many fields, including bioinformatics, signal processing, and social network analysis. However, identification of common (shared) and individual (unshared) structures across multiple data sets remains a major challenge in data fusion studies. In many disciplines, data from multiple sources are acquired and jointly analyzed for enhanced knowledge discovery. Multimodal imaging typically requires specialized postprocessing tools to merge data from the modalities because the "space" and "time" of meaningful qualities imaged is mostly different for each modality. Joint analysis of multimodal data sets has the aim to combine the complementary aspects (either in terms of underlying physiologic processes or in spatiotemporal resolution) of each modality in such a way that there is an added benefit compared with analyzing and interpreting each data set separately. Data fusion can be asymmetrical where weighting is given to one data set in particular over the others (these data may, for example, have greater certainty) or symmetrical with correlation between their spatiotemporal resolution [24].

In many disciplines, data from multiple sources are acquired and jointly analyzed for enhanced knowledge discovery. For instance, in metabolomics, different analytical techniques are used to measure biological fluids in order to identify the chemicals related to certain diseases. It is widely known that some of these analytical methods (eg, LC-MS and NMR) provide complementary data sets and that their joint analysis may enable us to capture a larger proportion of the complete metabolome belonging to a specific biological system.

12.3.2 Data Management and Interoperability

Data curation and information architecture play a critical part in that research management infrastructure transformation [25–27]. A complex and necessary research data management capacity and infrastructure-building exercise, with careful consideration of the structures and processes, will enable precision medicine and meet the big data challenge associated with biomedicine. Most importantly, data sharing and interoperability are the means for new knowledge to emerge. Collective initiatives such as the Research Data Alliance and Global Alliance for Genomics and Health [28], as well as large international strategic ventures [29,30], are driving this change in biomedicine.

Principal research exploration generally occurs at a domain knowledge level, where new and/or shared scholarly pathways can be discovered. Increasingly, large volumes of research objects require substantial standardization efforts and appropriate interfaces to enhance search capability. New methods for structuring meta-data (eg, new standards, linked data, structured ontologies, and formal domain-specific vocabularies) have become a semantic backbone for data-intensive research [31,32]. Achievements in standardization are exemplified in heterogeneous research sets such as the contemporary Linked Data for Libraries [33], and the SAO/NASA Astrophysics Data System [34] integrated search interface [35] provides a striking example of the power of linking a range of research objects in astrophysics to support exploration of research collaborations, peer networks, related areas of research, and gaps or crossovers in research activity. This same shift is occurring with biomedicine to aid the exploration of biomedical literature and associated scholarly research objects. This is evidenced by the range of systems developed and integrated with PubMed to support scholarly resource searching [36]. The semantic requirements for capturing the domain-specific knowledge are significantly different from those supporting peer data sharing or data integration.

Global reference data sets are now significant features of the research data landscape. These references sets range in type, that is, human chemistry and genetics [37,38], human disease [39], and clinical methods, treatments, and outcomes [40]. These data sets arise directly out of biomedical research and in the related research fields of pharmaceutics, health, and medicine. The life sciences community started publishing data in the linked open data (LOD) cloud around 2009 [41], and today this remains a notable section of that vast semantic web resource. There are, however, problems with a lack of shared semantic structures to realize the full power of using LOD methods. With bioinformatics efforts such as Bio2RDF, the aims of sharing data using LOD techniques are steadily improving [42]. New moves to enable genetic data to be shared by enabling matching to occur, along with well-established data portals [43], have allowed some of the hardest cases to be solved [44]. This shift in biomedical infrastructure development reveals that data sharing and interoperability are key elements of major international collaborations [45] to advance research. The changes in data keeping practices are occurring at the macro-level and connected to the specialist research communities [46,47]. The manner and form in which data are brought together and made available are the means by which research data management transformation can support data- and technology-intensive research practices.

Enabling cross-disciplinary and collaborative or translational research poses complex data management, integration, and fusion challenges. Foundational research data management with semantic interoperability is the infrastructural context for data- and technology-intensive research. The infrastructural context of the research domain and the new research practice need to inform the development of infrastructure to support analytical and modeling and connecting health, medical, genetic, population, and social data. The complexity of the framework underpinning the Commons being established through the National Institutes of Health (NIH) Big Data to Knowledge [48] program in support of biomedical research reveals the scale and complexity of bringing all these data together. Ensuring that the arrangement of the data is meaningful and coherent presupposes interoperability, for discovery and to serve recombinatory efforts at the research level. Informatics experts and data specialists provide critical input to the design of data architecture and management platforms to increase capacity for data discovery and reuse. In the past, medical subject headings [49] maintained a basic level of discovery consistency across health and medical information. With data-driven and computational research, the need for semantic structures to support integrated research on disease and genetics [50] has grown rapidly. Systematic sharing and moving data around (or moving the questions around instead) are part of this broader change in bioinformatics and biomedicine that needs to be brought to bear when developing new research infrastructure [51].

Data discovery and reuse occur at the broader community level (a discoverable resource) and at the research activity level (the unit of observation), for example, through research centers, cross-sector collaborations, and within research domains. The management of research material—from its raw captured state, to general data management processes, through to fine-grained research observations—requires the use of diverse information standards and vocabularies that are fit for purpose. The broader informatics changes in biomedicine will have an impact on development of services. Visualization services may or may not be the endpoint in a research process. Visualization may occur through tools at any point in the data and research lifecycles. The development of pipelines for different data types for transfer to visualization services will require the appropriate repository types (whether temporary or permanent). For the simple purpose of enabling discovery, an application layer may need to be developed above the diverse data repositories to support retrieval and further analysis [52]. The virtual environments where the data types can be merged or reformatted may need to be co-located. Where it is not technically feasible to move data for visualization into a service, then bringing the derived data from diverse repositories within or across institutions or nations, may need to be part of the process of data curation in support of biomedical and personalized medicine. Where data are streaming, new methods of sampling may be employed, to support the real-time visualization and potentially curation processes [53].

The curation [27] and scientific research activity [54] lifecycles must align and overlap in research infrastructure design. This overlay of lifecycles will support the movement of meta-data and data flows effectively for the researcher—from the point of ingest, into research applications and display in systems, and onto archival and discovery systems [55,56]. The data-intensive research activity, including services, needs to be incorporated into a research data management framework [57]. The data curator and architects aid through the development of research support processes to reduce duplication of effort, standardize ingest and transformation processes, and prepare and arrange the data. Methods for data integration and fusion may require different approaches developed by data curators and architects to meet the visual analytical needs of students [58] and researchers [59]. Where data are sensitive and contain personal information, some level of abstraction of the data or a mediation of view within storage and browser systems needs to be implemented to address privacy concerns.

Genome browser interface designs, however, reveal the limits of detailed analysis constrained by a two-dimensional (2D) computer screen interface design [60,61]. The success of the graph-oriented tool Cytoscape [62] has revealed a demand for stronger visualization infrastructure support services. Annotating data with domain-specific vocabularies and ontological structures [63] aids search. Additional techniques such as entity mapping may be employed to forge links across different data types in the recombination of clinical and pathology data. Visualization experts need multilevel semantics for resource discovery and to explore units of analysis and data features. Building a metadata layer is a significant investment in critical semantics connections that support the rendering of biomedical visualizations [64].

12.4   IMMERSIVE INTERACTIVE VIRTUAL ENVIRONMENTS

The true challenges lie in visualizing multiple scans, at multiple points in time, and even across multiple patients and multiple modalities [5]

The perception is enhanced drastically when you step inside the data as opposed to looking at projections on a computer screen, … so, you can just realize what the data means much faster, and that is what I think is so exciting about this technology: that you can perceive the data faster than by looking at them on a flat screen or on a report [65].

12.4.1 Tunnel Vision Versus Omnidirectional Vision

Evaluating the usefulness of 3D data imaging in radiology and surgery varies greatly among contexts, but in most cases where data input space exceeds human cognition—and for communication between disciplines—3D image synthesis is essential [66]. More recently, histopathology experts described 3D reconstructions of tissue and examinations at a microscopic resolution as having "significant potential to enhance the study of normal and disease processes," especially in circumstances where structural changes or spatial relationships are important [67]. This research commented on the limitations of existing 3D imaging practices, including "optical projection tomography, 3D ultrasonography, microscopic MRI, X-ray micro-CT, confocal laser scanning, and serial block face imaging"; and introduced a new means of high-resolution 3D histologic image reconstruction for virtual slides; the limits of this introduced method lay in the size of the native resolution of the slides restricting the 3D visualization because of the memory and screen resolution of the desktop viewing platform; the solution to this issue, gigapixel 3D volumes, requires a high-resolution display [67]. In the increasingly large, dense, and complex data climate, it appears that 3D imaging will be intrinsically grounded in medical practice.

Recent visualization research remains constrained to 2D small screen–based analysis and advances interactive techniques of clicking, dragging, panning, and rotating. Furthermore, the number of pixels available to the user remains a critical limiting factor in human cognition of data visualization. Research into new modalities of visualizing data is essential for a world producing and consuming digital data at unprecedented rates [68]. Visualization tools have proliferated in systems biology, but as Gehlenborg et al. noted:

> ...in spite of these changes, and in spite of the development of new methods for visualizing and analyzing these data, we still use the same primary visual metaphor to communicate ideas about biological systems: namely, pathways (graphs that show overall changes in state) or, more generally, networks (graphs that do not necessarily show state changes) [10].

Researchers who are limited to this form of virtual navigation suffer significant temporal and cognitive constraints, and this requires the user to spend time and effort moving back and forth between different parts of the information space in order to compare trends and look for outliers. Context and focus techniques may reduce the need for virtual navigation in some tasks but are less useful when the relevant information is distant [69]. Faced with this predicament, a user's natural response would be to reduce the amount of virtual navigation. This accommodating behavior may save precious time and reduce cognitive workload. Unfortunately, it may also contribute to a "tunnel vision" phenomenon, where analysis is focused on, and limited to, a small fraction of the available data [69].

Therefore, there is an increasing trend toward research requiring "unlimited" screen resolution has resulted in the recent growth of gigapixel displays. The high visual acuity of large-scale high-resolution visualization displays has already enabled new possibilities for data visualization in many fields of research. As the scale and complexity of data continue to grow, large-scale visualization instruments such as display walls and immersive environments become increasingly essential to researchers, letting them transform raw data into discovery. In particular, immersive systems are an attractive option for exploring 3D spatial data such as molecules, MRI, and other imaging data sets. On the other hand, multi-tile display walls with their high resolution help interpret large data sets, offering both overview and resolution, or can be used to lay out a variety of correlated visual data and documents for collaborative analysis [70]. As Andrew Johnson, Director of Research at the Electronic Visualization Laboratory (EVL), University of Illinois at Chicago (UIC), describes: "These kinds of environments are lenses to look at data—the modern equivalent of the microscope or telescope" [71].

Compared with data sets typically analyzed on traditional desktop computer monitors, the resulting visualizations as 2D slices and 3D volumes are substantially larger [72]. To optimize the processing of these data for present and future resolution conditions, the extraction and visualization of data must be adaptable and scalable; additionally, with the push toward Personalized Medicine, these tools will be required to aid medical practitioners making decisions in real time [72].

The next-generation immersive virtual reality (VR) systems emerging from systems such as the CAVE Automated Virtual Environments (CAVE, University of Illinois at Chicago [UIC]) and Advanced Visualisation Interaction Environment (AVIE, University of New South Wales [UNSW]) include the StarCAVE (University of California, San Diego [UCSD]), CAVE2™ (UIC), the Allosphere (University of California, Santa Barbara [UCSB]), and a new environment in production at UNSW, Omniform co-located with the full-dome visualization facility DomeLab. The design of these immersive interactive environments enhance cognitive exploration and interrogation of high-dimensional data, essential in big data bioinformatics while also providing powerful spaces akin to the real world for simulation. The immersive framing helps resolve problems of data occlusion and distribute the mass of data analysis in networked sequences revealing patterns, hierarchies, and interconnectedness, and algorithmic processing underpins data fusion techniques. These hybrid approaches to data representation also allow for the development of audio strategies to help augment the interpretation of the results.

I think there's a huge upside in having new ways of visualizing large data sets ...If we have innovative ways of being able to visualize that data and understand that data, hopefully that will lead to better discoveries [73].

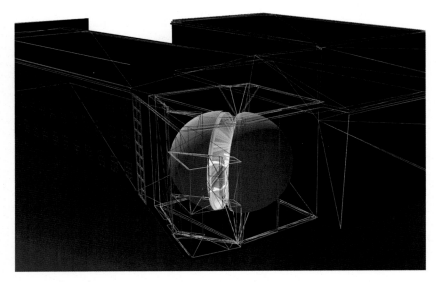

FIGURE 12.1   Graphics rendering of the AlloSphere immersive instrument created by composer JoAnn Kuchera-Morin, AlloSphere Director, California Nanosystems Institute University of California, Santa Barbara. *Image by Haru Ji.*

12.4.2 Allosphere

The AlloSphere was developed in 2007 by the UCSB as the first spherical immersive VR system used for medical data visualization (Fig 12.1). The system features 24 million pixel, 10 m diameter spherical projection screen using 14 projectors suspended in a three-story near-anechoic room equipped with camera tracking, audio recognition, and sensors offering immersive projection, interactivity, and unique sonification potential for up to 20 users [74]. Coded specifically for the AlloSphere, *allocore* software was developed to interpret and visualize medical data in 3D. "The result is something far beyond other VR systems such as a CAVE Automatic Virtual Environment (CAVE) or a planetarium: 360 degrees of sounds and images in a chamber large enough to hold 30 or more researchers at once..." writes Julian Smith for *Scientific American*. It is a place where you can use all of your senses to find new patterns in data, according to JoAnn Kuchera-Morin, the AlloSphere's director. "You can almost say researchers are shrunk down to the size of their data, immersed at a perceptual level" [75].

12.4.3 Allobrain

AlloSphere's first visualization in 2008 written in *allocore* titled Allobrain situates experimental medical data visualization in the field of aesthetic investigation. The Allobrain project offers a stereographic visualization model on the cusp of medical data and aesthetic visualization, using fMRI brain data to construct an interactive, immersive, and multimodal art installation [74]. The scale of neurologic investigation in the Allobrain makes use of the AlloSphere's spherical structure to represent the brain's two hemispheres. In digitizing the physical brain, viewers are positioned within the virtual fMRI data space and play out exploratory data mining narratives. Artificial and semi-autonomous mobile agents explore, measuring blood flow and sonifying data; a second set of agents responds to these signals by clustering in formations. These interactions create a complex and evolving 3D soundscape and data visualization, engaging pattern recognition of both the system and the user. The Allobrain was built to demonstrate the potential of engaging scientific data with immersive, interactive 3D environments rather than to interpret data in the conventional sense; in this capacity, the project established a new data visualization aesthetic.

12.4.4   Nanomed/MRI

Since the launching of the Allobrain project in 2008, the AlloSphere has facilitated research in nanomedicine, led by Professor Jamey Marth. Drawn from MRI data of a living human body, the AlloSphere's highly detailed simulated human is being used to observe the behavior of cell components (proteins, lipids, and glycans) and organs [71]. Current research is modeling the transport of chemotherapy directly and exclusively to pancreatic and renal cancers (avoiding healthy tissue), with the ambition of collaborating with materials scientists to design nanoparticles capable of safely traveling through the body to only affect diseased cells. Simulated fluid dynamics of arterial and venous blood flow are being integrated into the VH to then precisely track the organic and artificial nanoscale particles capacity to target cancerous cells. The AlloSphere's scale, clarity, and processing capacity are enabling the development of virtual tests on

FIGURE 12.2   AlloSphere Director JoAnn Kuchera-Morin immersed in anatomically correct human body data from an MRI. Researchers use wireless devices and tablets to mine various parameters of the data.

the nanoscale (Fig 12.2). As Professor Jamey Marth, Director of the Center for Nanomedicine at the UCSB states:

> *Right now, we use MRIs and PET scans to visualize these processes, but other imaging approaches are needed — and that's where the AlloSphere comes in. This is the breeding ground for the next generation of solutions in medicine [71].*

12.4.5 CAVE-Automated Virtual Environments

12.4.5.1 The Original CAVE

The "classic" CAVE is a cube-shaped VR room, typically $3\,m \times 3\,m \times 3\,m$ in size, whose walls, floor, and occasionally ceiling are entirely made of 3D computer-projected screens. The CAVE was conceived and designed in 1991 by Tom DeFanti and Dan Sandin, who were then professors and co-directors of the EVL at the UIC. All participants wear active stereo glasses to see and interact with complex 3D objects. One participant wears a six-degree-of-freedom location and orientation sensor, called a *tracker*, so that when he or she moves within the CAVE, correct viewer-centered perspective and surround stereo projections are produced quickly enough to give a strong sense of 3D visual immersion. The original CAVE was named not only for its obvious grotto-like qualities, but also after Plato's Cave, the allegorical place where shadows become reality for those inside [76]. Early CAVEs had less than one megapixel per wall per eye; newer ones have up to 15 times as many pixels.

In 2007, Harel Weinstein et al. at the Weill Cornell Medical College in New York examined how the addictive drug cocaine interacts with its target, a protein that normally pumps the neurotransmitter dopamine out of the synaptic clefts between neurons for reuse. When viewing a simulation of the subtle, yet dynamic, flexing and bending of the dopamine transporter, the researchers noticed that just above where cocaine bound sat two helical structures that oscillated in close proximity to each other—a feature that had not been obvious from looking at the interactions on a 2D desktop computer. Knowing this, the researchers devised a way to reversibly cross-link these two helices in an engineered version of the dopamine transporter. Using this molecular clamp to lock in or lock out cocaine molecules in tissue culture cells expressing the engineered protein, they showed that the cocaine bound to the transporter at precisely the place where the computer modeling had predicted. Without identifying the helices in the CAVE, "this idea would never have come to us... That was a revelation," says Weinstein [73]. Importantly, the work also confirmed that cocaine uses the same binding pocket on the transporter as dopamine. Efforts to find a drug therapy for cocaine addiction would now have to steer clear of simply looking for an inhibitor of cocaine binding, lest it also should disrupt vital dopamine function.

Robbert Creton, a developmental biologist at Brown University in Providence, Rhode Island, used his institution's CAVE to look at confocal microscopy images of a developing zebra fish embryo. Creton was trying to determine whether a sac-like organ, known as a Kupffer vesicle, which develops around 12 hours after fertilization, could be responsible for establishing left–right asymmetry in the growing fish embryo. Creton hypothesized that the mechanism could be related to the distribution of tiny hair-like cilia that project into the fluid-filled sac. On his computer monitor, it was hard for Creton to distinguish the cilia on one surface of the vesicle from those on the opposite surface, and quantifying the differences in density between the two surfaces was impossible. What he saw in the CAVE, but had failed to see on a flat screen, was that the cilia were, indeed, unevenly distributed throughout the vesicle. This explained how they could control the flow of fluid to establish a left–right chemical gradient that affected the development of the zebra fish [73].

12.4.5.2   StarCAVE

StarCAVE, a third-generation tiled CAVE, was completed in July 2007 at Calit2 at the UCSD (Calit2 has been renamed the Qualcomm Institute). The StarCAVE uses liquid crystal display (LCD) projectors with integrated circular polarization coupled with passive lightweight circularly polarized glasses. It has a resolution of 68 million pixels, 34 projectors, and 5.1 surround sound. User interaction is provided via a wand and a multi-camera, wireless tracking system. The StarCAVE offers 20/40 vision in a fully horizontally enclosed space with a diameter of 3 m and a height of 3.5 m (Fig. 12.3) [76]. Uses of the StarCAVE for medical applications are diverse and include neuroscience, bioengineering [77], obesity [78], dentistry [79], drug design [80], and biomedicine [81].

The Protein Viewer (Fig. 12.4) application, for example, connects the StarCAVE to the program database (PDB) data bank server [82] to download one of 50,000 protein structures and display it in 3D using PyMOL [83] to convert the PDB files to the Virtual Reality Modeling Language (VRML) format, which can be read by the visualization software. The user can choose between

FIGURE 12.3   Mechanical engineering professor Alison Marsden and a student explore blood flow data in a visualization facility at the StarCAVE Qualcomm Institute, University of California, San Diego. *Image courtesy of Qualcomm Institute, UC San Diego.*

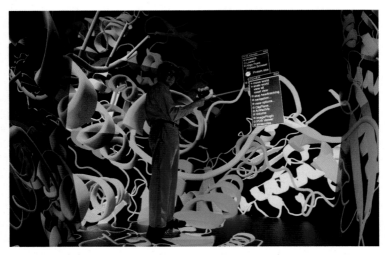

FIGURE 12.4   Professor Tom De Fanti inside the StarCAVE looking into an RNA protein rendering. *Image courtesy of Qualcomm Institute, UC San Diego.*

different visualization modes (cartoon, surface, sticks), load multiple proteins, align proteins, or display the corresponding amino acid sequence. The high resolution of the StarCAVE and its surround display capability allows the user to view a large number of protein structures at once, and scientists use this visualization to find similarities between proteins.

The NexCAVE takes a different approach to the design by using LCD panels of varying arrangements; for example, at CALIT2 the system comprises 10 panels (a 3 × 3 array with an additional display at the bottom of the middle column). It is a passive stereo system that uses simple circularly polarized glasses. The panels are installed such that the top and bottom tiles are tilted inward to help preserve polarization (since these displays have a limited vertical angle of view) [76]. The NexCAVE has inspired the development of the latest-generation CAVE2.

The VE-HuNT (Virtual Environment Human Navigation Task) system is another research endeavor of UCSB [84]. The system immerses test subjects in a human-scale, interactive, virtual reality–based environment to explore cognitive decline. As researchers in the field agree, early intervention is crucial for stemming the ravages of dementia; however, witnessing adults "in the act" of getting lost or disoriented is a challenging research problem that can be expensive to solve. Users interact through a computer interface device, similar to a steering wheel and an accelerator pedal, to navigate a virtual environment (created in a 3D portable version of the NexCAVE environment). The subjects are asked to perform a series of increasingly difficult navigational tasks such as finding a colored tile on the floor with and without navigational cues.

Current diagnostic tools such as volumetric MRI or PET scans are very expensive and would not be feasible for assessing all patients in their 50 second (or younger) when a diagnosis of risk would be most useful. The VE-HuNT system is made with off-the-shelf consumer components and is compact enough to be housed in any neurology or clinical facility and used as a quantitative assessment of patients along with other cognitive tests. "The use of VR head-mounted displays (HMDs) such as Oculus Rift has also been proposed," he adds, "but HMDs have the negative property that the user feels disembodied and cannot see herself, which in older persons often leads to vertigo or other vestibular problems. Perhaps more importantly, HMDs are characterized by a significant loss of peripheral vision, which is known to be very important in human navigation, particularly in seniors" [85].

12.4.6   CAVE2

CAVE2 is a 320-degree panoramic room, with over 70 3D LCDs with a range of resolutions from 37 to 42 megapixels (Figs. 12.5 and 12.6), which is about the limit of human 20/20 visual acuity. Proponents of these "CAVE automatic virtual environments" say they could be a boon for basic research scientists wanting to visualize complex 3D structures—from elaborate molecules to whole organs to networks of gene or protein interactions. CAVE2 "will push some science forward… It's a matter of finding out the right way to apply it," says David Barnes, manager of CAVE2 at Monash University in Australia [73].

Olusola Ajilore, a neuropsychiatrist at UIC, is using CAVE2 to study whether impairments in white matter integrity underlie depression in older

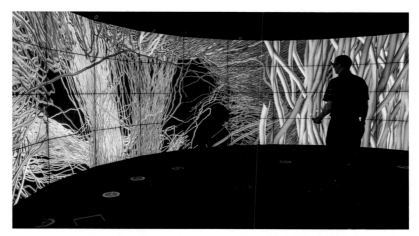

FIGURE 12.5   Dr. David Barnes immersed in the connectivity map of a human brain, in the Monash CAVE2 facility. *Data, visualization and photograph courtesy of M. Eager, O. Kaluza, D.G. Barnes and Monash University.*

FIGURE 12.6 Connectome data in CAVE2 where fiber tracts are color coded by their primary direction (green: front–back, red: left–right, blue: up–down). *Image courtesy of University of Illinois at Chicago (UIC) Electronic Visualization Laboratory (EVL) and Department of Psychiatry. Photo by Lance Long, UIC/EVL.*

adults (Fig 12.6). Within the immersive 3D laboratory, he can virtually step into and walk through the brains of his research participants or, more precisely, view their diffusion tensor images. The bundles of neural fiber tracts in this brain visualization are color coded by their primary direction (green: front–back; red: left–right; blue: up–down). Scientists examine this array of colors for faulty wiring in the neural connections in white matter, which takes up half the brain. It serves as a neural switchboard with millions of communication cables that operate like telephone trunk lines connecting different regions of the brain. Previous research suggests that damage to these cables in white matter is associated with depression. "We're hoping this 3-D environment will help us spot differences that aren't easily detectable in two dimensions on a flat screen," says Ajilore. "If we get better at mapping the brain areas responsible, it will lead to more precision in the use of technologies that may repair these damaged connections, like deep brain stimulation" [73].

Developed in 2004, SAGE (Scalable Adaptive Graphics Environment), an open-source middleware, provides a framework for large-scale collaborative scientific visualization spaces, allowing diverse data in varied resolutions, scales, and formats and from multiple sources to be viewed simultaneously. In 2013, SAGE2 launched a redesign of this system, making use of web-browser and cloud-based technologies to optimize the sourcing and co-located viewing of disparate data sets. This software architecture is widely used in the CAVE2 systems.

12.4.7   AVIE, Omniform, DOMELAB

The Advanced Visualisation Interaction Environment (AVIE) at UNSW's iCinema Research Centre is a landmark 360-degree stereoscopic interactive visualization environment designed in 2004 (Fig 12.7). The base configuration comprises a cylindrical 3D projection screen, 4.5 m in height and 10 m in diameter; a 12-channel stereoscopic projection system; and a 7.1 surround sound audio system. AVIE's immersive mixed-reality capability articulates an embodied interactive relationship between the viewers and the projected information spaces. It uses an active stereo projection solution and camera tracking, wands, and tablet interface interaction [86]. The omnidirectional interface prioritizes "users in the loop" in an egocentric model while also supporting embodied spherical (allocentric) relations to the respective data sets [68].

Building on the research developments embedded in AVIE, the UNSW group is embarking on a new omnispatial and omnidirectional immersive interactive environment, called Omniform (Fig. 12.8), which is located at the Expanded Perception and Interaction Centre (EPICentre). This new system resolves the challenges of both fidelity and resolution inherent to AVIE, to create a 120-megapixel, 3D interactive, 320-degree screen display, specifically for medical health visualization. Using rear-projected video "cubes," this display system also resolves the challenges of CAVE2 by reducing the bevel between screens (or cubes in this case) close to zero, maintaining resolution under polarization (in CAVE2, for example, resolution is halved when polarization is used to create the 3D effect), and minimizing the cross-reflection from opposite screens through the use of nonreflective black glass in front of each cube.

FIGURE 12.7   The AVIE system from iCinema, the application is T_Visionarium, visualizing the linkages among 25,000 videos. *Image courtesy of: iCinema, UNSW.*

FIGURE 12.8 An artist's impression of Omniform, a 120-million pixels 3D rear-projected visualization environment under development at UNSW, Australia. *Image courtesy of: EPICentre, UNSW.*

12.4.7.1 Operating Theater of the Future

Projects in development for Omniform come under the umbrella of UNSW's Medical Innovation Through Immersive Visualisation project and includes a series of collaborations between UNSW and Imperial College London.

1. *Simulation and data fusion design*: A simulated operating theater environment that remodels existing real-time data, including outputs from tools found in operating theater interfaces (eg, i-Knife) [65], extracts these data onto an ultra-high-definition surround screen environment, and adds multiple clinical, "-omics," and imaging data sets, perhaps on the same patient or from the same study—to create a new form of integrated and linked visualization framework for use during operations (eg, i-Knife telemetry linking to 3D ultrasound will enable optimal knife trajectory to the pathology with the i-knife giving chemically orientated ablation of diseased tissue).

2. *Bioinformatics and stratified medicine*: Combining genomic and metabolic profiling information (data fusion) to augment clinical diagnostics (eg, for cancer, cardiovascular disease) and to enable patient stratification and monitoring of interventional pathways (patient journeys). This would involve mathematical analysis of genomic and metabolic data sets in relation to patient clinical data and visualization of these data in simplified formats to aid medical decision making.

3. *Chemoinformatics and advanced metabolic pathway visualization*: 3D navigation through metabolic and genetic spaces to explore connectivity patterns of genes, proteins, and metabolites in health and disease. This will use an

extended "metabonetworks" approach to optimally reconstruct metabolic pathway structures and activities that reflect disease processes—this could also be linked to MS topographical chemistry imaging for tissues in molecular pathology studies.

12.4.7.2   Visualization of Complex MRI Data Sets

Modern noninvasive methods for mapping brain structure and function have been developed to the point where the amount of information produced is difficult to visualize, particularly in 3D space. The problem is confounded when different data obtained using different scanning techniques are also included. A typical examination could include structural (morphologic) information, fiber streamlines, functional connectivity, and blood flow. At UNSW, researchers have spent considerable time making sure that their data collection and analysis methods are as robust and reliable as possible. The UNSW group now has, for example, exciting new methods for calculating, on individual subjects, robust measures of network functional connectivity that are superior to other current methods, as well as robust methods for tracking structural white matter connections and calculating the errors associated with this [87]. Visualizing these data in 2D is difficult, but when multiple imaging modalities are considered, the problem becomes next to impossible. The project intends using the high computing and projection power of the DomeLab to be able to visualize, for the first time, complex data sets derived from a single subject in order to begin to understand the individualized organization of the brain and how this relates to function (Figs. 12.9 and 12.10).

12.4.8   Multi-tile Displays

Tiled LCD screen systems emerged in the mid-1990s as increased resolution visualization tools within research environments. Display systems that include tiled LCD panels also incorporate high-resolution, stereoscopic displays, which can be used to juxtapose large, heterogeneous data sets while providing a range of naturalistic interaction schemes. They, thus, empower designers to construct integrative visualizations that more effectively "mash up" 2D, 3D, temporal, and multivariate data sets.

12.4.9   The Virtual Microscope

The Virtual Microscope at the University of Leeds is being used to explore the potential of a large-scale high-resolution display for visual analysis of imaging data. Since 2003, the Leeds Digital Pathology project has developed research, software, and infrastructure for the analysis of virtual slides and development of 3D histopathology. The Leeds Virtual Microscope (LVM) engages two 50-megapixel PowerWalls and two 6-megapixel monitors and is used to view virtual slides at 200,000 dpi (Fig 12.11). In a study carried out in 2014, the

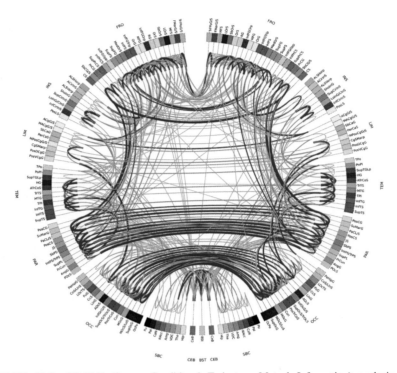

FIGURE 12.9 SCoTMI (Sparse Conditional Trajectory Mutual Information) analysis of Connectome data. Data are shown using a circos plot with the anatomical parcellations of the brain shown around the circumference (connection repeatability across 80 data sets). Connections detected more often are thicker, and subcortical connections are shown in orange. *Image courtesy of Ben Cassidy, Dept of Biostatistics, Columbia University / Caroline Rae, NeuRA - Neuroscience Research Australia, UNSW.*

usefulness of virtual slides to pathologists' systematic search tasks found that the increased resolution offered faster, easier initial identification and location of cancer in axillary lymph nodes samples. The times to diagnosis via traditional light microscope and the LVM were very similar, but diagnosis of digitized slides on a desktop display took 60% longer than that with either of these microscopes [2]. For the purposes of comprehensive search of an individual slide, the increased resolution did not offer any additional benefit [88]; however, a follow-up study found that when surveying multiple slides on the LVM, pathologists spent more time looking at the slides and revisited slides more often compared with a traditional light microscope [89]. These lingering behaviors were attributed to the speed and intuitive navigation of the LVM within and between slides. These interface benefits extended to the teaching context, where collaborative interaction was noted as a benefit [90].

FIGURE 12.10   Schematic view of connectdome data in the full dome (*DomeLab*). *Image courtesy of: Sarah Kenderdine, UNSW.*

FIGURE 12.11   The Leeds Virtual Microscope showing a gigapixel cancer slide on a 54-million-pixel "Powerwall" display. *Image courtesy of WELMEC, a Centre of Excellence in Medical Engineering funded by the Wellcome Trust and EPSRC (WT 088908/Z/09/Z).*

12.4.10   Bacterial Gene Neighborhood Investigation Environment

The rapid developments in genome sequencing technology and complete genome sequencing in the past decade have generated an unprecedented volume of data. The collection of complete sequences, particularly in bacterial genomics, is opening possibilities for comparative gene neighborhood analysis. The large volumes of complete sequences in bacterial genomics could contribute valuable information on "function and pathway membership of novel genes" [91]. To describe the neighborhood surrounding gene orthologs (highly similar sequences), visualizations capable of expressing nuances in pattern variation and relationships are required. The Bacterial Gene Neighborhood Investigation Environment, BactoGeNIE, was the first visualization approach to enable to large-scale comparisons of hundreds of gene neighborhoods in a large high-resolution display (Fig 12.12). Developed in the CyberCommons, BactoGeNIE allows the simultaneous viewing of up to 600 related bacterial genome sequences and employs analysis tools, including clustering techniques and ortholog clustering targeting functionality [91]. Previous genome neighborhood visualizations with low information density designed desktop displays do not scale up spatially. As Jillian Aurisano et al. state:

> …*existing techniques are largely not designed to accommodate comparative tasks across large collections of complete genome sequences. In particular, no visual tools exist for comparing gene neighborhoods that scale beyond small stretches of genes in 2–9 genomes. Even if the approaches in existing tools could be scaled to larger collections of genomes, our domain experts found that the fundamental designs did not scale visually or perceptually to allow for large-scale comparative tasks. This scalability issue limits analysis through visualization in comparative bacterial genomics, in particular in the case of comparative tasks across large collections (dozens to thousands) of bacterial genome sequences [92].*

The BactoGeNIE uses high-density visualization to produce a compressed view that encodes context sequences linearly. Coding sequences are represented

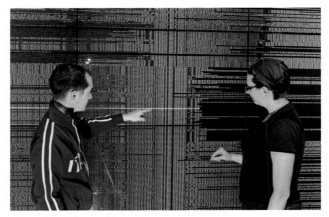

FIGURE 12.12 BactoGeNIE visualizes the gene neighborhood around a hypothetical protein in 600 draft *Escherichia coli* genomes and is displayed on the Cyber-Commons wall at the Electronic Visualization Laboratory (EVL) at the University of Illinois at Chicago [1]. BactoGeNIE was developed by Jillian Aurisano (UIC/EVL). *Image courtesy of UIC/EVL. Photo by Lance Long, UIC/EVL.*

as 2D arrows, while their orientations indicate the directions of transcription and the coding strand. Addressing the rapidly changing climate of data visualization in the late 2000s, CyberCommons was developed in SAGE as an experimental collaborative learning space. The room features a PowerWall array of high-density thin-bezel LCD screens with a combined resolution of 19 megapixels and allows participants to run networked video conferencing and share and collaboratively work on files with local and remote participants [93]. The micro-polarization layer on the screens enables 3D stereoscopic content when viewed through polarized 3D glasses. Studies of visual analysis using the CyberCommons show the improved cognitive efficiency for hypothesis validation [93]. Compared with conventional displays, where visual analysis processes were observed to involve "cycling back and forth between a series of temporally separated views," the scale of the CyberCommons enabled "quick glancing" at contextual data trajectories already visible in the single display [93] to observe patterns. Rather than comparing pairs of trajectories within the tested a data set to others, the analysis engaged multiple trajectories at once. The 3D stereoscopic encoding of the data and "the ability to see temporal pattern unraveled in space" contributed to the user finding it "easier to think visually than in Matlab" [93]. Commenting more broadly on large high-resolution displays (LHDs) (CAVE2 and CyberCommons), the studies found that users of higher-pixel display systems spent more time in exploratory processes; transitioned into and remained within an "insight generating" mental state for a longer period; and postulated hypotheses in a "broader and more integrative" way [93].

12.4.11   OmegaDesk

Software frameworks such as Omegalib offer high-resolution environments, dynamic configurability of the display environment, and co-located collaborative research. The OmegaDesk was successfully used in clinical research requiring high-resolution imaging. Developed by the EVL at the UIC, the OmegaDesk integrates touch-sensitive surfaces with ultra-high-resolution computer-aided workspaces and VR environments. Using a digital light microscope (×20 magnification) to digitize images and a 32-megapixel charge-coupled device and image creation software to capture the images, the OmegaDesk was used to generate 3D images to measure the extent of lateral tissue damage caused by Holium and Biolase dental lasers. With these 3D models, researchers were able to observe the lasers' effectiveness in bleeding control and coagulation and make recommendations for clinical and surgical applications [94].

CONCLUSIONS

Since big data problems frequently require the combined efforts of many individuals from disparate fields, the next generation of data intensive visualization and interaction environments will need to enable collaboration and group work [95].

Although visualization has long been key in helping understand biological systems such as metabolism, signaling, and the regulation of gene expression [10], conceivably the biggest hurdle yet to overcome for large-scale visualization facilities is one of awareness—"people not knowing what is possible," as Andrew Johnson, head of research at EVL, has said. Johnson does not see this problem as insurmountable, though. "Once people see an example that's close enough to what they want to do, it starts to click and they start getting ideas," he says [73]. Nowadays, visualization is used to support the entire research process—with each "view" prompting new questions and ways to sort and combine data sets. Integrating large-scale immersive and interactive systems into this research process is one of the major challenges for the future. The necessity for paradigm shifts in user interfaces for large-scale immersive environments involves the introduction of physical mechanisms and unique application design for data manipulation [96]. Researchers increasingly need to coordinate more data and multiple views in larger-format displays. And despite the ambition for collaborative dynamic, open-source software, the software written for large-scale immersive environments remains a custom-built undertaking and another challenge to the ubiquity of the powerful solutions that these systems and their applications offer. The future for advancing this field of research will be multidisciplinary where data curation, design-led approaches to HCI, data analytics, and software and hardware engineering intersect.

REFERENCES

[1] Reda K, Febretti A, Knoll A, Aurisano J, Leigh J, Johnson AE, et al. Visualizing large, heterogeneous data in hybrid-reality environments. IEEE Comput Graph Appl 2013;33(4): 38–48.

[2] Ruddle RA, Fateen W, Treanor D, Sondergeld P, Ouirke P, editors. Leveraging wall-sized high-resolution displays for comparative genomics analyses of copy number variation. Biol Data Vis (BioVis) 2013 IEEE Symposium on; 2013: IEEE.

[3] Holzinger A, Dehmer M, Jurisica I. Knowledge discovery and interactive data mining in bioinformatics-state-of-the-art, future challenges and research directions. BMC Bioinformatics 2014;15:I1.

[4] Botha CP, Preim B, Kaufman A, Takahashi S, Ynnerman A. From individual to population: challenges in Medical Visualization 2012.

[5] Blaas J. Visual analysis of multi-field data [PhD]. Delft: Delft University of Technology; 2010.

[6] Ackerman MJ. The visible human project. Proc IEEE 1998;86(3):504–11.

[7] Tang L, Chung MS, Liu Q, Shin DS. Advanced features of whole body sectioned images: virtual chinese human. Clin Anat 2010;23(5):523–9.

[8] VPH- FET Roadmap. European Commission Seventh Framework Programme; 2011 Contract No.: FP7-ICT-258087.

[9] Viceconti M, Hunter P, McCormack K, Henney A, Omholt SW, Graf N, et al. Big data, big knowledge: big data for personalised healthcare. Belgium: Virtual Physiological Human Insitute for Integrative Biomedical Research; 2014.

[10] Gehlenborg N, O'Donoghue SI, Baliga NS, Goesmann A, Hibbs MA, Kitano H, et al. Visualization of omics data for systems biology. Nat Methods 2010;7(3):S56–68.

[11] Nicholson JK, Holmes E, Kinross JM, Darzi AW, Takats Z, Lindon JC. Metabolic phenotyping in clinical and surgical environments. Nature 2012;491(7424):384–92.

[12] Lindon JC, Nicholson JK. Spectroscopic and statistical techniques for information recovery in metabonomics and metabolomics. Annu Rev Anal Chem 2008;1:45–69.

[13] Beckonert O, Keun HC, Ebbels TM, Bundy J, Holmes E, Lindon JC, et al. Metabolic profiling, metabolomic and metabonomic procedures for NMR spectroscopy of urine, plasma, serum and tissue extracts. Nat Protoc 2007;2(11):2692–703.

[14] Wong A, Jiménez B, Li X, Holmes E, Nicholson JK, Lindon JC, et al. Evaluation of high resolution magic-angle coil spinning NMR spectroscopy for metabolic profiling of nanoliter tissue biopsies. Anal Chem 2012;84(8):3843–8.

[15] Ramautar R, Mayboroda OA, Somsen GW, de Jong GJ. CE-MS for metabolomics: developments and applications in the period 2008–2010. Electrophoresis 2011;32(1):52–65.

[16] Trygg J, Holmes E, Lundstedt T. Chemometrics in metabonomics. J. Proteome Res 2007;6(2):469–79.

[17] Lindon JC, Nicholson JK. The emergent role of metabolic phenotyping in dynamic patient stratification. Expert Opin Drug Metab Toxicol 2014;10(7):915–9.

[18] Crockford DJ, Lindon JC, Cloarec O, Plumb RS, Bruce SJ, Zirah S, et al. Statistical search space reduction and two-dimensional data display approaches for UPLC-MS in biomarker discovery and pathway analysis. Anal Chem 2006;78(13):4398–408.

[19] Gerbig S, Golf O, Balog J, Denes J, Baranyai Z, Zarand A, et al. Analysis of colorectal adenocarcinoma tissue by desorption electrospray ionization mass spectrometric imaging. Anal Bioanal Chem 2012;403(8):2315–25.

[20] Schäfer K-C, Balog J, Szaniszló T, Szalay D, Mezey G, Dénes J, et al. Real time analysis of brain tissue by direct combination of ultrasonic surgical aspiration and sonic spray mass spectrometry. Anal Chem 2011;83(20):7729–35.

[21] McDonnell LA, Heeren R. Imaging mass spectrometry. Mass Spectrom Rev 2007;26(4):606–43.

[22] Fonville JM, Carter C, Cloarec O, Nicholson JK, Lindon JC, Bunch J, et al. Robust data processing and normalization strategy for MALDI mass spectrometric imaging. Anal Chem 2012;84(3):1310–9.

[23] Raghupathi W, Raghupathi V. Big data analytics in healthcare: promise and potential. Health Inf Sci Syst 2014;2(1):3.

[24] Uludağ K, Roebroeck A. General overview on the merits of multimodal neuroimaging data fusion. Neuro Image 2014;102:3–10.

[25] Data Archiving and Networked Services (DANS): KNAW & NWO. Available from: <http://www.dans.knaw.nl/en>; 2015 [cited May 18, 2015].

[26] Australian National Data Service: Australian National Data Service. Available from: <http://www.ands.org.au>; 2015 [cited May 18, 2015].

[27] Digital Curation Centre (DCC) DCC. Available from: <http://www.dcc.ac.uk>; 2015 [cited May 18, 2015].

[28] Research Data Sharing without Barriers: Research Data Alliance. Available from: <https://rd-alliance.org>; 2015 [cited May 18, 2015].

[29] Elixir Cambridgeshire UK: Wellcome Trust Genome Campus. Available from: <https://http://www.elixir-europe.org>; 2015 [cited May 18, 2015].

[30] Human Variome Project: Human Variome Project International. Available from: <http://www.humanvariomeproject.org>; 2015 [cited May 18, 2015].

[31] Tilahun B, Kauppinen T, Keßler C, Fritz F. Design and development of a linked open data-based health information representation and visualization system: potentials and preliminary evaluation. JMIR Med Inform 2014;2(2).

[32] Hasnain A, Sana e Zainab S, Kamdar MR, Mehmood Q, Warren Jr CN, Fatimah QA, et al. A roadmap for navigating the life sciences linked open data cloud. In: semantic technology: 4th joint international conference, JIST 2014, Chiang Mai, Thailand, November 9–11, 2014 Revised Selected Papers: Springer; 2015. p. 97–112.

[33] Linked Data for Libraries: Linked data for libraries. Available from: <https://http://www.ld4l. org>; 2014 [cited May 18, 2015].

[34] SAO/NASA Astrophysics Data System: Smithsonian Astrophysical Observatory. Available from: <http://adswww.harvard.edu/>; unknown [cited May 18, 2015].

[35] SAO/NASA Astrophysics Data System Labs: Smithsonian Astrophysical Observatory. Available from: <http://labs.adsabs.harvard.edu/adsabs/>; 2015 [cited May 18, 2015].

[36] Lu Z. PubMed and beyond: a survey of web tools for searching biomedical literature. Database. 2011;2011:baq036.

[37] European Bioinformatics Institute (EMBL-EBI) Cambridgeshire, UK: Wellcome Trust Genome Campus. Available from: <http://www.ebi.ac.uk>; 2015 [cited May 18, 2015].

[38] Protein Data Bank in Europe Cambridgeshire, UK: EMBL-EBI, Wellcome Trust Genome Campus. Available from: <http://www.ebi.ac.uk/pdbe/node/1>; [cited May 18, 2015].

[39] Decipher: Wellcome Trust Sanger Institute. Available from: <https://decipher.sanger.ac.uk>; 2015 [cited May 18, 2015].

[40] Accelerating Medicines Partnership—Alzheimer's Disease United States of America: National Institute on Aging. Available from: <http://www.nia.nih.gov/alzheimers/amp-ad>; 2015 [cited March 23, 2015].

[41] Cyganiak R, Jentzsch A. The linking open data—cloud diagram: can I get older versions?: linking open data. Available from: <http://lod-cloud.net/-history>; 2014 [cited May 18, 2015].

[42] BIO2RDF.

[43] The Cancer Genome Atlas: National Human Genome Research Institute, National Cancer Institute at NIH. Available from: <http://cancergenome.nih.gov>; 2015 [updated March 2015; cited May 23, 2015].

[44] Matchmaker Exchange Matchmaker Exchange. Available from: <http://matchmakerexchange.org>; 2015 [cited May 18, 2015].

[45] International Gene Consortium: ICGC. Available from: <https://icgc.org>; 2015 [cited May 18, 2015].

[46] Australian & New Zealand Neonatal Network (ANZNN): UNSW Australia. Available from: <https://npesu.unsw.edu.au/data-collection/australian-new-zealand-neonatal-network-anznn>; 2015 [cited May 18, 2015].

[47] International Network for Evaluation of Outcomes of Neonates: iNEO. Available from: <https://icgc.org>; 2012 [cited May 18, 2015].

[48] Big Data to Knowledge (BD2K): U.S. Department of Health and Human Services. Available from: <http://datascience.nih.gov/bd2k>; 2015 [cited May 18, 2015].

[49] Medical Subject Headings: U.S. National Library of Medicine. Available from: <https://http://www.nlm.nih.gov/mesh>; 1999–2015 [cited May 18, 2015].

[50] Gene Ontology Consortium (GOC): The Gene Ontology. Available from: <http://geneontology.org>; 1999–2014 [cited May 18, 2015].

[51] Regalado A. Internet of DNA. MIT Technology Review [Internet]. Available from: <http://www.technologyreview.com/featuredstory/535016/internet-of-dna/>; [cited May 18, 2015].

[52] Kamdar MR, Zeginis D, Hasnain A, Decker S, Deus HF. ReVeaLD: a user-driven domain-specific interactive search platform for biomedical research. J Biomed Inform 2014;47: 112–30.

[53] Keim DA, Mansmann F, Stoffel A, Ziegler H. Visual analytics Liu L, Özsu MT, editors. Encyclopedia of database systems. Springer; 2009. p. 3341–6.

[54] I2S2 Idealised Scientific Research Activity Lifecycle Model: UKOLN. Available from: <http://www.ukoln.ac.uk/projects/I2S2/documents/I2S2-ResearchActivityLifecycleMo del-110407.pdf>; 2010 [cited May 18, 2015].

[55] Goble C, Stevens R, Hull D, Wolstencroft K, Lopez R. Data curation+ process curation= data integration+ science. Brief Bioinform 2008;9(6):506–17.

[56] Rysavy SJ, Bromley D, Daggett V. DIVE: a graph-based visual-analytics framework for big data. Comput Graph Appl IEEE 2014;34(2):26–37.

[57] Bechhofer S, Buchan I, De Roure D, Missier P, Ainsworth J, Bhagat J, et al. Why linked data is not enough for scientists. Future Gener Comp Sy 2013;29(2):599–611.

[58] Wang X, Dou W, Lee S-w, Ribarsky W, Chang R, editors. Integrating visual analysis with ontological knowledge structure. IEEE Visualization Workshop on Knowledge Assisted Visualization; 2008.

[59] Huang D, Tory M, Aseniero B, Bartram L, Bateman S, Carpendale S, et al. Personal visualization and personal visual analytics; 2014.

[60] BioViz: UNC Charlotte. Available from: <http://bioviz.org/igb/index.html>; 2015 [cited May 18, 2015].

[61] Ensembl: WellcomeTrust Sanger Institute & EMBL-EBI. Available from: <http://asia. ensembl.org>; 2015 [cited May 18, 2015].

[62] Cytoscape: National Institute of General Medical Sciences of the National Institutes of Health. Available from: <http://www.cytoscape.org>; 2001–2015 [cited May 18, 2015].

[63] BioPortal: The National Center for Biomedical Ontology. Available from: <http://bioportal. bioontology.org>; 2005–2015 [cited May 18, 2015].

[64] Sackman JE, Kuchenreuther M. Marrying big data with personalized medicine. BioPharm Int 2014;27(8):36–8.

[65] Quraishi A-h. Cave2. Available from <http://chicagotonight.wttw.com/2013/04/23/cave2>; 2015 [cited May 18 2015].

[66] Kainz B, Portugaller RH, Seider D, Moche M, Stiegler P, Schmalstieg D. Volume visualization in the clinical practice Augmented environments for computer-assisted interventions: Springer; 2012.74.84

[67] Roberts N, Magee D, Song Y, Brabazon K, Shires M, Crellin D, et al. Toward routine use of 3D histopathology as a research tool. Am J Pathol 2012;180:1835–42.

[68] Kenderdine S, Au O-C, Shaw J, editors. Cultural data sculpting: omnispatial visualization for cultural datasets. Information Visualisation (IV), 2011 15th International Conference on: IEEE; 2011.

[69] Reda K, Offord C, Johnson AE, Leigh J, editors. Expanding the porthole: leveraging large, high-resolution displays in exploratory visual analysis. In: CHI'14 extended abstracts on human factors in computing systems: ACM; 2014.

[70] Febretti A, Nishimoto A, Mateevitsi V, Renambot L, Johnson A, Leigh J, editors. Omegalib: a multi-view application framework for hybrid reality display environments. Virtual Reality (VR), 2014 IEEE: IEEE; 2014.

[71] Marsa L. Diving into the data, literally. Discover 2014(July 29,).

[72] Goodyer C, Hodrien J, Wood J, Kohl P, Brodlie K. Using high-resolution displays for high-resolution cardiac data. Philos Trans Royal Soc A Math Phys Eng Sci 2009;367(1898):2667–77.

[73] Lewis D. The CAVE artists. Nat Med 2014;20(3):228–30.

[74] Thompson J, Kuchera-Morin J, Novak M, Overholt D, Putnam L, Wakefield G, et al. The allobrain: an interactive, stereographic, 3D audio, immersive virtual world. Int J Hum-Comput Stud 2009;67(11):934–46.

[75] Smith J. A 360-Degree virtual reality chamber brings researchers face to face with their data. Sci Am 2009.

[76] DeFanti T, Acevedo D, Ainsworth R, Brown M, Cutchin S, Dawe G, et al. The future of the CAVE. CEJE 2011;1(1):16–37.

[77] Ahsan K, Shahzad M. Visualizing protein structures in virtual interactive interface. J Med Biol Eng 2015;4:5.

[78] Cornick JE, Blascovich J. Are virtual environments the new frontier in obesity management? Soc Personal Psychol Compass 2014;8(11):650–8.

[79] Dias DRC, Brega JRF, Trevelin LC, Neto MP, Gnecco BB, de Paiva Guimaraes M, editors. Design and evaluation of an advanced virtual reality system for visualization of dentistry structures. Virtual Systems and Multimedia (VSMM), 2012 18th International Conference on: IEEE; 2012.

[80] Dalkas GA, Vlachakis D, Tsagkrasoulis D, Kastania A, Kossida S. State-of-the-art technology in modern computer-aided drug design. Brief Bioinform 2012:bbs063.

[81] Long G, Kim HS, Marsden A, Bazilevs Y, Schulze JP, editors. Immersive volume rendering of blood vessels. IS&T/SPIE Electronic Imaging: International Society for Optics and Photonics; 2012.

[82] Berman HM, Westbrook J, Feng Z, Gilliland G, Bhat T, Weissig H, et al. The protein data bank. Nucleic Acids Res 2000;28(1):235–42.

[83] DeLano WL. PyMOL. San Carlos, CA: DeLano Scientific; 2002.700.

[84] Edelstein EA, Macagno E. Form follows function: bridging neuroscience and architecture Sustainable environmental design in architecture: Springer; 2012.27.41.

[85] Fox T. New Virtual Reality Navigation System to Help Diagnose Cognitive Defects; 2014.

[86] McGinity M, Shaw J, Kuchelmeister V, Hardjono A, Favero DD, editors. AVIE: a versatile multi-user stereo 360 interactive VR theatre. In: Proceedings of the 2007 workshop on Emerging displays technologies: images and beyond: the future of displays and interacton: ACM; 2007.

[87] Cassidy B, Rae C, Solo V. Brain activity: connectivity, sparsity, and mutual information; 2014.

[88] Randell R, Ambepitiya T, Mello-Thoms C, Ruddle RA, Brettle D, Thomas RG, et al. Effect of display resolution on time to diagnosis with virtual pathology slides in a systematic search task. J Digit Imaging 2014;28:68–76.

[89] Randell R, Ruddle RA, Thomas RG, Mello-Thoms C, Treanor D. Diagnosis of major cancer resection specimens with virtual slides: impact of a novel digital pathology workstation. Hum Pathol 2014;45:2101–6.

[90] Using a high-resolution wall-sized virtual microscope to teach undergraduate medical studentsRandell R, Hutchins G, Sandars J, Ambepitiya T, Treanor D, Thomas R, editors. CHI'12 extended abstracts on human factors in computing systems. : ACM; 2012.

[91] Aurisano J, Reda K, Johnson A, Leigh J, editors. Bacterial gene neighborhood investigation environment: a large-scale genome visualization for big displays. Large Data Analysis and Visualization (LDAV), 2014 IEEE 4th Symposium on: IEEE; 2014.

[92] Aurisano J, Reda K, Johnson A, Marai GE, Leigh J. BactoGeNIE: a large-scale comparative genome visualization for big displays. BMC Bioinform J 2015.

[93] Reda MK. Exploratory visual analysis in large high-resolution display environments; 2014.

[94] Bassiony MA, Vesper BJ, Mateevitsi VA, Elseth KM, Colvard MD, Garcia KD, et al. Immunohistochemical evaluation of bleeding control induced by holium laser and biolase dental laser as coagulating devices of incisional wounds. In: Proceedings of the UIC College of Dentristry Clinic and Research Day 2014; Chicago, IL; 2014.

[95] Marrinan T, Aurisano J, Nishimoto A, Bharadwaj K, Mateevitsi V, Renambot L, et al., editors. SAGE2: a new approach for data intensive collaboration using Scalable Resolution Shared Displays. Collaborative Computing: Networking, Applications and Worksharing (CollaborateCom), 2014 International Conference on: IEEE; 2014.

[96] Moreland K, editor Redirecting research in large-format displays for visualization. Large Data Analysis and Visualization (LDAV), 2012 IEEE Symposium on: IEEE; 2012.

Chapter 13

Future Visions for Clinical Metabolic Phenotyping: Prospects and Challenges

John C. Lindon[1], Jeremy K. Nicholson[1], Elaine Holmes[1] and Ara W. Darzi[1,2]

[1]Department of Surgery and Cancer, Imperial College London, London, UK [2]Institute of Global Health Innovation, Imperial College London, London, UK

Chapter Outline

13.1 DEVELOPMENT OF METABOLIC PHENOTYPING

Metabolic phenotyping has now come of age. It has been a long journey from the initial studies of spectroscopic analysis of biofluids in the early 1980s, mainly in terms of small exploratory studies on human biofluids [1–3] and studies of xenobiotic toxicity [4,5]. Subsequent application of multivariate statistics to aid interpretation of data [6–9] has brought the field to the position it

E. Holmes, J.K. Nicholson, A.W. Darzi & J.C. Lindon (Eds): Metabolic Phenotyping in Personalized and Public Healthcare. DOI: http://dx.doi.org/10.1016/B978-0-12-800344-2.00013-6

currently commands with dedicated phenotyping centers being set up around the world capable of analyzing thousands of samples per study and hundreds of thousands of samples a year. Despite the formal definition of the topic being published in 1999, the development of metabolic phenotyping has taken close to 30 years to reach the strong position it holds today [10]. The reason has been a combination of factors: technologic advances in analytical chemistry, mainly in nuclear magnetic resonance (NMR) spectroscopy, in separations science and in mass spectrometry (MS); new developments in chemometrics and bioinformatics; increases in computational power; a willingness of scientists to embrace a new paradigm; an embracing of the approach by clinical professionals; and the advent of specialist teaching courses and a consequent increase in the numbered of well-trained scientists.

Modern metabolic phenotyping makes extensive use of multiple MS and NMR spectroscopic platforms. However, in the 1980s, most human and animal metabolic profiling studies heavily utilized ^{1}H NMR spectroscopy typically at an observation frequency of 400 MHz [1–4]. This had limited sensitivity and resolving power, and as time progressed, higher observation frequencies became available with a concomitant increase in sensitivity, with the highest-frequency commercially produced instrument now operating at 1000 MHz. Nevertheless, most modern studies have standardized the frequency at 600 MHz because this represents a good compromise between sensitivity and capital cost [11]. The development of NMR detectors (probes) operating at cryogenic temperatures resulted in a large reduction in electronic noise in the spectrometer systems, a substantial improvement in sensitivity, and consequently a lowering of detection limits [12]. With improvements in magnet technology and lowering of production costs, it is likely that higher-field NMR systems will come into routine use for large-scale metabolic phenotyping.

In parallel, improvements in chromatography meant that MS-based metabolic profiling became more effective and is now more widely used than NMR spectroscopy, although there remain issues of data and batch alignment for large-scale studies. Initially, studies used either gas chromatography (GC) [13,14] or high-performance liquid chromatography (HPLC) [15,16]. GC-based analysis suffers from the general need to chemically modify many metabolites so that they are volatile enough to enter the gas phase, and this can introduce unwanted extra variability in sample composition. Nevertheless, comprehensive protocols for GC-MS have been published [17], and it is a robust tool, particularly for quantifying specific groups of molecules such as short-chain fatty acids (SCFAs). HPLC has limited chromatographic resolution such that often several metabolites in complex mixtures of thousands of compounds co-elute, and this can cause problems in MS detection, where one metabolite can take sufficient ion current such that others become undetectable despite being above the nominal detection level of the instrument (so-called ion suppression). This problem has been partially alleviated by the development of ultra-performance liquid chromatography (UPLC), which uses a much higher pressure in the column and a very small particle size for the column packing material such that the chromatographic resolution is much improved (by up to a factor of 10 in time, depending on the method used) [18]. More recently, capillary electrophoresis (CE) has found a niche application, since it is targeted toward polar or charged metabolites [19]. Supercritical fluid chromatography (SFC)-MS is one of the emerging tools, and its use of carbon dioxide as the solvent (with possible addition of organic solvent modifiers) makes it greener in terms of amount and type of solvent used. Although this technique has been available for many years, only in the last few years have highly reproducible analytical systems been developed. Supercritical fluids operate at the interface of the point at which a substance can no longer exist as a liquid, regardless of any increase in pressure, and the point at which a substance can no longer exist as a gas, no matter how much the temperature is increased. The ability to apply strong solvent gradients with SFC-MS, combined with the low viscosity of the solvent ensures faster separations than typically achieved with LC-MS. Although it has many applications, one of the most successful has been in achieving separation of chiral compounds [20]. The MS-based approaches have also formed the basis for targeted assays aimed at specific families of metabolites such as bile acids, eicosanoids, amino acids, and so on. All of the various techniques and approaches can now be used in combinations in large-scale metabolic phenotyping studies to achieve a more comprehensive description of an individual's molecular phenotype.

In parallel with the analytical chemistry advances have come major increases in computer power such that calculations on large cohorts of spectra data are now possible. In addition, new ways of analyzing and visualizing complex data have been published. To make sense of the complex types of variations in a spectral data set, multivariate chemometrics was first applied about 25–30 years ago [6–9], and these methods have been refined and extended over the years [20]. One highly useful extension is statistical spectroscopy, which takes advantage of known internal peak intensity correlation in a spectral data set to identify features that can arise from the same molecule or from a set of molecules in a given biochemical pathway [21]. This methodology has been applied to NMR data sets [22], to MS data sets [23] and for correlation between NMR and MS data sets [24].

Alongside these developments have been advances in bioinformatics, data storage, databases, and data repositories. Thus, several software packages exist, both academic and commercial, for linking assigned metabolites to biochemical pathways [25,26] and for linking metabolic data to genome and proteome data, and this has opened new avenues for exploring interactions of small molecules, proteins, and genes. This facilitates a better understanding of how an individual functions at all levels of molecular biology and has opened up channels for the new discipline of Systems Medicine. Central data repositories are now being established, principally at the European Bioinformatics Institute (EBI) near Cambridge, in the United Kingdom, and at the University of California San Diego, funded by the US National Institutes of Health (NIH). Specifically,

MetaboLights, a database, has been developed as a conduit for holding biofluid metabolic spectra, molecular structures, and information about the biological significance of those molecules as well as archiving information relating to the analytical experimental conditions [27].

In terms of applications relevant to Personalized Medicine, the early studies were mostly in the area of assessing preclinical toxicology of candidate drugs and in small exploratory disease studies, mainly separating disease and healthy profiles and finding the underlying biochemical signatures that achieve this. However, with the increased power of the methodology and development of new technologies such as surgical metabonomics (see chapter "Precision Surgery and Surgical Spectroscopy"), there is now real potential, but this is tempered by the need for extensive product development and validation, to incorporate the technology into mainstream clinical practice.

13.2 CURRENT STATUS OF METABOLIC PHENOTYPING

Small-scale exploratory clinical studies on many diseases continue to be published and the metabolic biomarkers of class separation determined. It remains to be seen how robust such biomarkers turn out to be, given the often rather low statistical power of the studies because of the small cohort sizes and the failure to eliminate the effects of confounding variables or co-morbidities. For example, typical of many published studies that use statistics to separate a diseased class from some sort of control group is the fact that the two groups can have very different lifestyles, diets, and drug regimens as a consequence of the disease situation. Nevertheless, the main push in clinical studies using metabolic phenotyping continues, but it has moved in two different directions.

At one end of the scale, large cohorts of samples that arise from epidemiologic studies are examined in conjunction with the collected meta-data to identify the metabolic markers of disease risk [28–33]. Out of this body of analysis has arisen the concept of metabolome-wide association studies [28], which derive associations between metabolic traits and disease risk or prevalence, as described in chapter "Population Screening for Biological and Environmental Properties of the Human Metabolic Phenotype: Implications for Personalized Medicine." These metabolites can also be linked to gene information if the epidemiologic study also carries out genomics on the subjects [34,35]. Such studies can easily encompass 5000–10,000 samples and typically use both NMR spectroscopy and UPLC-MS in an untargeted fashion. Other studies use a targeted approach and simply measure a predefined set of analytes using commercially available test kits [36]. Indeed, often both approaches are combined, that is, untargeted profiling by NMR and MS plus targeted assays for different chemical classes such as acylcarnitines and bile acids by MS, and lipoprotein subclasses by NMR spectroscopy.

At the other end of the spectrum, it is now increasingly being seen that measuring metabolic information from a single patient as he or she embarks on a clinical journey can give the medical staff early insight into how effective a treatment regime is and allows the clinician to tailor treatment to the individual [37,38]. There has been a corresponding push in bioinformatic research to develop methods to accommodate the dynamic aspects of modeling a patient through time, and several methods for handling time-series data have been published [39,40], although many of these methods are based around fluxes in cellular systems. Data from similar individuals undergoing a given treatment regime can be combined to give average metabolic trajectories over the treatment period that can be used in a prognostic fashion to optimize treatment in future patients. This approach relies on the concept of pharmacometabonomics [41–46] (see chapter "Pharmacometabonomics and Predictive Metabonomics: New Tools for Personalized Medicine").

In a move toward real-time metabolic profiling, the use of MS with desorption electrospray ionization has been pioneered in the form of the intelligent knife (i-Knife), a topic covered in chapter "Precision Surgery and Surgical Spectroscopy." This uses the smoke from a surgical diathermy tool to identify the tissue being operated on and, thus, allow the surgeon to make a decision, in essentially real time, on exactly where to cut (eg, cut around a tumor or clear tumor margins) [47].

One area that enjoyed some popularity for a time but now seems to have declined somewhat is the measurement of metabolic concentrations in a system (usually a cell-based system, but sometimes in vivo) over time to access the fluxes of metabolites in certain biochemical pathways, a topic sometimes termed *flux-omics*. Often, these experiments used stable isotope versions of the substrates and NMR spectroscopy or MS to follow the time course of their biochemical transformations [48,49]. The technique has recently been used to profile time-dependent metabolic fluxes following hemorrhagic shock in a rodent model [50], and flux analysis still enjoys some prominence in the field of cancer medicine [51]. Research groups such as those of Zamboni at ETH Zurich [52] and Hellerstein at Berkley, California [53], with the NIH-funded center directed by Nair and colleagues at the Mayo Clinic core metabolomics resource center provide facilities for metabolic flux analysis. In reality, it is effectively impossible to apply stable isotope labeling studies at scale in routine patient profiling or hospital environments. In fact, any diagnostic procedure that generates significant extra work in the clinical diagnostic environment has little chance of long-term translation to practice unless it offers overwhelming improvements in optimization of the patient care pathway. Therefore, methods that can be attached to current patient biofluid or biopsy sampling procedures will always be favored. As a good case in point, the i-Knife (see chapter "Precision Surgery and Surgical Spectroscopy"), the new diagnostic tool that utilizes a waste product (smoke) from the standard surgical device, generates new real-time diagnostic information. This is congruent with existing clinical practice, and so the barrier to wider translation is low.

Collectively, these complementary analytical approaches have led directly to the establishment of dedicated phenome centers around the world, where scaling of the operation and strict adherence to predefined operating procedures allow uniform data to be acquired over large sample numbers and over extended periods of time (see chapter "Phenome Centers and Global Harmonization").

13.3 TECHNOLOGIC CHALLENGES FOR INTEGRATION OF METABOLIC PHENOTYPING INTO MEDICINE

Before metabolic phenotyping can finally be regarded as a stable approach for understanding and elucidating mechanisms of disease and disease risk, there remain significant challenges in the Technology area.

13.3.1 Standardization and Validation of Protocols

First, it will be necessary to create standard and well-validated protocols for metabolic analytical technology methods. Some considerable progress has been made in this respect. NMR spectroscopy enjoys a stable technology and detailed protocols have been published to allow high-quality data sets to be acquired [54]. Despite the recent creation of several journals dedicated to metabolic profiling, such as *Metabolomics*, and calls for standardization of metabolic profiling technology, there are still no well-defined and universally accepted protocols, particularly for MS-based studies; the primary reason is the availability of multiple and constantly changing analytical platforms for measurement of metabolic content of biological samples. Development of standardized protocols is particularly necessary for the so-called untargeted profiles, where analysis of data in terms of biological endpoints is undertaken prior to metabolite identification. In targeted studies where each analyte is known, changes in technology are less important so long as the method used is quantitative. MS has seen large changes in technology over the past decades such that many of the approaches used for ionization and detection are no longer commercially available. In this respect, comparison of MS data features (rather than identified metabolites) acquired recently and in the past is more difficult. Nevertheless, several protocols for UPLC-MS studies have been published, but as mentioned above, there has been little attempt at standardization [17].

It is envisaged that NMR spectroscopy data will be highly reproducible if the already published protocol is followed, but for GC-MS and UPLC-MS, despite their widespread use, there is as yet no consensus on the protocols to be used, and for other separation modalities such as CE, or SFC, most studies have been exploratory in nature, with no attempt to define standard methods.

Policing this will be a difficult process and will require setting up a group to establish and oversee standards. Journals can take a lead in this process in that studies should only be published if they adhere to an accepted practice (along the lines already defined for genomics experiments). This approach could be based on the original Standard Metabolic Reporting Structure (SMRS) Group [55], later taken over as the Metabolomics Standards Initiative [56]. Other initiative such as the MetaboLights database at the EBI and the Metabolomics

Workbench (an NIH-funded core facility) are also creating structures for the standardization of terminology, data structure, and analysis pipelines [57,58].

There is a need to ensure that data collected in the future will be compatible with data collected previously. This is not an insignificant challenge, given the rate at which technical developments occur in analytical chemistry and associated computer hardware and software. Ensuring retrospective compatibility should be a major goal. This is particularly an acute need for untargeted studies, where the identities of the significant metabolites are not known at the outset. For targeted studies aimed at specific metabolites, the problem is much less, as the output is simply a table of metabolite identities and their concentration (no matter by which technology they have been derived).

Ultimately, at the clinical deployment level, technologies must all be standardized, validated, and regulated, and the concept of an International Phenome Centre Network is a powerful driver toward this goal. From the point of view of comparative epidemiology and pathology studies, standardization also allows the comparative biology of disease to be studied worldwide without confounding variations in analytical chemistry. Such initiatives will, therefore, be essential to the mission of large-scale population phenotyping.

13.3.2   Data Analysis, Integration, and Big Data

In addition to the need for continued development of analytical technologies, there still have to be advances in chemometrics, bioinformatics, and data visualization (see chapter "Modeling People and Populations: Exploring Medical Visualization Through Immersive Interactive Virtual Environments"). A wide range of methods already exist for classification of samples by using chemometrics methods, and these are heavily used, ranging from approaches that investigate spectral features individually (univariate statistics taking into account variation caused by confounder traits [28]), linear combinations of features (multivariate methods such as principal components analysis or partial-least-squares analysis and its variants [59]), through to nonlinear methods such as random forest, machine learning, neural nets, and others [59]. Moreover, one of the main bottlenecks that current untargeted metabolic phenotyping studies are encountering is the need to identify the metabolites responsible for the significant spectral features. The stage where automated annotation of signals is possible has not yet been reached, since there is both overlap of signals (ultimately requiring improved resolution) and, more importantly, the fact that several signals can be attributed to the same parent molecule corresponding to different chemical environments in NMR spectroscopy, or to adducts or daughter ions in MS. Although software that correlates different features in the data into one metabolite and hence aids identification [21] is available, new tools are certainly needed. Not least, there is a requirement for databases of the spectral features arising from known metabolites and the need for intelligent matching software. Although some progress has been made in this area, much remains to be done.

The deposition of metabolic phenotyping data into open repositories for scientists to carry out their own analyses is in its infancy. Some journals now require data to be deposited on their web sites, and initial consolidation attempts have been made, particularly the "MetaboLights" system at the EBI (Cambridge, United Kingdom) [57], and the NIH-funded "Metabolic Workbench" at the University of California, San Diego [58]. It remains to be seen how useful the data will be, given the current lack of standardization of data collection protocols and, indeed, sample preparation protocols. In addition, well-validated databases of the spectral properties of known metabolites need to be established. There are several publicly available comprehensive databases, including the Human Metabolome Database [60], the Biological Magnetic Resonance Bank [61] system at Madison, Wisconsin, and the METLIN database from Scripps [62]. The entries, however, are not without errors. Moreover, some attempts have been made to assign all the peaks in NMR spectra and in the GC-MS and UPLC-MS profiles of human urine and serum to attempt to define the "complete metabolomes" of these fluids [63–65].

It is increasingly obvious that metabolic phenotyping data should not be regarded in isolation and that the need for integration and fusion of such data with genome, transcriptome, epigenome, and proteome data will allow greater understanding of the systems biology.

An individual does not function in isolated units of the genome, proteome, epigenome, or metabolome and is designed such that cellular units interact in a coherent manner to form biologically coordinated tissues and organs that communicate chemically and electrically with each other. Thus, analysis of biological or pathologic processes should attempt to embrace a holistic approach if the goal is to truly understand the inner workings of the human being. The integration of the results and knowledge for all of the levels of systems biology leads to an overall description of human biology and requires new analytical tools to accommodate this systems approach. As technology advances and provides the capacity for capturing an individual's phenome at a deeper level, the amount of raw, and potentially processed, data increases enormously. Given the explosion of data at each of the bio-organizational levels, the field is rapidly moving into the realm of "big data," whereby the data sets created are so large and complex that existing statistical and software tools are inadequate. The term *interactomics* describes an attempt at defining the complete set of molecular interactions within cells and uses networks to visualize how genes, proteins, and metabolites are associated with respect to biological function. Moving beyond the cellular level, trans-omic interactions present a greater challenge, given that there are spatiotemporal constraints, with events at the gene, protein, and molecule levels operating on different timescales [66]. However, in reality, it is not possible to determine *all* of the interacting metabolic, physiologic, transporter parameters in any in vivo system; therefore, by definition, the interactome must always be incomplete. Notwithstanding this, several promising studies have been carried out in this area, notably the use of genome-wide data on subjects who also have had metabolic profiles measured from their biofluids in epidemiologic studies. Linkage of the two data sets can lead to genes whose functions can be linked with certain metabolites; hence this, by a reverse-engineering approach, can help identify unannotated spectral features with the biological correlates [34,35]. Similarly, metabolic–proteomic correlations have proved useful for identifying the linked metabolites and proteins involved in a mouse xenograft model of human prostate tumors [67]. Certainly, this will be an area of future expansion. One of the major challenges will be the mathematical integration and visualization of all of these data to create new metrics of human health and disease. Again, attempts are being made in this respect (see Chapter 12 "Modeling People and Populations: Exploring Medical Visualization Through Immersive Interactive Virtual Environments"), and the integration of genomic, transcriptomic, and metabolic data into biochemical pathways is but one paradigm where many advances have been made.

13.4 THE HUMAN BEING AS A SUPRA-ORGANISM

There has been increasing recognition that health is not solely determined by human genetics and that humans harbor a significant pool of "alien" DNA arising from symbiotic and pathogenic bacteria, fungi, and viruses. The gut microbiome, in particular, typically contributing 1–1.5 kg to the weight of an adult, constitutes a huge metabolically active reservoir of cells that transform endogenous, dietary, and drug-related chemicals and thus represents a significant force in determining health outcomes. It should, therefore, not be forgotten that data fusion and systems medicine approaches should include information on the gut microbiome. The role or impact of the gut microbiome has been a recurrent theme in several chapters in this book. In the literature, there are numerous examples of how the gut bacteria interact with the mammalian system to influence chemical signaling pathways [68,69] and inflammatory processes, and there is even some suggestion that the host genome influences the colonization of the gut bacteria [70]. The influence of the microbiome in diseases as varied as cancer, inflammatory bowel syndrome, asthma, metabolic syndrome, arthritis, and neurobehavioral conditions such as autism [69] is under review, and it is widely accepted that bacteria have a role to play in these conditions. However, in most instances, it is unclear whether this relationship is causal, and their precise roles have yet to be deciphered. It is known that horizontal transfer of genes between extrinsic and internal gut bacteria can confer new metabolic capacity to the human host, such as the transfer of bacteria evolved to degrade algal carbohydrates, which results in the conferred ability to digest such materials [7]. Unlike genomics or proteomics, which focus only on the human, metabolic phenotyping delivers a more holistic picture of the metabolic landscape within the human and captures information on chemicals deriving from exogenous origin. Thus, metabolic profiles are rich in information relating to the activity of the gut microbiota. For this reason, metabolic phenotyping has been used as a tool in functional metagenomics to identify chemicals derived from bacteria that may have signaling roles in communication across the human body. For example, 4-cresyl sulfate and 4-cresyl glucuronide originate from bacterially derived cresol in the intestine, which subsequently undergoes phase II conjugation to sulfate and glucuronide in the liver. Cresols are known to be produced by *Clostridia* and related organisms. The composition of *Clostridia* is known to be altered in children with autism, and correspondingly these children manifest a higher excretion of 4-cresyl sulfate, as detected by NMR spectroscopy [2]. Other bacterial metabolites such as SCFAs are more ubiquitous and are produced by many bacterial species. Perturbation of SCFA metabolism is part of the signature of many diseases, including inflammatory bowel disease and cancers [71,72], and their profile in fecal content and tissues is easily measured with the use of targeted GC and GC-MS methods. Although, as yet, there has been no serious strategy for modeling the trans-genomic interactions that define the relationship between humans and resident bacteria, there have been a few examples of mapping metabolic interactions between two or more genomes, mainly in the plant pathogen area [73], but also in mammalian systems [74,75].

One particularly promising new approach is the application of rapid evaporation ionization MS (REIMS) as a means of detection and identification of bacteria [76]. Recently, it has been shown that application of this method together with multivariate modeling can classify cultured bacteria at the level of 95.9%, 97.8%, and 100% at the species, genus, and Gram-stain levels, respectively. Within one species, seven different *Escherichia coli* strains were subjected to different culture conditions and were distinguishable with 88% accuracy. In addition, the technique was able to distinguish five pathogenic *Candida* species with 98.8% accuracy. Several avenues for the potential use of this new technology include applying it in cases of sepsis treated in an intensive care setting to establish which pathogenic bacteria are present, rather than having to wait to culture bacteria from blood or tissue samples taken from the patient. Use of this in the clinic, however, relies on the construction of databases of bacterial phenotypes, which, although under way, requires additional work and external validation.

13.5 PROSPECTS FOR METABOLIC PHENOTYPING IN CLINICAL MEDICINE AND EPIDEMIOLOGY

Metabolic phenotyping as a driver in molecular epidemiology poses large challenges in terms of how to derive valuable information from thousands of samples taken usually in historical collections and to use this to better understand population-based factors in health and disease. The difficulty of dealing with large numbers of samples, analytical instrument performance drift over time and intercohort variability is complicated by the fact that people are highly variable in lifestyle and diet and often do not provide accurate meta-data information either unintentionally or deliberately. Furthermore, the presence of drugs and their metabolites and other environmentally present materials cause even more variability. A further increase in complexity is caused by the high variability among individuals in their gut microflora and the complex, two-way biochemical relationship between the microflora and the host. Nevertheless, metabolic phenotyping of epidemiology cohort biofluid samples has been successful with the derivation of population-wide risk factors and the comparison of such outcomes with information from genome-wide association studies. Examples include studies from the INTERMAP cohort [28,29,75,77] and the EC-funded COMBI-BIO project, which combined cohorts of samples from three countries [78]. Furthermore other large cohorts have been studied [31–33], including a pioneering genome–metabolome correlation that aided the identification of metabolites based on variations in the genomic data and by knowing the functions of interesting genes [34].

This approach is one way to improved understanding of gene–environment interactions that underpin disease risks. This is particularly important in two areas: (1) discovering biomarkers that help to define a healthy person as age progresses in an aging society and (2) using these biomarkers to define a lifestyle or pharmaceutical therapy regime to prolong and preserve that healthy status. As mentioned above, understanding the metabolic interplay between the host metabolism and that of the gut microbiome is becoming increasingly important because manipulation of gut microbial activity (or, indeed, species) can have an impact on recovered calories (and hence obesity), and considering the gut microbes as drug targets could lead to new therapies or to useful modification of existing drug pharmacokinetics.

To date, there have been many studies by metabolic phenotyping in small-scale clinical applications, usually in terms of achieving a differential diagnosis. These include cardiovascular diseases; arthritis and other inflammatory diseases; organ failure; cancers of various types; neurologic, neonatal, and immunologic disorders, and so on [79–81]. The challenge is to conduct sufficient numbers of such studies so that robust statistical significance can be attached to any derived metabolic biomarkers, something that has not always been achieved in the past.

One new approach that has arisen in the form of pharmacometabonomics, which describes the prediction of the outcome of a particular clinical intervention (eg, drugs or surgery, or nutritional or lifestyle modifications) based on a preinterventional metabolic profile [41–46]. It can be seen that this could be simply a predictive model of prognosis based on a characteristic preintervention baseline metabolic profile or a more dynamic or longitudinal approach, where changes in metabolic profiles over time can be monitored as a patient undergoes treatment, to identify early signs of efficacy or adverse effects [46]. This paradigm is ideally suited for the development of a new style of clinical trial—the phenotypically augmented clinical trial [82]. Thus, it is predicted that this approach will be a major driver for what has variously become known as *precision medicine, personalized medicine,* or *stratified medicine.*

The appearance of surgical metabonomics on the clinical horizon (see chapter "Precision Surgery and Surgical Spectroscopy") also heralds new opportunities in 21st century medicine. This is perhaps the crown of Personalized Medicine tools delivering near-real-time information to the surgeon in order to augment intraoperative decision making. The main focus of development of this technology has been in the area of oncology, and early studies have shown great promise with respect to improved definition of tumor margins.

In a parallel development, hyperspectral MS imaging technology [83–85] has also contributed in the area of oncology, providing a conduit for deriving high-resolution chemical images across tissue sections that are more detailed in structure compared with corresponding conventional histopathologic images. Together, these tools represent a powerful and unique approach to augmenting the current diagnostic potential but also hold the promise of unlocking new mechanistic information relating to tumor heterogeneity [85] and possible novel therapeutic targets.

The various factors that impact on the use of metabolic phenotyping in both population risk studies and in applications in stratified medicine are encapsulated in Fig. 13.1.

13.6 POTENTIAL NEW METABOLIC PHENOTYPING OUTPUTS

Two obvious linked questions have to be asked: (1) What new outputs will arise from metabolic phenotyping? (2) How will these be used to aid medicine in the future?

Clearly, metabolic biomarkers, either as single entities or as combination panels of biomarkers, could be used as evidence to help a clinician make a differential diagnosis of closely related diseases as symptomatically observed. This is most likely to be manifested as changes in the concentrations of the metabolites rather than the complete disappearance of any metabolite or the appearance of new substances, although both scenarios cannot be ruled out. It is thus possible that new commercially available diagnostic test kits based on such metabolic biomarker profiles will be developed.

Also, such metabolic biomarker profiles could be used in a prognostic fashion for more personalized or stratified health care when based either on preintervention metabolic phenotypes in biofluids or tissue biopsies or even on the time dependence of such metabolic fingerprints as the treatment progresses. This leads to the concept of a metabolic passport for each person that can be checked periodically, and any deviations from that person's normal range of values could be used as an early indication of a future health problem.

Related to this, of course, is the development of biomarkers of disease risk from large cohorts of samples from historical epidemiologic studies or the increasingly available national bio-banks of samples.

Finally, this should lead to new paradigms such as centralized metabolic phenotyping in phenome centers, novel partnership opportunities, for example,

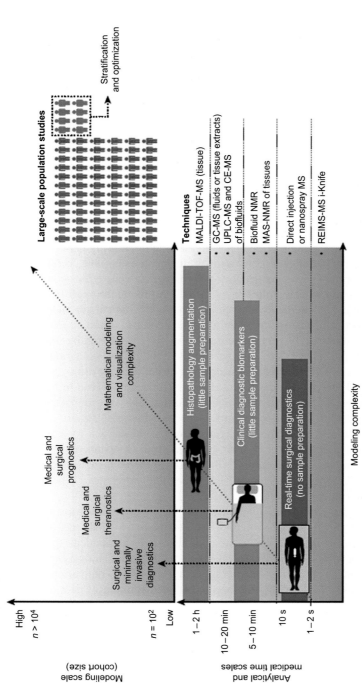

FIGURE 13.1 Technology platforms and analytical timescales for patient journey phenotyping, diagnostic and prognostic biomarker discovery, and population disease-risk biomarker modeling. Different analytical technologies can be applied to a variety of clinically derived bio-samples, and the choice of technology is dependent on the timescale for reaching a solution to the clinical problem, as well as the analytical performance characteristics of the technology. Thus, surgical problems require either real-time or near-real-time solutions for clinical decision making, whereas histopathologic augmentation has a multihour timescale. Epidemiologic problems, such as public health care, epidemiology, and identification of new biomarkers, as well as surgical risk stratification and preoperative optimization, involve the analysis of hundreds or thousands of samples from different populations, and the transfer of information can cause bottlenecks in data processing and total-cohort analyses. In the critical care or surgical setting, large-scale population studies can also be used to identify populations at risk of surgical morbidity or a poor outcome. *GC*, gas chromatography; *i-Knife*, intelligent knife; *MALDI-TOF-MS*, matrix-assisted laser desorption ionization time-of-flight mass spectrometry; *MAS-NMR*, magic-angle-spinning–nuclear-magnetic-resonance; *MS*, mass spectrometry; *REIMS*, rapid evaporative ionization mass spectrometry; *UPLC*, ultra-performance liquid chromatography. *Reprinted with permission from Nicholson JK, et al. Metabolic phenotyping in clinical and surgical environments. Nature 2012;491:384–92.*

between companies that promote pharmaceuticals, nutrition, and a healthy life-style and experts in metabolic phenotyping; in turn, this will lead to new funding routes for research and development in metabolic phenotyping.

13.7   POSSIBLE IMPACTS OF METABOLIC PHENOTYPING IN MEDICINE

Future directions involving metabolic phenotyping should have an impact in several areas. In a real-world scenario, it is possible that the new initiatives in the visualization of the so-called big data (see chapter "Modeling People and Populations: Exploring Medical Visualization Through Immersive Interactive Virtual Environments") will lead to improved decision making by medical professionals and will allow better harmonization of metabolic phenotyping and Systems Medicine, at both the national and international levels, in spite of the very different health care funding regimes around the world.

On a practical level, the integration of metabolic data in pathway networks and the identification of significant proteins via proteome–metabolome correlation studies and important genes via genome–metabolome connectivities could lead to the identification of new drug targets. Although this might be difficult in terms of mammalian biochemistry, the realization that the human gut microbiome is a major source of interaction with human metabolism opens up many new possible drug targets in the microbiome genome [83]. Moreover, by thus causing changes to the microbiome metabolism, this could have a favorable impact on the pharmacokinetics and pharmacodynamics of existing drugs.

On a policy level, the approaches discussed in this text should lead to new translatable strategies for stratified medicine and ultimately have impact on future health care policies. It is, of course, possible to conduct cost–benefit calculations on any new metabolite-based diagnostic or prognostic test for comparison of the approaches currently used around the world, independent of the health care funding model in any country. All of the factors that have been discussed in the above sections of this chapter have been integrated into a scheme that shows their interrelationships (see Fig. 13.2).

This book provides a brief introduction to the history, current status, and future potential of metabolic phenotyping in Personalized Medicine and Population Medicine. It draws from decades of research in chemical spectroscopy and outlines the necessary integration of the chemical analysis with advanced bioinformatics and computational modeling in order to deliver clinically relevant solutions, highlighting the necessity for integration of data at a systems level.

As we have described, many new and exciting technologies that are currently in their infancy have real potential to deliver improved diagnostic or prognostic solutions in a clinically relevant time frame. With any new medical technology or approach, the critical issue relates to the "clinical actionability" of the outputs. In addition to the rigorous validation criteria described previously, novel

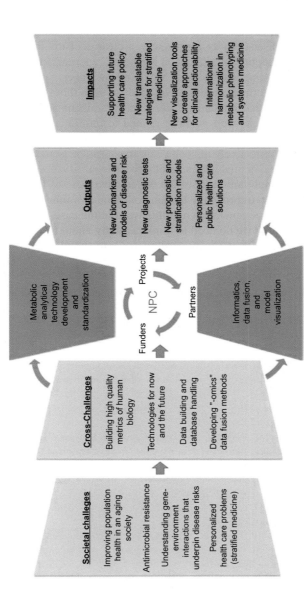

FIGURE 13.2 The various factors and their interactions for successful future use of metabolic phenotyping in Personalized Medicine and population disease risk screening. *NPC*, National Phenome Centre (see text for explanation).

diagnostic (ie, multiomic-platform) outputs must make sense to a practicing physician or surgeon in the clinical environment and result in a clear actionable decision that impacts favorably on the care pathway. In practice, this is a formidable challenge that incorporates social, medicolegal, and training issues, as well as the difficulties in taking complex mathematical modeling data and presenting them in a user-friendly format with a clear decision pathway. The consequences of the adoption of "Precision Medicine" approaches that encompass metabolic phenotyping and Systems Medicine have deep societal impacts [86], and the pathway to translation is at yet unclear, even though it is promising. What is certain is that humanity faces many imminent geopolitical and environmental problems, as well as those related to multidrug resistance, all of which will impact on future health care paradigms. Our ability to use the "-omics" technologies effectively in health care must also be geared to addressing these longer-term existential issues, as otherwise local advances in diagnostics will be lost. The important thing is that we do have these technologies and capabilities that can help improve the human condition at a time of great challenge; a century or even a generation ago this was not the case. Our prognosis is, therefore, positive, but we do need to think carefully about the allocation of the limited resources to deliver effective solutions for the whole of humankind. Complex interactions take place at the interface between genes and the environment, and the understanding of those interactions to take actions for human good is required. That is where metabolic profiling has its greatest potential impact.

REFERENCES

[1] Bales JR, Higham DP, Howe I, Nicholson JK, Sadler PJ. Use of high-resolution proton nuclear magnetic resonance spectroscopy for rapid multi-component analysis of urine. Clin Chem 1984;30:426–32.

[2] Nicholson JK, Buckingham MJ, Sadler PJ. High resolution ^{1}H n.m.r. studies of vertebrate blood and plasma. Biochem J 1983;211:605–15.

[3] Nicholson JK, O'Flynn MP, Sadler PJ, MacLeod AF, Juul SM, Sönksen PH. Proton-nuclear-magnetic-resonance studies of serum, plasma and urine from fasting normal and diabetic subjects. Biochem J 1984;217:365–75.

[4] Nicholson JK, Wilson ID. High resolution proton magnetic resonance spectroscopy of biological fluids. Prog NMR Spectrosc 1989;21:449–501.

[5] Lindon JC, Holmes E, Nicholson JK. Toxicological applications of magnetic resonance. Prog NMR Spectrosc 2004;45:109–43.

[6] Gartland KPR, Beddell CR, Lindon JC, Nicholson JK. A pattern recognition approach to the comparison of ^{1}H NMR and clinical chemical data for classification of nephrotoxicity. J Pharmaceut Biomed Anal 1990;8:963–73.

[7] Gartland KPR, Sanins S, Nicholson JK, Sweatman BC, Beddell CR, Lindon JC. Pattern recognition analysis of high resolution ^{1}H NMR spectra of urine. A nonlinear mapping approach to the classification of toxicological data. NMR Biomed 1990;3:166–72.

[8] Gartland KPR, Beddell CR, Lindon JC, Nicholson JK. Application of pattern recognition methods to the analysis and classification of toxicological data derived from proton NMR spectroscopy of urine. Mol Pharmacol 1991;39:629–42.

[9] Holmes E, Bonner FW, Sweatman BC, Lindon JC, Beddell CR, Rahr E, et al. NMR spectroscopy and pattern recognition analysis of the biochemical processes associated with the progression and recovery from nephrotoxic lesions in the rat induced by mercury(II)chloride and 2-bromo-ethanamine. Mol Pharmacol 1992;42:922–30.

[10] Nicholson JK, Lindon JC, Holmes E. "Metabonomics": understanding the metabolic responses of living systems to pathophysiological stimuli via multivariate statistical analysis of biological NMR spectroscopic data. Xenobiotica 1999;29:1181–9.

[11] Lindon JC, Holmes E, Nicholson JK. So what's the deal with metabonomics? Metabonomics measures the fingerprint of biochemical perturbations caused by disease, drugs and toxins. Anal Chem 2003;75:384A.8.

[12] Keun HC, Beckonert O, Griffin JL, Richter C, Moskau D, Lindon JC, et al. Cryogenic probe ^{13}C NMR spectroscopy of urine for metabonomic studies. Anal Chem 2002;74:4588–93.

[13] Garcia A, Barbas C. Gas chromatography-mass spectrometry (GC-MS)-based metabolomics. Methods Mol Biol 2011;708:191–204.

[14] Jonsson P, Gullberg J, Nordström A, Kusano M, Kowalczyk M, Sjöström M, et al. A strategy for identifying differences in large series of metabolomic samples analyzed by GC/MS. Anal Chem 2004;76:1738–45.

[15] Wilson ID, Plumb R, Granger J, Major H, Williams R, Lenz EM. HPLC-MS-based methods for the study of metabonomics. J Chromatogr B 2005;817:67–76.

[16] Gika HG, Theodoridis GA, Wingate JE, Wilson ID. Within-day reproducibility of an HPLC-MS-based method for metabonomic analysis: application to human urine. J Proteome Res 2007;6:3291–303.

[17] Dunn WB, Broadhurst D, Begley P, Zelena E, Francis-McIntyre S, Anderson N, et al. Procedures for large-scale metabolic profiling of serum and plasma using gas chromatography and liquid chromatography coupled to mass spectrometry. Nat Protoc 2011;6:1060–83.

[18] Want EJ, Wilson ID, Gika H, Theodoridis G, Plumb RS, Shockcor J, et al. Global metabolic profiling procedures for urine using UPLC-MS. Nat Protoc 2010;5:1005–18.

[19] Naz S, Moreira dos Santos DC, Garcia A, Barbas C. Analytical protocols based on LC-MS, GC-MS and CE-MS for non-targeted metabolomics of biological tissues. Bioanalysis 2014;6:1657–77.

[20] Garzotti M, Hamdan M. Supercritical fluid chromatography coupled to electrospray mass spectrometry: a powerful tool for the analysis of chiral mixtures. J Chromatogr B 2002;770:53–61.

[21] Cloarec O, Dumas ME, Craig A, Barton RH, Trygg J, Hudson J, et al. Statistical total correlation spectroscopy (STOCSY): a new approach for individual biomarker identification from metabonomic NMR datasets. Anal Chem 2005;77:1282–9.

[22] Holmes E, Loo RL, Cloarec O, Coen M, Tang H, Maibaum E, et al. Detection of urinary drug metabolite signatures in molecular epidemiology studies via statistical total correlation (NMR) spectroscopy. Anal Chem 2007;79:2629–40.

[23] Crockford DJ, Lindon JC, Cloarec O, Plumb RS, Bruce SJ, Zirah S, et al. Statistical search space reduction and two-dimensional data display approaches for UPLC-MS in biomarker discovery and pathway analysis. Anal Chem 2006;78:4398–408.

[24] Crockford DJ, Holmes E, Lindon JC, Plumb RS, Zirah S, Bruce SJ, et al. Statistical HeterospectroscopY (SHY), a new approach to the integrated analysis of NMR and UPLC-MS datasets: application in metabonomic toxicology studies. Anal Chem 2006;78:363–71.

[25] Posma JM, Robinette SL, Holmes E, Nicholson JK. MetaboNetworks, an interactive Matlab-based toolbox for creating, customizing and exploring sub-networks from KEGG. Bioinformatics 2014;30:893–5.

[26] Kaever A, Landesfeind M, Feussner K, Mosblech A, Heilmann I, Morgenstern B, et al. MarVis-Pathway: integrative and exploratory pathway analysis of non-targeted metabolomics data. Metabolomics 2015;11:764–77.

[27] Steinbeck C, Conesa P, Haug K, Mahendraker T, Willaims M, Maguire E, et al. MetaboLights: towards a new COSMOS of metabolomics data management. Metabolomics 2012;8:757–60.

[28] Chadeau-Hyam M, Ebbels TM, Brown IJ, Chan Q, Stamler J, Huang CC, et al. Metabolic profiling and the Metabolome-Wide Association Study: significance level for biomarker identification. J Proteome Res 2010;9:4620–7.

[29] Holmes E, Loo RY, Stamler J, Bictash M, Yap IKS, Chan Q, et al. Human metabolic phenotype diversity and its association with diet and blood pressure. Nature 2008;453:396–400.

[30] Fischer K, Kettunen J, Würtz P, Haller T, Havulina AS, Kangas AJ, et al. Biomarker profiling by nuclear magnetic resonance spectroscopy for the prediction of all-cause mortality: an observational study of 17,345 persons. PLoS One 2014;11:e1001606.

[31] Yu B, Zheng Y, Nettleton JA, Alexander D, Coresh J, Boerwinkle E. Serum metabolomics profiling and incident CKD among African Americans. Clin J Am Soc Nephrol 2014;9:1410–17.

[32] Würtz P, Havulinna AS, Soininen P, Tynkkynen T, Prieto-Merino D, Tillin T, et al. Metabolite profiling and cardiovascular event risk: a prospective study of 3 population-based cohorts. Circulation 2015;131:774–85.

[33] Altmaier E, Fobo G, Heier M, Thorand B, Meisinger C, Römisch-Margl W, et al. Metabolomics approach reveals effects of antihypertensives and lipid-lowering drugs on the human metabolism. Eur J Epidemiol 2014;29:325–36.

[34] Gieger C, Geistlinger L, Altmaier E, de Angelis MH, Kronenberg F, Meitlinger T, et al. Genetics meets metabolomics: a Genome-Wide Association Study of metabolite profiles in human serum. PLoS Genet 2008;4:e1000282.

[35] Klenø TG, Kiehr B, Baunsgaard D, Sidelmann UG. Combination of "omics" data to investigate the mechanism(s) of hydrazine-induced hepatotoxicity in rats and to identify potential biomarkers. Biomarkers 2004;9:116–38.

[36] Adamski J, Suhre K. Metabolomics platforms for genome wide association studies—linking the genome to the metabolome. Curr Opinion Biotechnol 2013;24:39–47.

[37] Foxall PJ, Mellotte GJ, Bending MR, Lindon JC, Nicholson JK. NMR spectroscopy as a novel approach to the monitoring of renal transplant function. Kidney Int 1993;43:234–45.

[38] Kinross JM, Holmes E, Darzi AW, Nicholson JK. Metabolic phenotyping for monitoring surgical patients. Lancet 2011;377:1817–19.

[39] Omranian N, Klie S, Mueller-Roeber B, Nikoloski Z. Network-based segmentation of biological multivariate time series. PloS One 2013;8:e62974.

[40] Netzer M, Weinberger KM, Handler M, Seger M, Fang X, Kugler KG, et al. Profiling the human response to physical exercise: a computational strategy for the identification and kinetic analysis of metabolic biomarkers. J Clin Bioinform 2011;1:34.

[41] Clayton TA, Lindon JC, Cloarec O, Antti H, Chareul C, Hanton G, et al. Pharmacometabonomic phenotyping and personalised drug treatment. Nature 2006;440:1073–7.

[42] Clayton TA, Baker D, Lindon JC, Everett JR, Nicholson JK. Pharmacometabonomic identification of a significant host-microbiome metabolic interaction affecting human drug metabolism. Proc Natl Acad Sci USA 2009;106:14728–33.

[43] Backshall A, Sharma R, Clarke SJ, Keun HC. Pharmacometabonomic profiling as a predictor of toxicity in patients with inoperable colorectal cancer treated with capecitabine. Clin Cancer Res 2011;17:3019–28.

[44] Phapale PB, Kim S-D, Lee HW, Lim M, Kale DD, Kim Y-L, et al. An integrative approach for identifying a metabolic phenotype predictive of pharmacokinetics of Tacrolimus. Nat Clin Pharmacol Therap 2010;87:426–36.

[45] Everett JR. Pharmacometabonomics in humans: a new tool for personalized medicine. Pharmacogenomics 2015;16:737–54.

[46] Lindon JC, Nicholson JK. The emergent role of metabolic phenotyping in dynamic patient stratification. Expert Opin Drug Metab Toxicol 2014;10:915–19.

[47] Balog J, Sasi-Szabó L, Kinross J, Lewis MR, Muirhead LJ, Veselkov K, et al. Intraoperative tissue identification using rapid evaporative ionization mass spectrometry. Sci Trans Med 2013;5:194ra93.

[48] Cascante M, Marin S. Metabolomics and fluxomics approaches. Essays Biochem 2008;45:67–81.

[49] Winter G, Krömer JO. Fluxomics—connecting 'omics analysis and phenotypes. Environ Microbiol 2013;15:1901–16.

[50] D'Alessandro A, Slaughter AL, Peltz ED, Moore EE, Silliman CC, Wither M, et al. Trauma/hemorrhagic shock instigates aberrant metabolic flux through glycolytic pathways, as revealed by preliminary (13)C-glucose labeling metabolomics. J Transl Med 2015;13:253.

[51] Vaitheesvaran B, Xu J, Yee Q-Y, Lu V, Go L, Xiao GG, et al. The Warburg effect: a balance of flux analysis. Metabolomics 2015;11:787–96.

[52] Zamboni N, Fendt SM, Rühl M, Sauer U. (13)C-based metabolic flux analysis. Nat Protoc 2009;4:878–92.

[53] Hellerstein M, Turner S. Reverse cholesterol transport fluxes. Curr Opin Lipidol 2014;25:40–7.

[54] Dona AC, Jimenez B, Schaefer H, Humpfer E, Spraul M, Lewis MR, et al. Precision high-throughput proton NMR spectroscopy of human urine, serum, and plasma for large-scale metabolic phenotyping. Anal Chem 2014;86:9887–94.

[55] Lindon JC, Nicholson JK, Holmes E, Keun HC, Craig A, Pearce JT, et al. Summary recommendations for standardization and reporting of metabolic analyses. Nat Biotech 2005;23:833–8.

[56] Sansone S-A, Fan T, Goodacre R, Griffin JL, Hardy NW, Kaddurah-Douk R. The metabolomics standards initiative. Nat Biotech 2007;25:846–8.

[57] MetaboLights database. Available at: http://www.ebi.ac.uk/metabolights

[58] Metabolomics Workbench. Available at: http://www.metabolomicsworkbench.org

[59] Lindon JC, Holmes E, Nicholson JK. Pattern recognition methods and applications in biomedical magnetic resonance. Prog NMR Spectrosc 2001;39:1–40.

[60] Human Metabolome Database. Available at: http://www.hmdb.ca

[61] Biological Magnetic Resonance Data Bank. Available at: http://www.bmrb.wisc.edu

[62] Metlin database. Available at: https://metlin.scripps.edu/index.php

[63] Psychogios N, Hau DD, Peng J, Guo AC, Mandal R, Bouatra S, et al. The human serum metabolome. PLoS One 2011;6:e16957.

[64] Bouatra S, Aziat F, Mandal R, Guo AC, Wilson MR, Knox C, et al. The human urine metabolome. PLoS One 2013;8:e73076.

[65] Dunn WB, Lin W, Broadhurst D, Begley P, Brown M, Zelena E, et al. Molecular phenotyping of a UK population: defining the human serum metabolome. Metabolomics 2015;11:9–26.

[66] Nicholson JK, Connelly J, Lindon JC, Holmes E. Metabonomics: a platform for studying drug toxicity and gene function. Nat Rev Drug Discov 2002;1:153–61.

[67] Rantalainen M, Cloarec O, Beckonert O, Wilson ID, Jackson D, Tonge R, et al. Statistically integrated metabonomic-proteomic studies on a human prostate cancer xenograft model in mice. J Proteome Res 2006;5:2642–55.

[68] Redinbo MR. The microbiota, chemical symbiosis, and human disease. J Mol Biol 2014;426:3877–91.

[69] Nicholson JK, Holmes E, Kinross J, Burcelin R, Gibson G, Jia W, et al. Host-gut microbiota metabolic interactions. Science 2012;336:1262–7.

[70] Cenit MC, Olivares M, Codoñer-Franch P, Sanz Y. Intestinal microbiota and celiac disease: cause, consequence or co-evolution? Nutrients 2015;7:6900–23.

[71] Hester CM, Jala VR, Langille MG, Umar S, Greiner KA, Haribabu B. Fecal microbes, short chain fatty acids, and colorectal cancer across racial/ethnic groups. World J Gastroenterol 2015;21:2759–69.

[72] Mortensen PB, Anderen JR, Arffmann S, Krag E. Short-chain fatty acids and the irritable bowel syndrome: the effect of wheat bran. Scand J Gastroenterol 1987;22:185–92.

[73] Sahu SS, Weirick T, Kaundal R. Predicting genome-scale Arabidopsis-Pseudomonas syringae interactome using domain and interolog-based approaches. BMC Bioinform 2014;15(Suppl. 11):S13.

[74] Ghosh S, Prasad KV, Vishveshwara S, Chandra N. Rule-based modelling of iron homeostasis in tuberculosis. Mol Biosyst 2011;7:2750–68.

[75] Elliott P, Posma JM, Chan Q, Garcia-Perez I, Wijeyesekera A, Bictash M, et al. Urinary metabolic signatures of human adiposity. Sci Transl Med 2015;7:285ra62.

[76] Strittmatter N, Rebec M, Jones EA, Golf O, Abdolrasouli A, Balog J, et al. Characterization and identification of clinically relevant microorganisms using rapid evaporative ionization mass spectrometry. Anal Chem 2014;86:6555–62.

[77] Stamler J, Elliott P, Dennis B, Dyer AR, Kesteloot H, Liu K, et al. INTERMAP: background, aims, designs, methods and descriptive statistics. J Hum Hypertens 2003;17:591–608.

[78] CombiBio homepage. Available at: http://www.combi-bio.eu

[79] Williamson MP, Humm G, Crisp AJ. 1H nuclear magnetic resonance investigation of synovial fluid components in osteoarthritis, rheumatoid arthritis and traumatic effusions. Br J Rheumatol 1989;28:23–7.

[80] Niewczas MA, Sirich T, Mathew AV, Skupien J, Mohney RP, Warram JH, et al. Uremic solutes and risk of end-stage renal disease in type-2 diabetes: metabolomics study. Kidney Int 2014;85:1214–24.

[81] Zordoky BN, Sung MM, Ezekowitz J, Mandal R, Han B, Bjorndahl TC, et al. Metabolomic fingerprint of heart failure with preserved ejection fraction. PloS One 2015;10:e124844.

[82] Nicholson JK, Holmes E, Kinross JM, Darzi AW, Takats Z, Lindon JC. Metabolic phenotyping in clinical and surgical environments. Nature 2012;491:384–92.

[83] Golf O, Strittmatter N, Karancsi T, Pringle SD, Speller AV, Mroz A, et al. Rapid evaporative ionization mass spectrometry imaging platform for direct mapping from bulk tissue and bacterial growth media. Anal Chem 2015;87:2527–34.

[84] Veselkov KA, Mirnezami R, Strittmatter N, Goldin RD, Kinross J, Speller AV, et al. Chemoinformatic strategy for imaging mass spectrometry-based hyperspectral profiling of lipid signatures in colorectal cancer. Proc Natl Acad Sci USA 2014;111:1216–21.

[85] Guenther S, Muirhead LJ, Speller AV, Golf O, Strittmatter N, Ramakrishnan R, et al. Spatially resolved metabolic phenotyping of breast cancer by desorption electrospray ionization mass spectrometry. Cancer Res 2015;75:1828–37.

[86] Mirnezami R, Nicholson JK, Darzi A. Preparing for precision medicine. N Engl J Med 2012;366:489–91.

Index

Note: Page numbers followed by "*b*," "*f*," and "*t*" refer to boxes, figures, and tables, respectively.